Duxbury

D0025444

Student Solutions Manual

for

Johnson and Kuby's

Elementary Statistics

Tenth Edition

Patricia Kuby
Monroe Community College

THOMSON
— ✳ ™
BROOKS/COLE
Australia • Brazil • Canada • Mexico • Singapore • Spain • United Kingdom • United States

Printed in the United States of America
1 2 3 4 5 6 7   10 09 08 07 06

Printer: Thomson/West

0-495-10531-7
Cover Image: Getty Images

For more information about our products,
contact us at:
Thomson Learning Academic Resource Center
1-800-423-0563

For permission to use material from this text or
product, submit a request online at
http://www.thomsonrights.com.
Any additional questions about permissions can be
submitted by email to **thomsonrights@thomson.com.**

Thomson Higher Education
10 Davis Drive
Belmont, CA 94002-3098
USA

CONTENTS

Preface   iv

Solutions and Student Annotations
Chapter 1   Statistics   1
Chapter 2   Descriptive Analysis and Presentation
                 of Single-Variable Data   11
Chapter 3   Descriptive Analysis and Presentation
                 of Bivariate Data   98
Chapter 4   Probability   142
Chapter 5   Probability Distribution
                 (Discrete Variables)   172
Chapter 6   Normal Probability Distributions   204
Chapter 7   Sample Variability   233
Chapter 8   Introduction to Statistical Inferences   252
Chapter 9   Inferences Involving One Population   298
Chapter 10   Inferences Involving Two Populations   358
Chapter 11   Applications of Chi-Square   417
Chapter 12   Analysis of Variance   452
Chapter 13   Linear Correlation and Regression
                 Analysis   479
Chapter 14   Elements of Nonparametric Statistics   516

Introductory Concepts
                 Summation Notation   552
                 Using the Random Number Table   560
                 Round-Off Procedure   562
Review Lessons
                 The Coordinate-Axis System and the
                 Equation of a Straight Line   565
                 Tree Diagrams   573
                 Venn Diagrams   579
                 The Use of Factorial Notation   582
Answers to Introductory Concepts and Review Lessons Exercises
                 Summation Notation   584
                 Using the Random Number Table   584
                 Round-Off Procedure   585
                 The Coordinate-Axis System and the
                 Equation of a Straight Line   585
                 Tree Diagrams   589
                 Venn Diagrams   594
                 The Use of Factorial Notation   596

PREFACE

This Student Solutions Manual contains solutions for the odd
exercises in Elementary Statistics, 10th edition, as well as
information and assistance. Included at the end of the manual are
sections covering Introductory Concepts and Review Lessons on
various algebraic and/or basic statistical concepts.
For each chapter, the following are provided:
1) Chapter Preview
2) Chapter Solutions for Odd-Numbered Exercises
3) Student Annotations.

Student annotations are printed inside a border and
placed before related exercises.
There are several ways for a student to use this manual. One
possibility that has proved to be beneficial contains the following
steps. Begin by reading the exercise under consideration in the
text and attempting to answer the question. Locate the exercise
number in the manual. If the exercise was answered, compare
solutions. If solutions are not similar or difficulty has been
encountered, read the boxed annotations before the exercise in
detail.
Try the exercise again. If difficulty remains at this point,
review the boxed annotations and the solution. Ask the instructor
for aid or verification to get needed additional information.
In several exercises, a concluding question is asked to determine
if the overall concept is comprehended. These questions are marked
with an asterisk *. Locations of the corresponding answers are
given and marked with an * and the exercise number.

CHAPTER 1 ∇ STATISTICS

Chapter Preview

The purpose of Chapter 1 is to present:
1. an initial image of statistics that includes both the key role statistics has in the technical aspects of life as well as its everyday applicability,
2. its basic vocabulary and definitions,
3. basic ideas and concerns about the processes used to obtain sample data.

SECTION 1.1 EXERCISES

1.1 a. Americans, based on the title of the section 'Americans, Here's Looking at You'
  b. The communication method workers preferred was obtained.
  c. 63% of those people surveyed said they would like to live to see 100.
  d. Answers will vary.
  e. There are 7.2 fatal crash involvements per 100 million miles traveled for 19 year-olds.

1.3 a. Answers will vary.
  b. It does not appear, based on the list of averages given, that Java professionals work a 40-hour week.
  c. Only if long work hours are desirable.

SECTION 1.2 EXERCISES

> The article in Applied Example 1.1 gives information about the sample (the number of people surveyed). Be watchful of articles that do not give any of this information. Sometimes not knowing something about the sample or survey size causes a question of credibility.

1.5 a. Answers will vary.

> Descriptive Statistics - refers to the techniques and methods for organizing and summarizing the information obtained from the sample.
>
> Inferential Statistics - refers to the techniques of interpreting and generalizing about the population based on the information obtained from the sample.

1.7 a. descriptive
b. inferential

1.9 a. American heads of households   b. 1000
c. hardest household place to clean
d. $1000(0.12) = 120$
e. actual percentage could be 5% lower or 5% higher than quoted
f. Between 30% and 40% of all adults think that Venetian blinds are the the hardest to clean.

A variable is the characteristic of interest (ex. height), where data is a value for the variable (ex. 5'5"). A variable varies (heights vary), that is, heights can take on different values. Data (singular) such as 5'5" (one person's height) is constant; it does not change in value for a specific subject.

An attribute variable can take on any qualitative or "numerical" qualitative information (ex. kinds of fruit, types of music, religious preference, model year - most answers are in words, although model year would have "numerical" answers such as "2003"). An attribute variable can be nominal (description or name) or ordinal (ordered position or rank; first, second,…).

A numerical variable can take on any quantitative information. This includes any count-type and measurable-type data (ex. number of children in a family, amount of time, age, height, area, volume, miles per gallon). A numerical variable can be discrete or continuous. The domain of a discrete variable has gaps between the possible values; there are numerical values that cannot occur. Theoretically, the domain of a continuous variable has no gaps since all numerical values are possible. Do not be confused by data that has been rounded due to scale being used or for convenience reasons.

1.11 a. It could be any one of these: All Americans that file taxes, or all Americans that file taxes and receive refunds, or all Americans that receive refunds.
b. Users of TurboTax.
c. Action taken upon receipt of tax refund.
d. Majority of people plan to pay bills with their tax refund. The majority shows as the largest portion of the dollar bill.

1.13 a. Yes, if the rate increases from 4% to 6% that is a 50% increase in the rate: $(6-4)/4 = 2/4 = 0.50 = 50\%$. As a percent alone; the 50% is meaningless, it does not give the actual size of the numbers involved.
b. The phrase "50% jump" works much more effectively at getting people's attention than does "2% increase."

> Population - the collection of all individuals, objects, or scores whose properties are under consideration.
>
> Parameter - a number calculated from the population of values.
>
> Sample - that part of the population from which the data values or information is obtained.
>
> Statistic - a number calculated from the sample values.
>
> **NOTE:**
>   Parameters are calculated from populations; both begin with *p*.
>   Statistics are calculated from samples; both begin with *s*.

1.15 a. The population is all US adults.
  b. A sample is the 1200 randomly selected adults.
  c. The variable is "allergy status" for each adult.
  d. The statistic is the 33.2% based on the sampled adults.
  e. The parameter is the percent of all US adults who have an allergy, in this case, 36%.

1.17 A football jersey number is a categorical variable. It is attribute information that can identify something about the position played by a player [for example; 60's & 70's are numbers for lineman and they are not eligible to catch passes, other number groups have similar restrictions], but does not give any measurable information about that player.

1.19 a. Nominal possibilities: marital status, gender, ZIP code
  b. Ordinal possibilities: highest level of education, ranking of department preferences, rating for first impression of store

1.21 a. Score can only be whole numbers (scores are counted).
  b. Number of minutes (a length of time) can be any value (time is measured); its accuracy depends on the precision with which it is measured.

1.23 a. Satisfaction level with summer schedule
b. Ordinal

1.25 a. All individuals who have hypertension and use prescription drugs to control it (a very large group)
b. The 5,000 people in the study
c. The proportion of the population for which the drug is effective
d. The proportion of the sample for which the drug is effective, 80%
e. No, but it is estimated to be approximately 80%

1.27 a. All assembled parts from the assembly line
b. infinite   c. the parts checked
d. attribute, attribute (it identifies the assembler), numerical

1.29 a. The population being studied is composed of all people suffering from seasonal allergies.
b. The sample is the 679 people given the dose.
c. The characteristics of interest are relief status and side effects.
c. The data being collected are qualitative (relief status and side effects).

1.31 a. numerical   b. attribute   c. numerical
d. attribute   e. numerical   f. numerical

1.33 a. The population contains all objects of interest, while the sample contains only those actually studied.
b. convenience, availability, practicality

SECTION 1.3 EXERCISES

1.35 Group 2, the football players, because their weights cover a wider range of values, probably 175 to 300+, while the cheerleaders probably all weigh between 110 and 150.

1.37 By using a standard weight or measure in conjunction with money, prices between competing product brands can be more easily compared, irrespective of purchase quantity. There is a great deal of variability in container sizes between brands and even within brands of the same product. Problems associated with this variability are simplified by showing the standard unit price in addition to the cash register amount at the point of sale.

1.39 Answers will vary but there is no way to differentiate between the students if everybody attains the same grade. If all students received a 100%, then the test is too easy. If all students received a 0%, then the test was too hard. If the scores are between 40 to 95%, you can distinguish among the students' knowledge about the subject.

SECTION 1.4 EXERCISES

A <u>convenience sample</u> or <u>volunteer sample</u>, as indicated by their very names, can often result in <u>biased samples</u>.

Data collection can be accomplished with <u>experiments</u> (the environment is controlled) or <u>observational studies</u> (environment is not controlled). <u>Surveys</u> fall under observational studies.

Sample designs can be categorized as <u>judgement samples</u> (believed to be typical) or <u>probability samples</u> (certain chance of being selected is given to each data value in the population).

The <u>random sample</u> (each data value has the same chance) is the most common probability sample.
Methods (simply defined) to obtain a random sample include:
  Single-Stage Methods:
    1. <u>Random Number Table</u> - (see Introductory Concepts in the Student Solutions Manual)
    2. <u>Systematic</u> - every kth element is chosen

(continued)

> Multistage Methods:
>   1. <u>Stratified</u> - fixed number of elements from each strata
>   (group)
>   2. <u>Proportional (Quota)</u> - number of elements from each strata is
>   determined by its size
>   3. <u>Cluster</u> - fixed number or all elements from certain strata.

1.41 Volunteer; yes, only those with strong opinions will respond.

1.43 Answers will vary but Landers' survey was a volunteer survey, therefore there is a bias - mostly, only those with strong opinions will respond.

1.45 convenience sampling

1.47 a. (1,1), (1,2), (1,3), (1,4)
           (2,1), (2,2), (2,3), (2,4)
           (3,1), (3,2), (3,3), (3,4)
           (4,1), (4,2), (4,3), (4,4)

   b. (1,1,1), (1,1,2), (1,1,3)
      (1,2,1), (1,2,2), (1,2,3)
      (1,3,1), (1,3,2), (1,3,3)

      (2,1,1), (2,1,2), (2,1,3)
      (2,2,1), (2,2,2), (2,2,3)
      (2,3,1), (2,3,2), (2,3,3)

      (3,1,1), (3,1,2), (3,1,3)
      (3,2,1), (3,2,2), (3,2,3)
      (3,3,1), (3,3,2), (3,3,3)

1.49 probability samples

1.51 Statistical methods presume the use of random samples.

1.53 Randomly select an integer between 1 and 25 $(100/x = 100/4 = 25)$. Locate the first item by this integer value. Select every 25th data thereafter until the sample is complete.

1.55 A proportional sample would work best since the area is already divided into 35 (different size) listening areas. The size of the listening area determines the size of the subsample. The total for all subsamples would be 2500.

1.57 Only people with telephones and listed phone numbers will be considered.

SECTION 1.5 EXERCISES

> <u>Statistics</u> - allows you to make inferences or generalizations about the population based on a sample.
>
> <u>Probability</u> - the chance that something will happen when you know **all** the possibilities (you know all the possibilities when you have the population).

1.59 a. probability   b. statistics

1.61 a. statistics   b. probability
       c. statistics   d. probability

SECTION 1.6 EXERCISES

1.63 Several large comprehensive computer programs (called statistical packages) have been developed that perform many of the computations and tests you will study in this text. In order to have the statistical package perform the computations, you simply enter the data into the computer and the computer does the rest on command. Quickly and easily

1.65 Calculators only do the calculations they are directed to perform. The results are only as good as the operator is with precise entries.

1.67 Each student's answers will be different. A few possibilities are:
    a. color of hair, major, gender, marital status
    b. number of courses taken, number of credit hours, number of jobs, height, weight, distance from hometown to college, cost of textbooks

1.69 a. T = 3 is a data value - a value from one person.
    b. What is the average number of times per week the people in the sample went shopping?
    c. What is the average number of times per week that people (all people) go shopping?

1.71 a. US adults
    b. flu vaccine status, location where vaccine was obtained, precaution status, type of precaution
    c. all variables are attribute

1.73 a. All US public schools   b. 1000 principals
    c. Probability sample   d. Stratified

1.75 a. With population and drivers parallel it means that both are increasing at the same rate. Not all of the population is in the category of drivers. The number of non-drivers appears to remain constant. If population and drivers were not parallel, then one would be increasing at a faster rate than the other.
    b. When the lines cross for drivers and motor vehicles, it shows that there were more drivers than vehicles before 1971, then more vehicles than drivers after 1973. From 1971 to 1973, the number of drivers and vehicles was about the same.

1.77 a. Observational study
    b. Percent of helmet use in children.
    c. Proportion of sample that wore a helmet, 41%
    d. activity, gender, helmet use - attribute; age - numberical

1.79 a. All Americans
b. physical activities and nutrition
c. age - numerical; leisure activity status - attribute; nutrition status - attribute

1.81 a. Number of new prescriptions
b. Women, by 1.5 new prescriptions
c. 75 and older, 13 new prescriptions
d. Yes, geography matters. Tennessee is the state with the highest usage and Alaska is the state with the lowest usage.

1.83 Each will have different examples.

1.85 Each will have different examples.

1.87 Answers will vary.

CHAPTER 2 ∇   DESCRIPTIVE ANALYSIS AND PRESENTATION OF SINGLE-VARIABLE DATA

Chapter Preview

Chapter 2 deals with the presentation of data that were obtained through the various sampling techniques discussed in Chapter 1. The four major areas for presentation and summary of the data are:

1. graphical displays,
2. measures of central tendency,
3. measures of variation, and
4. measures of position.

A study supported by the Stanford Institute for the Quantitative Study of Society on how people utilize the internet is presented in Section 1 of this chapter.

SECTION 2.1 EXERCISES

2.1   a.   Answers will vary. Possibilities might include: sort data, frequency of each value, average.

       b.   Answers will vary. Possibilities might include: location of my data value with respect to the other values, proximity to average.

SECTION 2.2 EXERCISES

MINITAB - Statistical software
Data is entered by use of a spreadsheet divided into columns and rows. Data for each particular problem is entered into its own column. Each column represents a different set of data. Be sure to name the columns in the space provided above the first row, so that you know where each data set is located. (C1 = Column 1). . .

EXCEL - Spreadsheet software
Data is entered by use of a spreadsheet divided into columns and rows. Data for each particular problem is entered into its own column. Each column represents a different set of data. If needed, use the first row of a column for a title. (A1 = 1st cell of column A)

TI-83/84 Plus - Graphing calculator
Data is entered into columns called lists. Data for each particular problem is entered into its own list. Each list represents a different set of data. If needed, use the space provided above the first row for a title. Lists are found under STAT > 1:Edit.
(L1 = List 1)

2.3 a. Answers will vary but both graphs show relative size with respect to the individual answers.
   b. Answers will vary. The circle graph does a better job of representing the relative proportions of the answers to the group as a whole.
   c. The bar graph is more dramatic in representing the relative proportions between the individual answers.

Partitioning the circle:

   1. Divide all quantities by the total sample size and turn them into percents.

   2. 1 circle   =   100%
      1/2 circle   =   50%
      1/4 circle   =   25%
      1/8 circle   =   12.5%

   3. Adjust other values accordingly.

   4. Be sure that percents add up to 100 (or close to 100, depending on rounding).

Computer and calculator commands to construct a Pie Chart can be found in ES10-p42.   . . .

The TI-83 program 'CIRCLE' and others can be downloaded from the textbook companion website at http://statistics.duxbury.com/jkes10e Select 'TI-83/84 Programs' from the book resource menu. Right-click on the zip file to save it to your computer and unzip the file using a zip utility, such as Winzip. Then download the programs to your calculator using TI-Graph Link Software. Further details on page 42 in ES10*.
*(ES10 denotes the textbook Elementary Statistics, 10th edition)

MINITAB commands to construct a bar graph can be found in the GRAPH > BAR CHART pull-down menu. With the categories in C1 and the corresp. freq. in C2, select 'Values from a table'; '1 col of values: Simple'.
EXCEL commands to construct a bar graph can be found under Chart Wizard. Input the categories into column A and the corresponding frequencies into column B.
TI-83/84 commands to construct a bar graph can be found using the STATPLOT command. Input the categories as numbers into list 1 and the corresponding frequencies into list 2. Adjust the x-scale in the WINDOW to 0.5 to allow for gaps between bars.

NOTE: Bar graphs may be vertical (as shown on the next page) or horizontal (as shown in the chapter opener Section 2.1). Information typically on the axes may also be printed inside the bars themselves. There is much flexibility in constructing a bar graph. Remember to leave a space between the bars. Be sure to label both axes and give a title to the graph.

2.5 a.

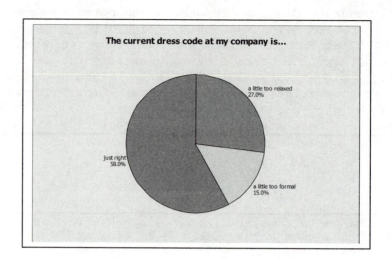

b.

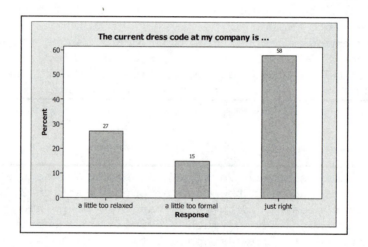

c. Answers will vary.

2.7 a.

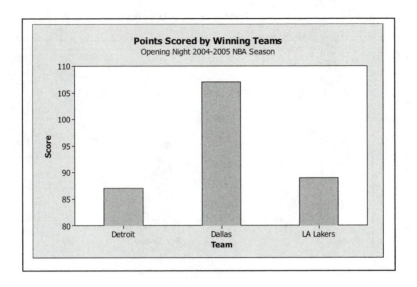

b.

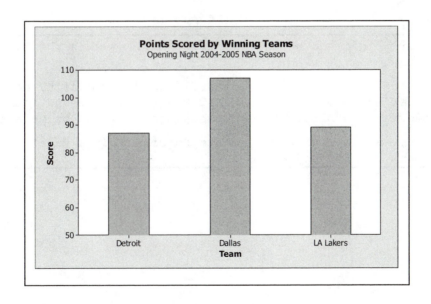

2.9

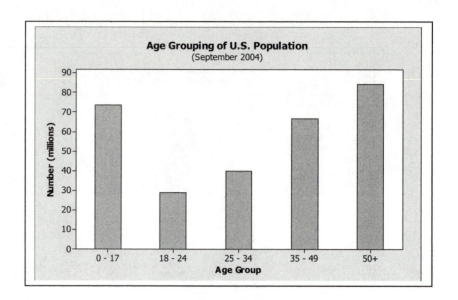

The Pareto command generates bars, starting with the largest category.

NOTE: Pareto diagrams are primarily used for quality control applications and therefore MINITAB's PARETO command identifies the categories as "Defects", even when they may not be defects.

Computer and calculator commands to construct a Pareto diagram can be found in ES10-pp43&44.

2.11

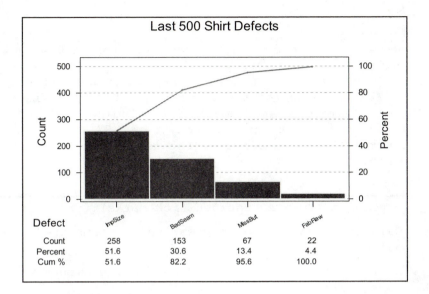

Last 500 Shirt Defects

Defect	ImpSize	BadSeam	MssBut	FabFlaw
Count	258	153	67	22
Percent	51.6	30.6	13.4	4.4
Cum %	51.6	82.2	95.6	100.0

2.13   a.

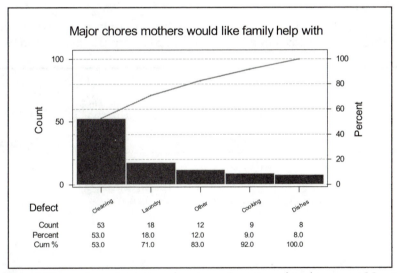

Major chores mothers would like family help with

Defect	Clearing	Laundry	Other	Cooking	Dishes
Count	53	18	12	9	8
Percent	53.0	18.0	12.0	9.0	8.0
Cum %	53.0	71.0	83.0	92.0	100.0

b. The "Other" category is too large, it is a collection of several answers and as such is larger than two of the categories. If it were broken down, then the Pareto diagram would have the categories in order of mother's wishes.

2.15 a. 150 defects
   b. percent of scratch = n(scratch)/150 = 45/150 = 0.30 or
      30.0%
   c. 90.7% = [37.3 + 30.0 + 15.3 + 8.0]% [round-off error]
      90.7% is the sum of the percentages for all defects that
      occurred more often than Bend, including Bend.
   d. Two defects, Blem and Scratch, total 67.3%. If they can
      control these two defects, the goal should be within
      reach.

Picking increments (spacing between tick marks) for a dot plot

   1. Calculate the spread (highest value minus the lowest
      value).

   2. Divide this value by the number of increments you wish
      to show (no more than 7 usually).   . . .
   3. Use this increment size or adjust to the nearest number
      that is easy to work with (5, 10, etc.).

Computer and calculator commands to construct a dotplot can be
found in ES10-p45. Commands for multiple dotplots are also in ES10-
pp49-50.

2.17 Points Scored per Game by Basketball Team

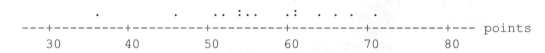

2.19 a.

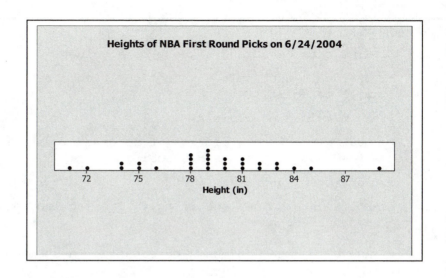

Heights of NBA First Round Picks on 6/24/2004

Height (in)

b. shortest – 71 inches, tallest – 89 inches
c. most common – 79 inches, 5 players share that height
d. most common height = tallest column of dots

2.21 Overall length of commutators

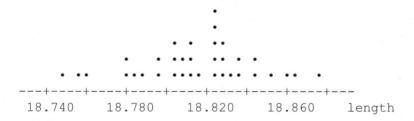

```
STEM-AND-LEAF DISPLAYS

    1. Find the lowest and highest data values.

    2. Decide in what "place value" position, the data values
       will be split.

    3. Stem = leading digit(s)

    4. Leaf = trailing digit(s) (if necessary, data is first
       "rounded" to the desired position)

    5. Sort stems and list.

    6. Split data values accordingly, listing leaves on the
       appropriate stem.

Computer and calculator commands to construct a stem-and-leaf
display can be found in ES10-p47.
```

2.23 Points scored per game

```
        3 | 6
        4 | 6
        5 | 6 4 5 4 2 1
        6 | 1 1 8 0 6 1 4
        7 | 1
```

2.25 a. Quik Delivery's delivery charges

```
2.  | 0
2.  | 9 8 8 9
3.  | 1 1
3.  | 5 8 8 5 8 6 6 8 7 7 8
4.  | 0 3 1 0 0
4.  | 5 5 9 6 8 6
5.  | 0 4 0 2 4
5.  | 6 7
6.  | 0 1
6.  | 8
7.  |
7.  | 8
```

b. The distribution is slightly skewed to the right; the bulk of the distribution is between 2.8 and 5.4, with only one smaller value and 6 larger values that create a longer right-hand tail.

2.27 a. The place value of the leaves is in the hundredths place; i.e., 59|7 is 5.97.

b. 16

c. 5.97, 6.01, 6.04, 6.08

d. Cumulative frequencies starting at the top and the bottom until it reaches the class that contains the median. The number in parentheses is the frequency for just the median class.

SECTION 2.3 EXERCISES

Frequency distributions can be either grouped or ungrouped. Ungrouped frequency distributions have single data values as x values. Grouped frequency distributions have intervals of x values, therefore, use the class midpoints (class marks) as the x values.

Histograms can be used to show either type of distribution graphically. Frequency or relative frequency is on the vertical axis. Be sure the bars touch each other (unlike bar graphs). Increments and widths of bars should all be equal. A title should also be given to the histogram.   . . .

Computer or calculator commands to construct a histogram can be found in ES10pp61-63. Note the two methods, depending on the form of your data.

2.29 a.

x	f
0	2
1	5
2	3
3	0
4	2
	12

   b. f is frequency, therefore value of 1 occurred 5 times.
   c. 2 + 5 + 3 + 0 + 2 = 12
   d. The sum represents the sum of all the frequencies, which is the number of data, or the sample size.

2.31 a. Bargraph, the player's name is qualitative.

   b.
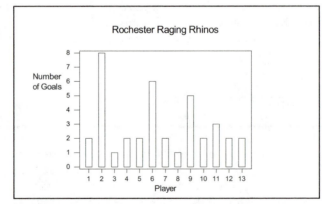

   c. Histogram, working only with numerical data, want to show sequence.

d.

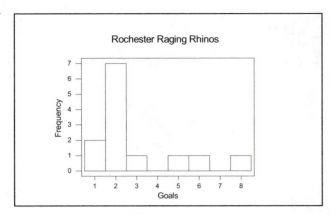

Rochester Raging Rhinos

2.33   a.

Height (in)	Frequency
64	2
65	4
66	3
67	3
68	2
69	2
71	2

b.

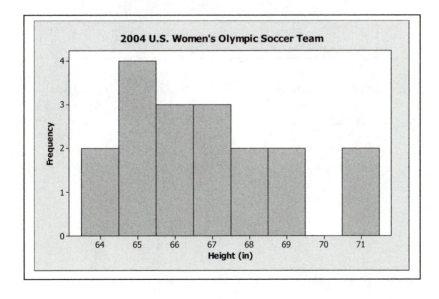

2004 U.S. Women's Olympic Soccer Team

c. Height
(in)	Rel. Freq.
64	0.111
65	0.222
66	0.167
67	0.167
68	0.111
69	0.111
71	0.111

d. 0.167 + 0.167 + 0.111 + 0.111 + 0.111 = 0.667 = 66.7%

2.35 a.

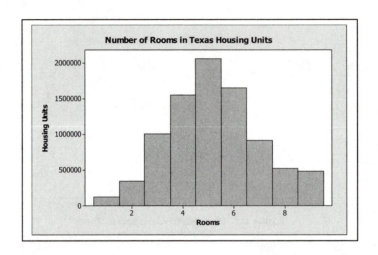

b. Mounded, truncated on right due to 9+ class

c. Centered on 5 rooms, there are 3 to 7 rooms for the most

2.37 a.

x	67	68	69	70	71	72	73	74	75	76	77	78	79	80	81	82	83
f	1	8	3	5	10	22	17	28	17	9	9	9	4	1	1	1	1

b.

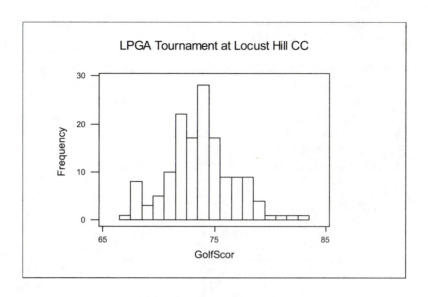

Parts of a grouped frequency distribution -

    class boundaries = the low and the high endpoints of the
                    interval

    class width = distance from any point in one class to the
               same position point in the next class or the
               difference between the upper and lower class
               boundaries

    class midpoint (mark) = (lower boundary + upper boundary)/2,
                       midpoint of the interval

Example: with respect to the second class interval

        30 - 40   form: $(40 \leq x < 50)$

        40 - 50   lower class boundary = 40

        50 - 60   upper class boundary = 50

                      class width = 50 - 40 = 10

                      class midpoint = (40 + 50)/2 = 45

2.39 a. 35-45

b. Values greater than or equal to 35 and also less than 45
   belong to the class 35-45.

c. Difference between upper and lower class boundaries.
   i. Subtracting the lower class boundary from the upper
      class boundary for any one class
   ii. Subtracting a lower class boundary from the next
      consecutive lower class boundary
   iii. Subtracting an upper class boundary from the next
      consecutive upper class boundary

d.

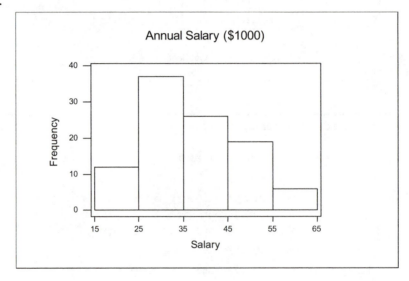

2.41 a. 12 and 16
b. 2, 6, 10, 14, 18, 22, 26
c. 4.0
d. 0.08, 0.16, 0.16, 0.40, 0.12, 0.06, 0.02

e.

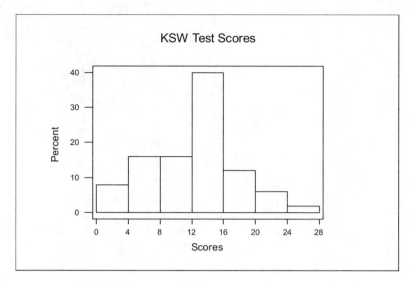

Refer to frequency distribution information before exercise 2.39 if necessary. Either class boundaries or class midpoints, may be used to determine increments along the horizontal axis for histograms of grouped frequency distributions.

2.43 a. <u>Class limits</u>   <u>frequency</u>   b. class width = <u>6</u>

Class limits	frequency
12 - 18	1
18 - 24	14
24 - 30	22
30 - 36	8
36 - 42	5
42 - 48	3
48 - 54	2

c. class midpoint = (24+30)/2
                  = <u>27</u>
   lower class
      boundary = <u>24</u>
   upper class
      boundary = <u>30</u>

d.

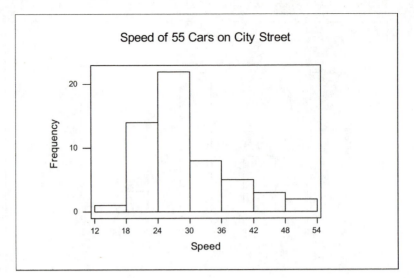

2.45 a.   Third Graders at Roth Elementary School

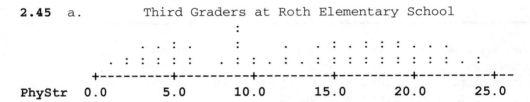

PhyStr   0.0   5.0   10.0   15.0   20.0   25.0

  b.

Class boundaries	frequency
1 – 4	6
4 – 7	10
7 – 10	7
10 – 13	6
13 – 16	8
16 – 19	11
19 – 22	10
22 – 25	6
	64

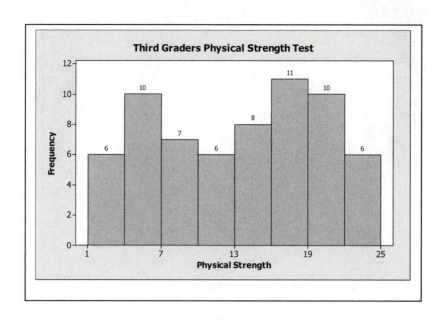

c. <u>Class boundaries</u>   <u>frequency</u>

```
 0 -  3          3
 3 -  6         10
 6 -  9          4
 9 - 12          9
12 - 15          7
15 - 18         11
18 - 21         11
21 - 24          7
24 - 27          2
                64
```

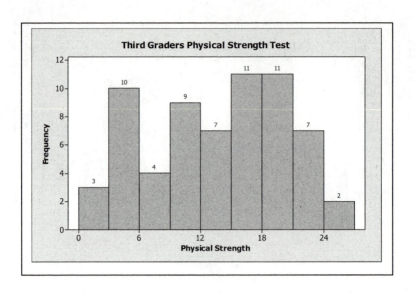

d.

Class boundaries	frequency
-2.5 - 2.5	3
2.5 - 7.5	13
7.5 - 12.5	13
12.5 - 17.5	15
17.5 - 22.5	17
22.5 - 27.5	3
	64

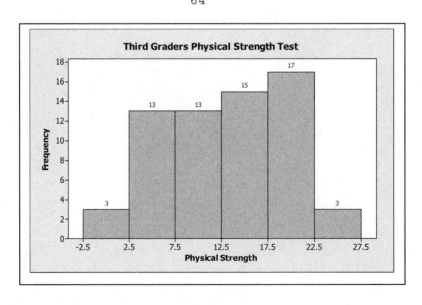

e. Answers will vary.

f. The histograms in parts b and c demonstrate a bimodal distribution whereas the distribution in part d is skewed left.

Dotplot shows mode to be 9, which is in the 7-10 class and a cluster centered around 17; while the histogram shows the two modal classes to be 4-7 and 16-22. The mode is not in either modal class.

g. Answers will vary but as the number of classes and the choice of class boundaries change, values will fall into various classes, thereby giving different appearances, all for the same set of data.

A guideline that can be used for selecting the number of classes is : n(classes) ≈ $\sqrt{n}$

2.47 a.

4 - 5	1
5 - 6	9
6 - 7	10
7 - 8	12
8 - 9	4

b. 1
c. 4.5, 5.5, 6.5, 7.5, 8.5,
d.

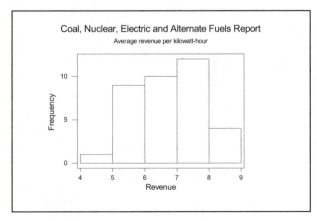

2.49 a. Symmetric: weight of dry cereal per box, breaking
        strength of certain type of string
   b. Uniform: result from rolling a die several hundred times
   c. Skewed Right: salaries, high school class sizes
   d. Skewed left: hour exam scores
   e. Bimodal: heights, weights for groups containing both male
     and female

An ogive is a line graph of a cumulative frequency or cumulative
relative frequency distribution. Start the line at zero for a
class below the smallest class. Plot the upper class boundary
points from the remaining values of the cumulative (relative)
frequency distribution. Connect all of the points with straight
line segments. The last point (class) is at the value of one
(vertically).

Computer and calculator commands to construct an ogive can be found
in ES10pp66&67.

2.51 a.

Class Boundaries	Cumulative Frequency
$15 \leq x < 25$	12
$25 \leq x < 35$	49
$35 \leq x < 45$	75
$45 \leq x < 55$	94
$55 \leq x \leq 65$	100

b.

Class Boundaries	Cum. Rel. Frequency
$15 \leq x < 25$	0.12
$25 \leq x < 35$	0.49
$35 \leq x < 45$	0.75
$45 \leq x < 55$	0.94
$55 \leq x \leq 65$	1.00

c.

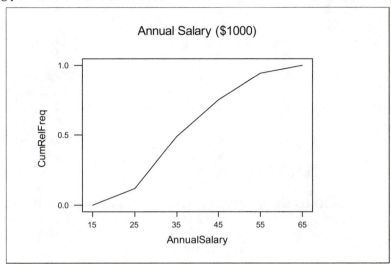

Annual Salary ($1000)

2.53 a.

Class limits	Cum.Rel.Freq.
0 - 4	0.08
4 - 8	0.24
8 - 12	0.40
12 - 16	0.80
16 - 20	0.92
20 - 24	0.98
24 - 28	1.00

(relative frequencies taken from ex. 2.41)

b.

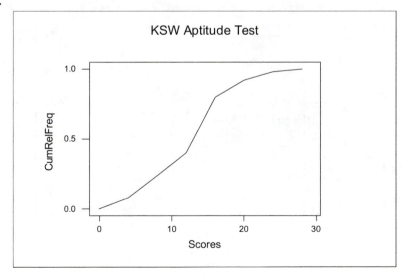

KSW Aptitude Test

2.55 a.

Class Boundar.	Class Midpoints	Freq.
-2.5 - 2.5	0	9
2.5 - 7.5	5	15
7.5 - 12.5	10	17
12.5 - 17.5	15	8
17.5 - 22.5	20	7
22.5 - 27.5	25	11
27.5 - 32.5	30	11
32.5 - 37.5	35	2
37.5 - 42.5	40	1
42.5 - 47.5	45	1

b&d.

Class Midpoints	Rel.Freq.	Cum. Rel. Freq.
0	0.110	0.110
5	0.183	0.293
10	0.207	0.500
15	0.098	0.598
20	0.085	0.683
25	0.134	0.817
30	0.134	0.951
35	0.024	0.975
40	0.012	0.987
45	0.012	0.999 • 1.000

c.

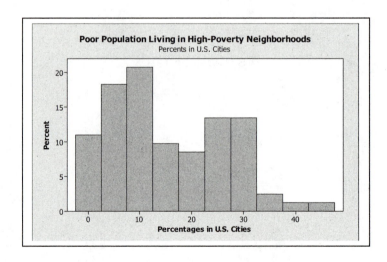

e.

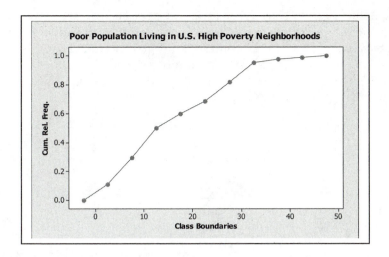

SECTION 2.4 EXERCISES

NOTE: A <u>measure of central tendency</u> is a value of the variable. It is that value which locates the "average" value for a set of data. The "average" value may indicate the "middle" or the "center" or the most popular data value.

NOTATION AND FORMULAS FOR MEASURES OF CENTRAL TENDENCY

$\sum x$ = sum of data values
n = # of data values in the sample
$\overline{x}$ = sample mean = $\sum x/n$
$\tilde{x}$ = sample median = middle data value
$d(\tilde{x})$ = depth or position of median = $(n + 1)/2$
mode = the data value that occurs most often
midrange = (highest value + lowest value)/2

NOTE: REMEMBER TO RANK THE DATA BEFORE FINDING THE MEDIAN.
   $d(\tilde{x})$ only gives the depth or position, not the value of the median. If n is even, $\tilde{x}$ is the average of the two middle values.

   See Introductory Concepts at the back of this manual for additional information about the Σ (summation) notation.

Computer and calculator commands to find the mean and median can be found in ES10-pp74&76 respectively.

2.57 The data resulting from a quantitative variable are numbers with which arithmetic (addition, subtraction, etc.) can be performed. The data resulting from a qualitative variable are 'category' type values such as color. It is not possible to add three colors together, and divide by 3, to obtain a value for the mean color.

2.59 a. 9   b. value = 0

2.61 a. $\overline{x} = \sum x/n = 123/36 = \underline{3.4}$
b. $\overline{x} = \sum x/n = 161/31 = \underline{5.2}$
c. $\overline{x} = \sum x/n = 252/43 = \underline{5.9}$
d. $\overline{x} = \sum x/n = 217/39 = \underline{5.6}$
e. $\overline{x} = \sum x/n = (123+161+252+217)/152 = 753/152 = \underline{4.95}$
        where $152 = (37+32+44+40) - 1 = 153 - 1$
f. $\overline{x} = \sum x/n = (3.4+5.2+5.9+5.6)/4 = 20.1/4 = \underline{5.025}$
g. Answers will vary. There are a different number of interchanges in each state, so the states do not weigh equally in finding the mean.

2.63 Ranked data: 4.15, 4.25, 4.25, 4.50, 4.60, 4.60, 4.75, 4.90
$d(\tilde{x}) = (n+1)/2 = (8+1)/2 = 4.5th; \quad \tilde{x} = (4.50+4.60)/2 = \underline{4.55}$

2.65 mode = $\underline{2}$

2.67 a. $\overline{x} = \sum x/n = (9+6+7+9+10+8)/6 = 49/6 = 8.166 = \underline{8.2}$
    Ranked data: 6, 7, 8, 9, 9, 10
    $d(\tilde{x}) = (n+1)/2 = (6+1)/2 = 3.5th; \quad \tilde{x} = \underline{8.5}$
    mode = $\underline{9}$
    midrange = $(L+H)/2 = (6+10)/2 = 16/2 = \underline{8.0}$
b. Answers will vary. All show centers.

2.69 a. $\overline{x} = \sum x/n = (3+5+6+7+7+8)/6 = 36/6 = \underline{6.0}$
b. $d(\tilde{x}) = (n+1)/2 = (6+1)/2 = 3.5th; \quad \tilde{x} = (6+7)/2 = \underline{6.5}$
c. mode = $\underline{7}$
d. midrange = $(H+L)/2 = (8+3)/2 = 11/2 = \underline{5.5}$

2.71   {28, 29, 33, 40, 41, 42, 44, 48, 48, 49}
a. $\bar{x} = \sum x/n = 402/10 = \underline{40.2}$
b. $d(\tilde{x}) = (n+1)/2 = (10+1)/2 = 5.5th; \tilde{x} = \underline{41.5}$
c. midrange = (H+L)/2 = (28+49)/2 = $\underline{38.5}$
d. mode = $\underline{48}$

2.73   a. $\bar{x} = \sum x/n = 2205.89/31 = 71.158 = \underline{71.16\%}$
b. $d(\tilde{x}) = (n+1)/2 = (31+1)/2 = 16th; \tilde{x} = \underline{72.66\%}$
c.

```
Stem-and-leaf of Percentage   N  = 31
Leaf Unit = 1.0


   2    5   89
   5    6   244
  10    6   57999
 (17)   7   00111223333444444
   4    7   578
   1    8   0
```

d.   The left-tail (or smaller value) tail causes the
     mean to be less in value than the median. The 2
     data, 58.60 and 59.25, are separate from the rest of
     the pack and have a reducing effect on the mean-
     value, but do not effect the value of the median.

2.75 a.   Third Graders at Roth Elementary School

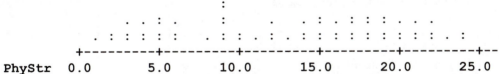

PhyStr 0.0   5.0   10.0   15.0   20.0   25.0

       b.   mode = $\underline{9}$

c.

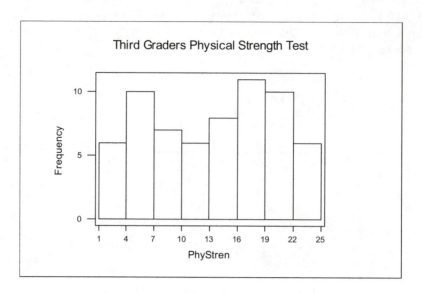

d. The distribution appears to be bimodal. Modal classes are 4-7, and 16-19.

e. Dotplot shows mode to be 9, which is in the 7-10 class; while the histogram shows the two modal classes to be 4-7 and 16-19. The mode is not in either modal class.

f. No. In an ungrouped distribution there is only one numerical value per class.

g. The mode is simply the single data value that occurs most often, while a modal class results from data tending to bunch-up forming a cluster of data values, not necessarily all of one value.

2.77 a. & b.

	Runs at Home	Runs Away	Difference
Mean	4.828	4.797	0.031
Median	4.870	4.860	0.01
Maximum	6.380	5.570	0.81
Minimum	3.630	3.430	0.20
Midrange	5.005	4.500	0.505

c. All five measures of central tendency are greater
   for runs scored at home than for the number of runs
   scored away. Conclusion, they score more runs at
   home.

d.

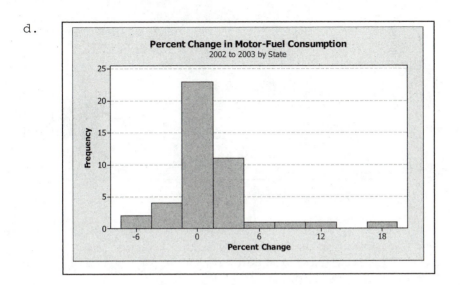

e. $\overline{x} = \sum x/n = 47.9/44 = 1.08864 = 1.089$

f. Part e is the mean of the 44 state values. It is not the
   mean for the whole country.

2.79 a. A quick look at the data listed suggests, Yes. The
   number of female licensed drivers is larger for most of
   the states states listed.

   b. Ratio M/F
         0.98520   0.95703   0.96727   1.01151   0.99234
      1.10043
         0.93043   1.01435   1.01596   0.98231   0.98466
      0.97655
         0.99562   1.03506   0.99592   0.98321   1.01944
      0.99244

   c. Near 1.0 means little difference.
      Greater than 1.0 means more male drivers.
      Less than 1.0 means more female drivers.

d.

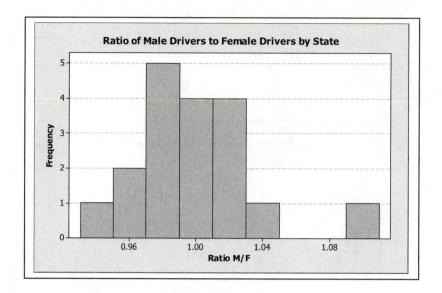

Ratio of Male Drivers to Female Drivers by State

e. The distribution of "Ratio M/F" is mounded and appears to be somewhat normal except for the one value that is considerably larger than the others, thus making the data skewed to the right.

f. $\bar{x} = \Sigma x/n = 17.9397/18 = 0.996651 = 0.997$

g. The value to the extreme right means that state has considerably more male drivers than female, approximately 10% more. The value to the extreme left means that state has fewer male drivers than female drivers and since the value is approximately 0.94, there are 6% fewer male drivers.

h. Answers will vary.

i.

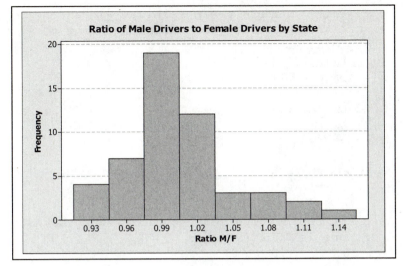

Ratio of Male Drivers to Female Drivers by State

$$\overline{x} \; = \; \Sigma x/n \; = \; = \; 51.1858/51 \; = \; 1.00364 \; = \; 1.004$$

j. The results are very similar.

k. Answers will vary.

2.81 a. 'Taxes per capita' means amount of taxes paid per person
          while 'percent of personal income' is the percent of
          personal income that is paid in taxes. A person in North
          Dakota pays a lesser amount of taxes per person but
          apparently they make so much less that the percentage of
          their income that goes for taxes is actually larger than
          in New Hampshire. $1283 is 4.8% of $26,729 while $1478
          is 4.4% of $33,590.

       b. The only 'average' that can be found is the midrange.
           Midrange = ($2748 + $1283)/2 = $2015.50

       c. Midrange = (9.6 + 4.4)/2 = 7.0%

       d. The only 'average' that is defined by only the extreme
           values of the variable is the midrange.

2.83 Many different answers are possible.

   a. Σx needs to be 500; therefore, need any three numbers that total 330.
     <u>100, 100, 130</u> [70, 100]
     $\bar{x} = \Sigma x/n = 500/5 = 100$, ck

   b. Need two numbers smaller than 70 and one larger.
     __, __, 70,___, 100: <u>50, 60, 80</u> [70, 100]
     $d(\tilde{x}) = (n+1)/2 = (5+1)/2 = 3rd$; $\tilde{x} = 70$, ck

   c. Need multiple 87's.   <u>87, 87, 87</u> [70, 100]
     mode = 87, ck

   d. Need any two numbers that total 140 for the extreme values where one is 100 or larger. _, _, _, 70, 100
     <u>40, 50, 60</u> [70, 100]
     midrange = (L+H)/2 = (40+100)/2 = 70, ck

   e. Need two numbers smaller than 70 and one larger than 70 so that their total is 330. _, _, 70, _, 100;
     <u>60, 60, 210</u> [70, 100]
     $\bar{x} = \Sigma x/n = 500/5 = 100$, ck
     $d(\tilde{x}) = (n+1)/2 = (5+1)/2 = 3rd$; $\tilde{x} = 70$, ck

   f. Need two numbers of 87 and a third number large enough so that the total of all five is 500.
     <u>87, 87, 156</u> [70, 100]
     $\bar{x} = \Sigma x/n = 500/5 = 100$, ck; mode = 87, ck

   g. Mean equal to 100 requires the five data to total 500 and the midrange of 70 requires the total of L and H to be 140;   40, _, 70, _, 100; that is a sum of 210, meaning the other two data must total 290. One of the last two numbers must be larger than 145, which would then become H and change the midrange. Impossible.

   h. There must be two 87's in order to have a mode, and there can only be two data larger than 70 in order for 70 to be the median. _, 70, 87, 87, 100; Impossible

SECTION 2.5 EXERCISES

NOTE: A <u>measure of dispersion</u> is a value of the variable. It is that value which describes the amount of variation or spread in a data set. A small measure of dispersion indicates data that are closely grouped, whereas, a large value indicates data that are more widely spread.

MEASURES OF DISPERSION - THE SPREAD OF THE DATA

<u>Range</u> = highest value - lowest value

<u>Standard Deviation</u> - s - the average distance a data value is from the mean

$$s = \sqrt{\Sigma (x - \overline{x})^2 / (n - 1)}$$

Variance - s^2 - the square of the standard deviation
(i.e., before taking the square root)

For exercises 2.89-2.93, be sure that the $\Sigma (x - \overline{x}) = 0$.

NOTE: Standard deviation and/or variance cannot be negative. This would indicate an error in sums or calculations.

See Introductory Concepts at the end of this manual for additional information about Rounding Off.

Computer and calculator commands to find the range and standard deviation can be found in ES10-pp88&89.
If using a non-graphing statistical calculator (one that lets you input the data points) to find the standard deviation of a sample, use the $\sigma(n-1)$ or s_x key. $\sigma(n)$ or σ_x would give the population standard deviation; that is, divide by "n" instead of "n-1".

2.85 a. range = H - L = $2748 - $1283 = $1465

b. range = H - L = 9.6% - 4.4% = 5.2%

2.87 The mean is the 'balance point' or 'center of gravity' to all the data values. Since the weights of the data values on each side of $\overline{x}$ are equal, $\Sigma(x - \overline{x})$ will give a positive amount and an equal negative amount, thereby canceling each other out.

Algebraically: $\Sigma(x - \overline{x}) = \Sigma x - n\overline{x} = \Sigma x - n\cdot(\Sigma x/n) = \Sigma x - \Sigma x = 0$

2.89 a. 1st: find mean, $\overline{x} = \Sigma x/n = 25/5 = 5$

$\underline{x}$	$\underline{x - \overline{x}}$	$\underline{(x - \overline{x})^2}$
1	-4	16
3	-2	4
5	0	0
6	1	1
10	5	25
Σ 25	0	46

$s^2 = \Sigma(x-\overline{x})^2/(n-1)$

$= 46/4 = \underline{11.5}$

b.

$\underline{x}$	$\underline{x^2}$
1	1
3	9
5	25
6	36
10	100
25	171

$SS(x) = \Sigma x^2 - ((\Sigma x)^2/n)$

$= 171 - ((25)^2/5)$

$= 171 - 125 = 46$

$s^2 = SS(x)/(n-1) = 46/4 = \underline{11.5}$

c. Both results are the same.

2.91 a. range = H - L = 8 - 3 = $\underline{5}$

b. 1st: find mean, $\overline{x} = \Sigma x/n = 36/6 = 6$

$\underline{x}$	$\underline{x - \overline{x}}$	$\underline{(x - \overline{x})^2}$
3	-3	9
5	-1	1
6	0	0
7	1	1
7	1	1
8	2	4
Σ 36	0	16

$s^2 = \Sigma(x-\overline{x})^2/(n-1)$

$= 16/5 = \underline{3.2}$

c. $s = \sqrt{s^2} = \sqrt{3.2} = 1.789 = \underline{1.8}$

2.93 a. 1st: find mean, $\bar{x} = \Sigma x/n = 104/15 = 6.9$

x	$x - \bar{x}$	$(x - \bar{x})^2$
4	-2.9	8.41
5	-1.9	3.61
5	-1.9	3.61
6	-0.9	0.81
6	-0.9	0.81
6	-0.9	0.81
7	0.1	0.01
7	0.1	0.01
7	0.1	0.01
7	0.1	0.01
8	1.1	1.21
8	1.1	1.21
8	1.1	1.21
9	2.1	4.41
11	4.1	16.81
Σ 104	+0.5*	42.95

$s^2 = \Sigma(x-\bar{x})^2/(n-1)$

$= 42.95/14$

$= 3.0679$

$= \underline{3.1}$

*The 0.5 is due to the round-off error introduced by
using $\bar{x} = 6.9$ instead of 6.933333.

An <u>easier</u> formula for <u>s</u> - <u>sample standard deviation</u>

1. Calculate "the sum of squares for x", SS(x):
$$SS(x) = \Sigma x^2 - ((\Sigma x)^2/n)$$

2. $s = \sqrt{SS(x)/(n-1)}$

This formula eliminates the problem of accumulating round-off errors.

NOTE: SS(x) is formed from the "sum of squared deviations from the mean", $\Sigma(x - \bar{x})^2$. Σx^2 is the "sum of the squared x's". SS(x) $\neq$ Σx^2.

b.

x	x²
4	16
5	25
5	25
6	36
6	36
6	36
7	49
7	49
7	49
7	49
8	64
8	64
8	64
9	81
11	121
$\sum$ 104	764

$$SS(x) = \sum x^2 - ((\sum x)^2/n)$$

$$= 764 - ((104)^2/15)$$

$$= 764 - 721.0667 = 42.93333$$

$$s^2 = SS(x)/(n-1)$$

$$= 42.9333/14 = 3.0667 = \underline{3.1}$$

c. $s = \sqrt{s^2} = \sqrt{3.0667} = 1.751 = \underline{1.8}$

2.95 a. Original data: n = 6, $\sum x = 37,116$, $\sum x^2 = 229,710,344$
$SS(x) = \sum x^2 - ((\sum x)^2/n) = 229,710,344 - (37,116^2/6) = 110,768.0$

$s^2 = SS(x)/(n-1) = 110,768.0/5 = \underline{22,153.6}$

b. Smaller numbers: n = 6, $\sum x = 1,116$, $\sum x^2 = 318,344$

$SS(x) = \sum x^2 - ((\sum x)^2/n) = 318,344 - (1,116^2/6) = 110,768.0$

$s^2 = SS(x)/(n-1) = 110,768.0/5 = \underline{22,153.6}$

Both sets of data have the same variance.

2.97 a.

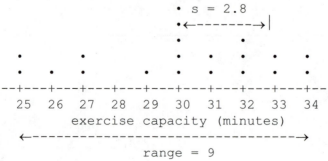

Police Recruits

range = 9

b. $\bar{x}$ = 601/20 = 30.05

c. range = H − L = 34 − 25 = <u>9</u>

d. n = 20, $\sum x$ = 601, $\sum x^2$ = 18,209
$SS(x) = \sum x^2 − ((\sum x)^2/n) = 18,209 − (601^2/20) = 148.95$
$s^2 = SS(x)/(n−1) = 148.95/19 = 7.83947 = \underline{7.8}$

e. $s = \sqrt{s^2} = \sqrt{7.83947} = 2.7999 = \underline{2.8}$

f. See graph in (a).

g. Except for the value x = 30, the distribution looks rectangular. Range is a little more than 3 standard deviations.

2.99 a. Range = H − L = 80.3 − 58.6 = 21.7;   s = 5.242
b.

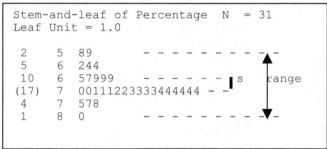

```
Stem-and-leaf of Percentage   N  = 31
Leaf Unit = 1.0

   2    5   89          - - - - - - - - -
   5    6   244
  10    6   57999       - - - - - -  s    range
 (17)   7   00111223333444444 - -
   4    7   578
   1    8   0           - - - - - - - -
```

c. The distribution of the data is skewed to the left. To accommodate the smaller values, the standard deviation is a bit high compared to the range. For a normal distribution, approximately 6 standard deviations equal the range.

2.101 The statement is incorrect. The standard deviation can never be negative. There has to be an error in the calculations or a typographical error in the statement.

2.103 Different answers will result depending on the relative size of the data making up the two sets will determine your answer.

<u>Larger</u>, if the data in one set are larger in value than the data values of the first set; the combined set is more dispersed.

Set I: {4,6,10,14,16}   and   Set II: {14,16,20,24,26}

Set I: $n = 5$, $\sum x = 50$, $\sum x^2 = 604$

$SS(x) = \sum x^2 - ((\sum x)^2/n) = 604 - (50^2/5) = 104$

$s^2 = SS(x)/(n-1) = 104/4 = 26$

$s = \sqrt{s^2} = \sqrt{26} = 5.099 = \underline{5.1}$

Set II: $n = 5$, $\sum x = 100$, $\sum x^2 = 2104$

$SS(x) = \sum x^2 - ((\sum x)^2/n) = 2104 - (100^2/5) = 104$

$s^2 = SS(x)/(n-1) = 104/4 = 26$

$s = \sqrt{s^2} = \sqrt{26} = 5.099 = \underline{5.1}$

Together, Set I and Set II;

$n = 10$, $\sum x = 150$, $\sum x^2 = 2708$

$SS(x) = \sum x^2 - ((\sum x)^2/n) = 2708 - (150^2/10) = 458$

$s^2 = SS(x)/(n-1) = 458/9 = 50.88888$

$s = \sqrt{s^2} = \sqrt{50.88888} = 7.133644 = \underline{7.13}$

SECTION 2.6 EXERCISES

NOTE: A <u>measure of position</u> is a value of the variable. It is that value which divides the set of data into two groups: those data smaller in value than the measure of position, and those larger in value than the measure of position.

To find any measure of position:

    1. Rank the data - <u>DATA MUST BE RANKED LOW TO HIGH</u>

    2. Determine the depth or position in two separate steps:
      a. Calculate $nk/100$, where n = sample size,
                              k = desired percentile
      b. Determine $d(P_k)$:
           If $nk/100$ = integer $\Rightarrow$ add .5
                  (value will be halfway between 2 integers)
           If $nk/100$ = decimal $\Rightarrow$ round up to the nearest
                                Whole number

    3. Locate the value of P_k

REMEMBER:
Q_1 = P_{25} = 1st quartile - 25% of the data lies below this value
Q_2 = P_{50} = $\tilde{x}$ = 2nd quartile - 50% of the data lies below this
                                   value
Q_3 = P_{75} = 3rd quartile - 75% of the data lies below this value

---

2.105 a. 91 is in the 44th position from the Low value of 39
        91 is in the 7th position from the High value of 98

    b. $nk/100$ = (50)(20)/100 = 10.0;
        therefore $d(P_{20})$ = 10.5th from L
    P_{20} = (64+64)/2 = <u>64</u>

    $nk/100$ = (50)(35)/100 = 17.5;
        therefore $d(P_{35})$ = 18th from L
    P_{35} = <u>70</u>

c. $nk/100 = (50)(20)/100 = 10.0;$
$\qquad$ therefore $d(P_{80}) = 10.5$th from H
$\qquad P_{80} = (88+89)/2 = \underline{88.5}$

$nk/100 = (50)(5)/100 = 2.5;$
$\qquad$ therefore $d(P_{95}) = 3$rd from H
$\qquad P_{95} = \underline{95}$

2.107 a.

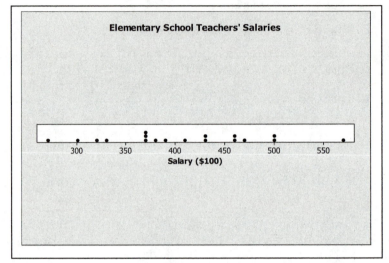

Ranked data:

269	295	317	326	367	367	371	376	391	413
433	434	455	458	471	495	501	574		

b. 2nd from the L, 17th from the H

c. $nk/100 = (18)(25)/100 = 4.5;$
$\qquad$ therefore $d(Q_1) = 5^{th}$ from L
$\qquad Q_1 = \underline{367 = \$36,700}$

d. $nk/100 = (18)(75)/100 = 13.5;$
$\qquad$ therefore $d(Q_3) = 14^{th}$ from L
$\qquad Q_3 = \underline{458 = \$45,800}$

2.109 Ranked data:

2.6   2.7   3.4   3.6   3.7   3.9   4.0   4.4   4.8   4.8
4.8   5.0   5.1   5.6   5.6   5.6   5.8   6.8   7.0   7.0

a. $nk/100 = (20)(25)/100 = 5.0$; therefore $d(P_{25}) = 5.5$th
$Q_1 = P_{25} = (3.7 + 3.9)/2 = \underline{3.8}$

$nk/100 = (20)(75)/100 = 15.0$; therefore $d(P_{75}) = 15.5$th
$Q_3 = P_{75} = (5.6 + 5.6)/2 = \underline{5.6}$

b. midquartile $= (Q_1 + Q_3)/2 = (3.8 + 5.6)/2 = \underline{4.7}$

c. $nk/100 = (20)(15)/100 = 3.0$; therefore $d(P_{15}) = 3.5$th
$P_{15} = (3.4+3.6)/2 = \underline{3.5}$

$nk/100 = (20)(33)/100 = 6.6$; therefore $d(P_{33}) = 7$th
$P_{33} = \underline{4.0}$

$nk/100 = (20)(90)/100 = 18.0$; therefore $d(P_{90}) = 18.5$th
$P_{90} = (6.8+7.0)/2 = \underline{6.9}$

Box-and-whisker displays may be drawn horizontal or vertical.
The Student Suite CD contains the Excel macro, Data Analysis Plus,
for constructing box-and-whisker displays.

2.111

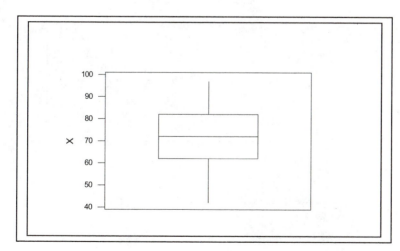

2.113 a.

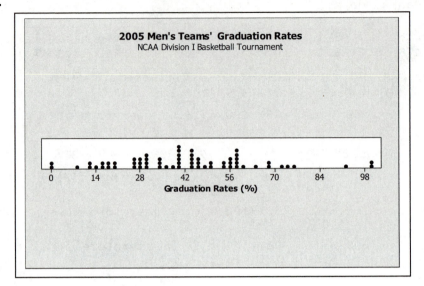

b.

```
Stem-and-leaf of Graduation Rates (%)   N  = 64
Leaf Unit = 1.0


  3    0    008
 11    1    11455779
 20    2    055577799
 27    3    0033368
(15)    4    000000334445557
 22    5    003455577888
 10    6    0477
  6    7    135
  3    8
  3    9    2
  2   10    00
```

c. 5-number summary: 0, 27, 40, 55, 100

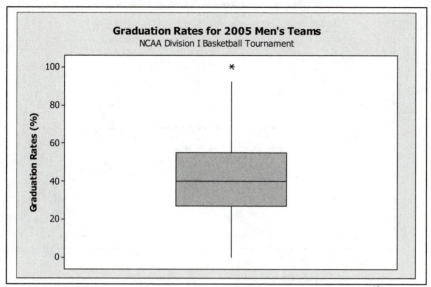

d. nk/100 = (64)(5)/100 = 3.2; therefore $d(P_5) = 4^{th}$
   P_5 =   11

   nk/100 = (64)(95)/100 = 60.8; therefore $d(P_{95}) = 61^{st}$
   P_{95} =   75

e. Skewed to the right, centered around 40% graduation rate. The
   two 100% values are distinctly separate from the other
   values.

f. The 92% and the two 100% are quite different from the rest.
   The next closest rate is a 75% graduation rate.

2.115 a.

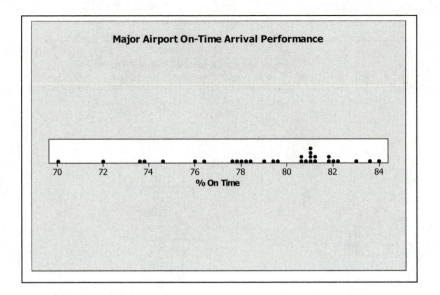

b.

```
Stem-and-leaf of % On Time   N  = 31
Leaf Unit = 0.10


   1    69   9
   1    70
   2    71   9
   2    72
   4    73   58
   5    74   6
   5    75
   7    76   04
  10    77   689
  13    78   239
  15    79   45
  (5)   80   55899
  11    81   0012779
   4    82   29
   2    83   69
```

c. 5-number summary: 69.96, 77.62, 80.50, 81.20, 83.93

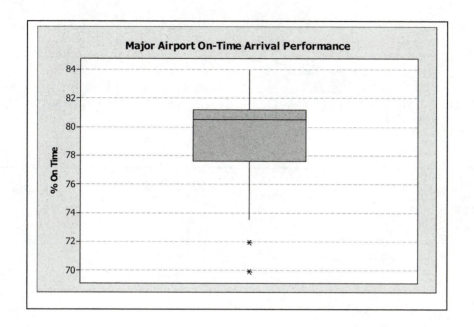

Major Airport On-Time Arrival Performance

Ranked Data:

69.96	71.96	73.55	73.85	74.60	76.00	76.45	77.62
77.82	77.94	78.26	78.38	78.92	79.42	79.50	80.50
80.51	80.85	80.90	80.91	81.04	81.08	81.10	81.20
81.71	81.73	81.90	82.28	82.99	83.60	83.93	

d. $nk/100 = (31)(10)/100 = 3.1$; therefore $d(P_{10}) = 4^{th}$
   $P_{10} = 73.85$

   $nk/100 = (31)(20)/100 = 6.2$; therefore $d(P_{20}) = 7^{th}$
   $P_{20} = 76.45$

e. The distribution is skewed to the left with the 2 airports
   having the lowest rate of on-time performance being quite
   separated for the others.

f. Travelers more interested in being on time and want the
   best on-time performance rate.

g. The airports with the lowest percentages are quite
   different from the rest of the airports. The two airports
   are EWR (Newark, NJ) and ORD (Chicago-O'Hare).

2.117 The distribution needs to be symmetric

---

z is a measure of position. It gives the number of standard deviations a piece of data is from the mean. It will be <u>positive</u> if x is to the <u>right of the mean</u> (larger than the mean) and <u>negative</u>, if x is to the <u>left of the mean</u> (smaller than the mean). Keep 2 decimal places. (hundredths)

$$z = (x - \text{mean})/\text{st. dev.} \qquad z = (x - \bar{x})/s$$

2.119 $z = (x - \text{mean})/\text{st.dev.}$

for x = 92,   z = (92 - 72)/12 = <u>1.67</u>
for x = 63,   z = (63 - 72)/12 = <u>-0.75</u>

2.121 $z = (x - \text{mean})/\text{st.dev.}$

a. for x = 54,   z = (54 - 74.2)/11.5 = <u>-1.76</u>
b. for x = 68,   z = (68 - 74.2)/11.5 = <u>-0.54</u>
c. for x = 79,   z = (79 - 74.2)/11.5 = <u>0.42</u>
d. for x = 93,   z = (93 - 74.2)/11.5 = <u>1.63</u>

2.123 If $z = (x - \text{mean})/\text{st.dev}$; then $x = (z)(\text{st.dev}) + \text{mean}$

a. for z = 0.0,   x = (0.0)(20.0) + 120 = <u>120</u>
b. for z = 1.2,   x = (1.2)(20.0) + 120 = <u>144.0</u>
c. for z = -1.4,   x = (-1.4)(20.0) + 120 = <u>92.0</u>
d. for z = 2.05,   x = (2.05)(20.0) + 120 = <u>161.0</u>

2.125 a. Ranked data:

0.03	0.05	0.05	0.06	0.07	0.10	0.13	0.14	0.14	0.14
0.14	0.14	0.14	0.16	0.16	0.16	0.16	0.17	0.17	0.17
0.19	0.20	0.20	0.20	0.20	0.20	0.21	0.21	0.21	0.22
0.22	0.23	0.23	0.24	0.25	0.29	0.29	0.30	0.30	0.31
0.31	0.32	0.32	0.34	0.35	0.36	0.37	0.39	0.39	0.55

b.

Lowest Value	First Quartile	Median	Third Quartile	Highest Value
0.03	0.14	0.20	0.30	0.55

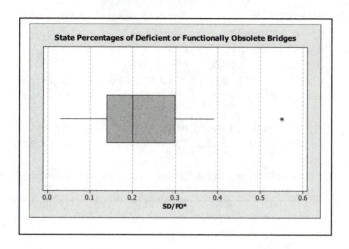

c. Midquartile = (0.14 + 0.30) / 2 = 0.22
   Interquartile range = 0.30 - 0.14 = 0.16

d.

	California	Hawaii	Nebraska	Oklahoma	Rhode Is.
z-score	-0.75	1.66	-1.42	0.22	3.20

2.127 for A: z = (85 - 72)/8 = 1.625
       for B: z = (93 - 87)/5 = 1.2
       Therefore, A has the higher relative position.

The empirical rule applies to a normal distribution.
    Approximately 68% of the data lies within 1 standard
    deviation of the mean.
    Approximately 95% of the data lies within 2 standard
    deviations of the mean.
    Approximately 99.7% of the data lies within 3 standard
    deviations of the mean.

Chebyshev's theorem applies to any shape distribution.
    At least 75% of the data lies within 2 standard deviations
    of the mean.
    At least 89% of the data lies within 3 standard deviations
    of the mean.

2.129 From 175 through 225 words, inclusive.

2.131 Nearly all of the data, 99.7%, lies within 3 standard
deviations of the mean.

2.133 a. 97.6 is 2 standard deviations above the mean
    $\{z = (97.6-84.0)/6.8 = 2.0\}$, therefore 2.5% of the time
    more than 97.6 hours will be required.

    b. 95% of the time the time to complete will fall within 2
    standard deviations of the mean, that is $84.0 \pm 2(6.8)$ or
    from 70.4 to 97.6 hours.

2.135 a. 50%   b. 0.50 – 0.34 = 0.16 = 16%
    c. 0.50 + 0.34 = 0.84 = 84%   d. 0.34 + 0.475 = 0.815 =
    81.5%

2.137 a. at least 75%   b. at least 89%

2.139 a. at most 11%   b. at most 6.25%

2.141 a.

class limits	x	f
1 - 4	2.5	6
4 - 7	5.5	9
7 - 10	8.5	8
10 - 13	11.5	10
13 - 16	14.5	6
16 - 19	17.5	4
19 - 22	20.5	4
22 - 25	23.5	2
25 - 28	26.5	1
	Σ	50

b. $\bar{x} = \Sigma x/n = 560/50 = \underline{11.2}$

$SS(x) = \Sigma x^2 - ((\Sigma x)^2/n) = 8184.5 - (560^2/50) = 1912.5$

$s = \sqrt{SS(x)/(n-1)} = \sqrt{1912.5/49} = \sqrt{39.0306} = 6.247 = \underline{6.2}$

c. $\bar{x} \pm 2s = 11.2 \pm 2(6.2) = 11.2 \pm 12.4$ or $\underline{-1.2}$ to $\underline{23.6}$
   96% of the data (48/50) is between -1.2 and 23.6.

2.143 a.

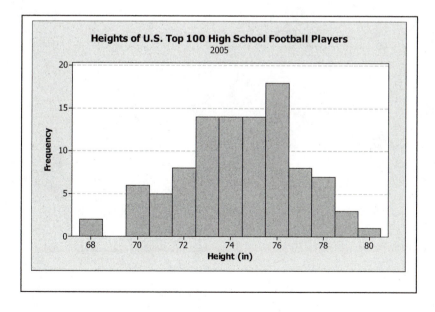

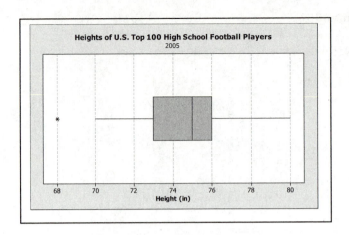

Heights of U.S. Top 100 High School Football Players
2005

Height (in)

b. $\bar{x} = \sum x/n = 7442/100 = 74.42 = \underline{74.4}$

$SS(x) = \sum x^2 - ((\sum x)^2/n) = 554456 - (7442^2/100) = 622.36$

$s = \sqrt{SS(x)/(n-1)} = \sqrt{622.36/99} = \sqrt{6.2864646}$
$= 2.507285 = \underline{2.5}$

c.
68	68	70	70	70	70	70	70	71	71	71	71	71
72	72	72	72	72	72	72	72	73	73	73	73	73
73	73	73	73	73	73	73	73	73	74	74	74	74
74	74	74	74	74	74	74	74	74	74	75	75	75
75	75	75	75	75	75	75	75	75	75	75	76	76
76	76	76	76	76	76	76	76	76	76	76	76	76
76	76	76	77	77	77	77	77	77	77	77	78	78
78	78	78	78	78	79	79	79	80				

d. $\bar{x} \pm 1s = 74.4 \pm (2.5)$ or $\underline{71.9}$ to $\underline{76.9}$
68% of the data (68/100) is between 71.9 and 76.9.

$\bar{x} \pm 2s = 74.4 \pm 2(2.5) = 74.4 \pm 5$ or $\underline{69.4}$ to $\underline{79.4}$
97% of the data (97/100) is between 69.4 and 79.4.

$\bar{x} \pm 3s = 74.4 \pm 3(2.5) = 74.4 \pm 7.5$ or $\underline{66.9}$ to $\underline{81.9}$
100% of the data (100/100) is between 66.9 and 81.9.

e. The empirical rule says approximately 68%, 95%, and 99.7% of the data are within one, two, and three standard deviations, respectively; the 68%, 97% and 100% do somewhat agree with the rule; based solely on this information, the distribution can be considered 'approximately' normal.

f. Chebyshev's theorem says at least 75%, and 89%, of the data are within two, and three standard deviations, respectively; 97%, and 100% both satisfy the theorem.

g. The graphs indicate a skewed right distribution, and therefore not normal. The boxplot shows an outlier to left and the right side of the box is much shorter than the left side indicating skewness; the histogram shows one value to the far left and modal class is to right within the center cluster, thus both graphs show a slight skewness to the left that the empirical rule alone cannot detect.

Helpful hint for use when expecting to count data on histogram:

Minitab: While on the Histogram dialogue box,
                    Select: **Labels > Data Labels...**
                    Select: **Use y-value labels**

This will direct the computer to print the frequency of each class above its corresponding bar.

Excel: Returns a frequency distribution with the histogram.
TI-83/84 Plus: Use the TRACE and arrow keys.

2.145 a.   Answers will vary.

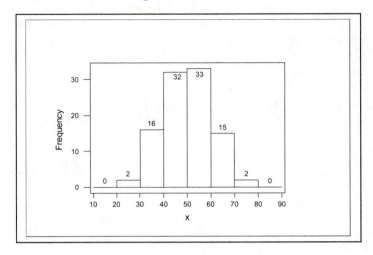

Within one standard deviation, 40 to 60, is 33 + 32 or 65 of the 100 data.   65%

Within two standard deviations, 30 to 70, is 16+32+33+15 or 96 of the 100 data.   96%

Within three standard deviations, 20 to 80, is all 100 of the data, or 100%.

The above results are extremely close to what the empirical rule claims will happen.

b, c, d.   Not all sets of 100 data will result in percentages this close.   However, expect very similar results to occur most of the time.

SECTION 2.8 EXERCISES

2.147 Yes, if all 8 employees earned $300 each, the mean would be $405.56 and if all 8 employees earned $350 each, the mean would be $450. $430 falls within this interval.

Or, if the mean of 9 employees is 430, then the total is 3870.   The 8 employees then would need to make 2620 (3870-1250) and their average earnings would be 2620/8 or 327.50, which is within the interval.

2.149 a. Here's a few of the more obvious ones. If you research
and find the original information, you'll find more.
1. The ranking information covers an 11-year period
while the tuition information covers 35 years. Yet
they are shown horizontally as being the same.
2. The units for tuition (share of median income) and
ranking (rank number) are totally different, yet
the vertical axis treats them as having common
units.
3. The ranking graph is placed below the tuition graph
creating the impression that cost exceeds quality.
Since the vertical scale is meaningless, either
line could have been ''on top''.
4. The sharp "drop" in the ranking graph actually
represents an improved ranking. A ranking of
''15th best'' is not better than a ranking of ''6th
best'', however the vertical scale used makes it
look like it is.

b.
1. The caption under the graph suggests that Cornell's
rank has been erratic by varying from 6th to 15th on
the national ranking over the 12 years reported.
With the hundreds of colleges and universities that
exist, to consistently hold a rank like this is
quite good.

2. The "upside-down" scale with the best ranking at
the bottom is totally misleading.

2.151

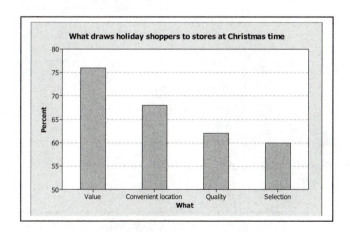

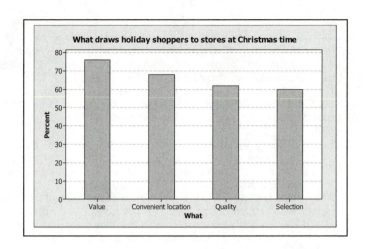

Answers will vary but the 50 to 80 scale gives a misleading
conclusion, namely that there is a significant difference
among the responses. Starting the percent scale at 0, gives
the true perspective of the relationship among the percents.

SECTION 2.9 EXERCISES

MEAN AND STANDARD DEVIATION OF FREQUENCY DISTRIBUTIONS

Mean - $\overline{x} = \sum xf / \sum f$

Standard Deviation - s

    1. $SS(x) = \sum x^2 f - ((\sum xf)^2 / \sum f)$

    2. $s^2 = SS(x) / (\sum f - 1)$

    3. $s = \sqrt{s^2}$

NOTE: in grouped frequency distributions, the calculated statistics
are approximations.   . . .

There are no* "mean of grouped data" nor "standard deviation of grouped data" computer or calculator commands. Therefore, the computer and calculator commands needed to form the xf and x^2f columns and to work through the formulas to calculate the mean and standard deviation are shown in ES10pp120&121.
* Grouped data techniques are for "hand" calculations. Usually a computer has all of the data and has no problem working with large sets of data.

2.153 a.

x	f	xf	x^2f
0	1	0	0
1	3	3	3
2	8	16	32
3	5	15	45
4	3	12	48
Σ	20	46	128

b. $\Sigma f = 20$; $\Sigma xf = 46$;

$\Sigma x^2f = 128$

c. x = 4; 4 is one of the possible data values
f = 8; 8 is the number of times an 'x' value occurred
Σf : sum of the frequencies = sample size
Σxf : sum of the products formed by multiplying a data
value by its frequency; the sum of the data

d. (i) Sum of the x-column has no meaning unless each value occurred only once.
(ii) Each data value is multiplied by how many times it occurred. Summing these products will give the same sum if all data values were listed individually.
Note: x   f   xf
        3   5   15   xf = 15   or   3+3+3+3+3=15

2.155

x	f	xf	x^2f
1	9	9	9
2	11	22	44
3	23	69	207
4	16	64	256
5	21	105	525
Σ	80	269	1041

a. $\bar{x} = \Sigma xf / \Sigma f$
   $= 269/80 = 3.3625 = \underline{3.4}$

b. $d(\tilde{x}) = (n+1)/2 = (80+1)/2 = 40.5$th; $\tilde{x} = \underline{3}$

c. $SS(x) = \sum x^2 f - ((\sum xf)^2/\sum f)$
$= 1041 - (269^2/80) = 136.4875$

$s^2 = SS(x)/(n-1) = 136.4875/79 = \underline{1.7277}$

$s = \sqrt{s^2} = \sqrt{1.7277} = 1.31442 = \underline{1.31}$

2.157

x	f	xf	$x^2 f$
12.5	2	25.0	312.5
12.7	6	76.2	967.74
13.0	22	286.0	3718.00
13.1	29	379.9	4976.69
13.2	12	158.4	2090.88
13.8	4	55.2	761.76
$\sum$	75	980.7	12,827.57

a. $\overline{x} = \sum xf/\sum f$
$= 980.7/75 = 13.076$
$= \underline{13.1}$

b. $SS(x) = \sum x^2 f - ((\sum xf)^2/\sum f)$
$= 12827.57 - (980.7^2/75) = 3.9368$

$s^2 = SS(x)/(\sum f-1) = 3.9368/74 = \underline{0.0532}$

c. $s = \sqrt{s^2} = \sqrt{0.0532} = 0.23065 = \underline{0.23}$

In exercises 2.159 through 2.163, the calculated means, variances, and standard deviations of the grouped frequency distributions will be approximations. This is due to the use of the class midpoints versus the actual data values.

For example, suppose the class limits for a particular class are 0-6, and that 5 data values fall in that class interval. The class midpoint of 3 would be used in the calculations, thereby treating all 5 data values as 3s, when they each could be any numbers from 0 through 6 (even all 0's).

2.159

Class limits	x	f	xf	x²f
3 - 6	4.5	2	9	40.50
6 - 9	7.5	10	75	562.50
9 - 12	10.5	12	126	1323.00
12 - 15	13.5	9	121.5	1640.25
15 - 18	16.5	7	115.5	1905.75
	$\sum$	40	447.0	5472.00

$\overline{x} = \sum xf/\sum f = 447.0/40 = 11.175 = \underline{11.2}$

$SS(x) = \sum x^2 f - ((\sum xf)^2/\sum f) = 5472 - (447^2/40) = 476.775$

$s^2 = 476.775/39 = 12.225 = \underline{12.2}$

$s = \sqrt{s^2} = \sqrt{12.225} = 3.496 = \underline{3.5}$

2.161

Class limits	x	f	xf	x²f
15.95-15.98	15.965	2	31.930	509.762450
15.98-16.01	15.995	4	63.980	1023.360100
16.01-16.04	16.025	15	240.375	3852.009375
16.04-16.07	16.055	3	48.165	773.289075
16.07-16.10	16.085	1	16.085	258.727225
	$\sum$	25	400.535	6417.148225

$\overline{x} = \sum xf/\sum f = 400.535/25 = 16.0214 = \underline{16.021}$

$SS(x) = \sum x^2 f - ((\sum xf)^2/\sum f) = 6417.148225 - (400.535^2/25)$
$= 0.016776$

$s^2 = SS(x)/(\sum f-1) = 0.016776/24 = 0.000699$

$s = \sqrt{s^2} = \sqrt{0.000699} = 0.026438 = \underline{0.026}$

2.163

Class limits	x	f	xf	x²f
0 - 10	5	37	185	925
10 - 20	15	24	360	5400
20 - 30	25	38	950	23750
30 - 40	35	32	1120	39200
40 - 50	45	27	1215	54675
$\sum$		158	3830	123950

$$\overline{x} = \sum xf/\sum f = 3830/158 = 24.240506 = \underline{24.2}$$

$$SS(x) = \sum x^2f - ((\sum xf)^2/\sum f) = 123950 - (3830^2/158)$$
$$= 31108.86076$$

$$s^2 = SS(x)/(\sum f-1) = 31108.86076/157 = 198.1456099$$
$$s = \sqrt{s^2} = \sqrt{198.1456099} = 14.0764 = \underline{14.1}$$

2.165 a.

Class limits	x	f	xf	x²f
-5 - +5	0	4	0	0
5 - 15	10	12	120	1200
15 - 25	20	9	180	3600
25 - 35	30	13	390	11700
35 - 45	40	13	520	20800
45 - 55	50	12	600	30000
55 - 65	60	5	300	18000
65 - 75	70	4	280	19600
75 - 85	80	2	160	12800
85 - 95	90	1	90	8100
$\sum$		75	2640	125800

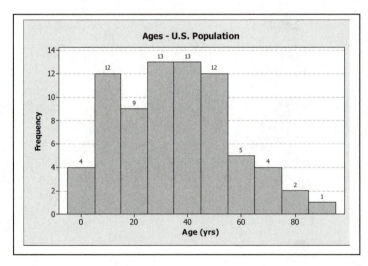

c. Graph appears skewed to the right.

d. $\overline{x} = \sum xf / \sum f = 2640/75 = \underline{35.2}$

e. $d(\tilde{x}) = (n+1)/2 = (75+1)/2 = 38$th; $\quad \tilde{x} = \underline{30}$

f. $R = H - L = 90 - 0 = 90$

g. $SS(x) = \sum x^2 f - ((\sum xf)^2 / \sum f) = 125800 - (2640^2/75)$
$\qquad = 32872$

$\quad s^2 = SS(x)/(\sum f - 1) = 32872/74 = 444.2162162$

$\quad s = \sqrt{s^2} = \sqrt{444.2162162} = 21.07643 = \underline{21.1}$

h.

	Ungrouped	Grouped	Percent Error
Mean	34.37	35.2	2.41%
Median	34	30	-11.76%
Range	90	90	0%
St. Deviation	20.95	21.08	0.62%

OTHER MEASURES OF CENTRAL TENDENCY FOR FREQUENCY DISTRIBUTIONS

NOTE: Data are already ranked.

Median - $\tilde{x}$ - find the depth and count down the frequency column until you include that position number. This is the median class. In an ungrouped frequency distribution, the median equals the x value of that class. In a grouped distribution, the data must be ranked in that particular class, then count to the appropriate position. If the original data are not given, use the class mark.

Mode - class midpoint of the highest frequency class

Modal Class - interval bounded by the class boundaries of the class with the highest frequency

Midrange = (highest value + lowest value)/2. If the original data is not given, use the lowest class boundary and the highest class boundary from the entire distribution, or the lowest and highest class midpoints.

2.167

Class limits	x	f	xf
0	0	44	0
1 - 2	1.5	31	46.5
3 - 4	3.5	19	66.5
5 - 6	5.5	14	77.0
7	7	11	77.0
8 or more	8.5	6	51.0
	Σ	125	318.0

a. $\Sigma f = 125$

b.

This distribution cannot (should not) be shown as a histogram since there are three different widths of classes within the distribution – 2 classes with width of 1, 3 classes with width of 2 and one that has an unspecified width. A bar chart would be the correct graph for this distribution.

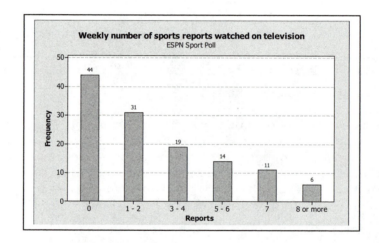

c. $\overline{x} = \Sigma xf / \Sigma f = 318/125 = 2.544 = 2.5$

d. $d(\tilde{x}) = (n+1)/2 = (125+1)/2 = 63^{rd}$; $\tilde{x} = \underline{1.5}$

e. mode = 0

2.169 a.   The first and last classes are not the same width as the
      others which are all 10 units wide. The first is
      'Younger than 5' therefore is 0 to 5 and 5 units wide.
      While the last class is '85 and older' and is of open
      width.

   b. The last class can be changed to '85 - 94 years' - the
      change should have little effect on the resulting
      statistics.
      The first class is not as easy since an interval 10 units
      wide includes numbers below zero, which do not apply to
      the variable. In part (c), use -5 - 5 as the class, In
      part (d) for the calculations, use class from 0 - 5.

   c.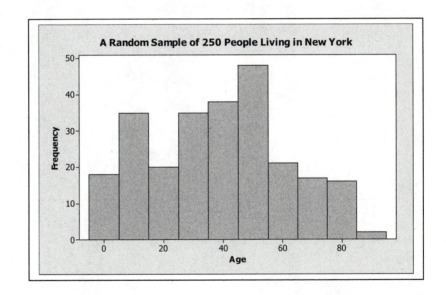

d.

Class limits	x	f	xf	x^2f
0 - 5	2.5	18	45	112.5
5 - 15	10	35	350	3500.0
15 - 25	20	20	400	8000.0
25 - 35	30	35	1050	31500.0
35 - 45	40	38	1520	60800.0
45 - 55	50	48	2400	120000.0
55 - 65	60	21	1260	75600.0
65 - 75	70	17	1190	83300.0
75 - 85	80	16	1280	102400.0
85 - 95	90	2	180	16200.0
Σ		250	9675	501412.5

$\bar{x} = \Sigma xf/\Sigma f = 9675/250 = \underline{38.}7$ years

e. $SS(x) = \Sigma x^2f - ((\Sigma xf)^2/\Sigma f) = 501412.5 - (9675^2/250)$
$= 126990.0$

$s^2 = SS(x)/(\Sigma f-1) = 126990.0/249 = 510.0$

$s = \sqrt{s^2} = \sqrt{510.0} = 22.583 = \underline{22.6\ years}$

CHAPTER EXERCISES

2.171 a.

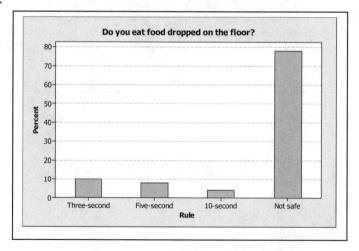

b. (300)(0.10) = 30 respond 'Three-second rule'
   (300)(0.08) = 24 respond 'Five-second rule'
   (300)(0.04) = 12 respond '10-second rule'
   (300)(0.78) = 234 respond 'Not safe'

2.173 a.

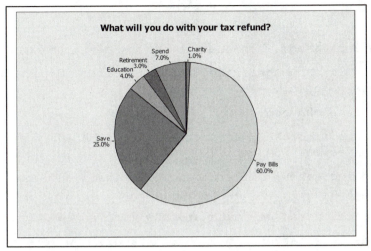

b. Answers may vary. Circle graph demonstrates the smaller percentages differences. It is easier to read.

2.175 a.

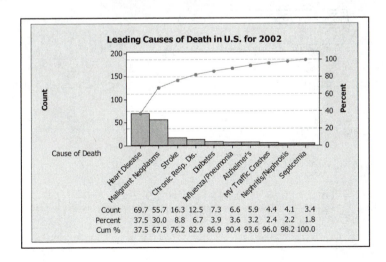

b. Answers will vary but the Pareto shows that the two leading causes of death, heart disease and malignant neoplasms account for nearly 70% of deaths.

2.177 a. numerical   b. attribute   c. numerical
       d. attribute   e. numerical

2.179 a. Mean increased; when one data increases, the sum increases.

   b. Median is unchanged; the median is affected only by the middle value(s).

   c. Mode is unchanged.

   d. Midrange increased; an increase in either extreme value increases the sum H+L.

   e. Range increased; difference between high and low values increased.

   f. Variance increased; data are now more spread out.

   g. Standard deviation increased; data are now more spread out.

2.181 Data summary: $n = 8$, $\sum x = 36.5$, $\sum x^2 = 179.11$

   a. $\bar{x} = \sum x/n = 36.5/8 = 4.5625 = \underline{4.56}$

   b. $s = \sqrt{[\sum x^2 - ((\sum x)^2/n)]/(n-1)}$
      $= \sqrt{[179.11 - (36.5^2/8)]/7}$
      $= \sqrt{1.79696} = 1.3405 = \underline{1.34}$

   c. These percentages seem to average very closely to 4%.

2.183 Data summary: $n = 118$, $\sum x = 2364$

   a. $\bar{x} = \sum x/n = 2364/118 = 20.034 = \underline{20.0}$

   b. $d(\tilde{x}) = (n+1)/2 = (118+1)/2 = 59.5\text{th}$;
      $\tilde{x} = (17+17)/2 = \underline{17}$

   c. mode $= \underline{16}$

d. $nk/100 = (118)(25)/100 = 29.5$; therefore $d(P_{25}) = 30th$
   $Q_1 = P_{25} = \underline{15}$

   $nk/100 = (118)(75)/100 = 88.5$; therefore $d(P_{75}) = 89th$
   $Q_3 = P_{75} = \underline{21}$

e. $nk/100 = (118)(10)/100 = 11.8$; therefore $d(P_{10}) = 12th$
   $P_{10} = \underline{14}$

   $nk/100 = (118)(95)/100 = 112.1$; therefore $d(P_{95}) = 113$
   $P_{95} = \underline{43}$

2.185 Data: 63 67 66 63 69   74 72 70 71 71
              72 70 75 85 84   85 85 86 94 91
              90 90 95 105 104

Data summary: $n = 25$, $\sum x = 1,997$, $\sum x^2 = 163,205$

$\bar{x} = \sum x/n = 1997/25 = 79.88 = \underline{79.9}$

$SS(x) = \sum x^2 - ((\sum x)^2/n)$
       $= 163,205 - (1,997^2/25) = 3684.64$

$s^2 = SS(x)/(n-1) = 3684.64/24 = 153.5267$

$s = \sqrt{s^2} = \sqrt{153.5267} = 12.3906 = \underline{12.4}$

2.187 a. The population is U.S. commercial airline industry; three
         variables are involved: number of reports, numbers of
         passengers, number of reports per 1000 airline passengers.
      b. Data; they are values of the variable, number of reports
         per 1000 airline passengers.
      c. Statistic; it summarizes the data for one month. It is
         used to estimate the parameter, the value for the whole
         population.
      d. No. The 19 airline values are a sample of the airline
         industry, not all are included here.

2.189 a. $\overline{x}$ = \$196,861   s = \$62,819

b. $\overline{x}$ - s = 196,861 - 62,819 = \$134,042   and

   $\overline{x}$ + s = 196,861 + 62,819 = \$259,680

c. 34, 34/50 = 0.68 = 68%

d. $\overline{x}$ - 2s = 196,861 - 2(62,819) = \$71,223   and

   $\overline{x}$ + 2s = 196,861 + 2(62,819) = \$322,499

e. 48, 48/50 = 0.96 = 96%

f. $\overline{x}$ - 3s = 196,861 - 3(62,819) = \$8,404   and
   $\overline{x}$ + 3s = 196,861 - 3(62,819) = \$385,318

g. 50/50 = 1.00 = 100%

h. They agree with Chebyshev's theorem, both percentages
   exceed the values cited.

i. 68%, 96% and 100% as a set are very close to the 68%,
   95% and 99.7% cited by the empirical rule. The
   results suggest the distribution might be
   approximately normal, but we need to look at an
   appropriate graph before making that decision.

j.

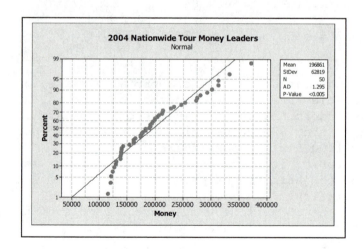

k. The normality test graph suggests that the distribution is not normal both by means of the points not following a straight line and the p-value being less than 0.05. By itself, the empirical rule is not enough to go by, one must also graph the distribution and/or use a standardized normality test. You might also construct a histogram.

2.191 Data summary: $n = 100$, $\sum x = 1315$

a. $\bar{x} = \sum x/n = 1315/100 = \underline{13.15}$
b. $d(\tilde{x}) = (n+1)/2 = 50.5\text{th}$; $\tilde{x} = \underline{13.85}$
c. mode = $\underline{15.0}$
d. midrange = $(H+L)/2 = (15.8+10.1)/2 = \underline{12.95}$
e. range = $H - L = 15.8 - 10.1 = \underline{5.7}$
f. $nk/100 = (100)(25)/100 = 25$; therefore $d(P_{25}) = 25.5\text{th}$
   $Q_1 = P_{25} = \underline{10.95}$

   $nk/100 = (100)(75)/100 = 75$; therefore $d(P_{75}) = 75.5\text{th}$
   $Q_3 = P_{75} = \underline{14.9}$
g. midquartile = $(Q_1 + Q_3)/2 = (10.95+14.9)/2 = \underline{12.925}$
h. $nk/100 = (100)(35)/100 = 35$; therefore $d(P_{35}) = 35.5\text{th}$
   $P_{35} = \underline{12.05}$

   $nk/100 = (100)(64)/100 = 64$; therefore $d(P_{64}) = 64.5\text{th}$
   $P_{64} = \underline{14.5}$

i.

Class limits	x	f	xf	x²f	rel.fr	cum.r
10.0 - 10.5	10.25	15	153.75	1575.94	0.15	0.15
10.5 - 11.0	10.75	10	107.50	1155.62	0.10	0.25
11.0 - 11.5	11.25	6	67.50	759.38	0.06	0.31
11.5 - 12.0	11.75	3	35.25	414.19	0.03	0.34
12.0 - 12.5	12.25	4	49.00	600.25	0.04	0.38
12.5 - 13.0	12.75	4	51.00	650.25	0.04	0.42
13.0 - 13.5	13.25	2	26.50	351.13	0.02	0.44
13.5 - 14.0	13.75	9	123.75	1701.56	0.09	0.53
14.0 - 14.5	14.25	12	171.00	2436.75	0.12	0.65
14.5 - 15.0	14.75	11	162.25	2393.19	0.11	0.76
15.0 - 15.5	15.25	23	350.75	5348.94	0.23	0.99
15.5 - 16.0	15.75	1	15.75	248.06	0.01	1.00
Σ		100	1314.00	17635.26		

j.

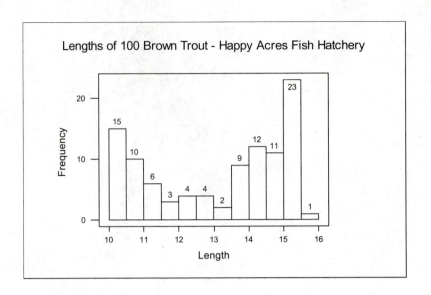

Lengths of 100 Brown Trout - Happy Acres Fish Hatchery

k. Shown in (i) above.

l.

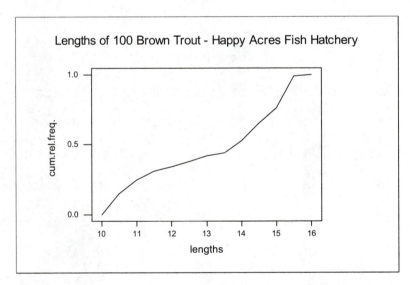

Lengths of 100 Brown Trout - Happy Acres Fish Hatchery

m. Summary: n = 100; Σxf = 1314, $\Sigma x^2 f$ = 17635.26

$\overline{x}$ = $\Sigma xf/\Sigma f$ = 1314/100 = <u>13.14</u>

n. SS(x)= $\Sigma x^2 f$ - $((\Sigma xf)^2/\Sigma f)$

= 17635.26 - $(1314^2/100)$ = 369.3

s^2 = SS(x)/(Σf-1) = 369.3/99 = 3.73030

$s = \sqrt{s^2} = \sqrt{3.73030}$ = 1.931399 = <u>1.93</u>

2.193 a. Answers will vary.

b. Answers will vary.

c. Answers will vary.

d. Sum of Area = 3022316

Sum of population = 279583437

Overall density = 279583437/3022316

= 92.506 = 93 people/sq. mile

Densities

5.04	6.13	9.09	9.80	14.95	15.49	18.08
22.16	26.30	32.68	35.28	38.38	41.26	45.04
49.35	50.34	51.99	58.52	59.62	63.35	74.79
78.00	80.29	86.17	86.43	92.10	95.52	100.06
129.19	132.82	134.68	139.05	152.70	167.55	170.72
173.43	213.44	220.20	270.90	272.92	275.41	380.94
382.78	554.79	679.89	768.94	863.52	1073.81	

e. Data summary: n = 48, Σx = 8503.88

$\overline{x}$ = $\Sigma x/n$ = 8503.88/48 = 177.164 = <u>177.2</u>

d($\tilde{x}$) = (n+1)/2 = 24.5th; $\tilde{x}$ = (86.17+86.43)/2 = <u>86.3</u>

no mode - no value repeats

midrange = (H+L)/2 = (1073.81+5.04)/2 = <u>539.425</u>

f.

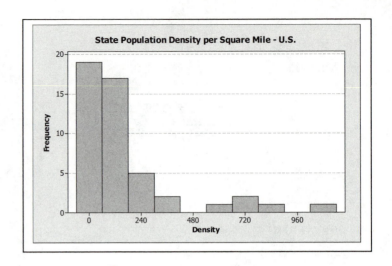

g. Highest: New Jersey, Rhode Island, Massachusetts, Connecticut, and Maryland

   Lowest: Wyoming, Montana, North Dakota, South Dakota, New Mexico

h. Answers will vary depending on answers in parts a – c.

2.195 a. Answers will vary.

b.

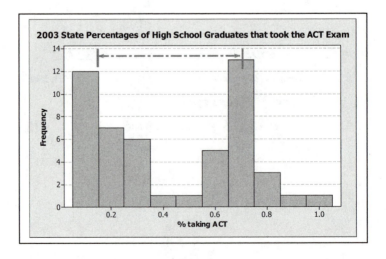

c. Distribution is bimodal. Large concentrations between 10%
   and 30% and then 60% and 80%.

d. Answers will vary.

e. $\overline{x}$ = 0.4308

f. Mean falls between the two high concentration areas. It
    is not representative of these data.

g. s = 0.2874

h. Percentage between:
    Looking at histogram: [7+6+1+1+5+(1/2)(13)]/50 = 53%
    Looking at data: 28/50 = 56%

i. The standard deviation is so large due to there being
    sofew data near the mean resulting in the data being
    quite wide spread. The lowest value is 0.05 and the
    highest is 1.0, that is from 5% to 100% with very few
    data between 0.35 and 0.55.

2.197 a.   Weight

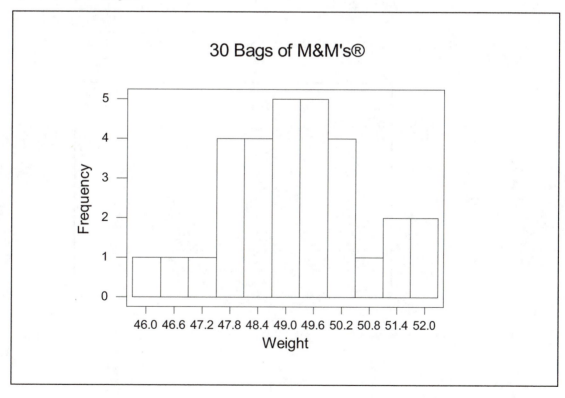

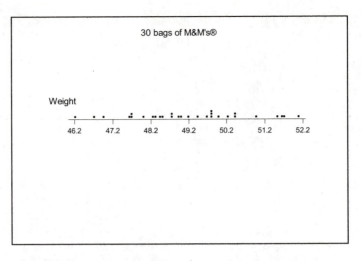

30 bags of M&M's®

b. $\bar{x}$ = 49.215, median = 49.07, s = 1.522,
       min = 46.22, max = 52.06

c. No, there does not seem to be any inconsistencies in the weight data.

d. Find number per bag: 58, 62, 59, etc.

e.

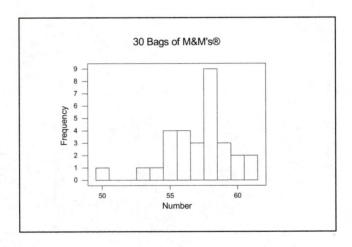

30 Bags of M&M's®

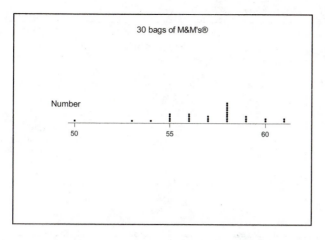

f. $\overline{x}$ = 57.1, median = 58, s = 2.383, min = 50, max = 61

g. One bag has "only 50" M&M's in it. This one value appears to be quite different from the rest of the values. Case 14 seems to have a total bag weight that is "typical" (see histogram below). However, its count number is approximately 10% smaller than the "typical" count (see histogram below) – this means that the individual M&M's would have to be 10% larger to make up the weight (see histogram below). Very suspicious!

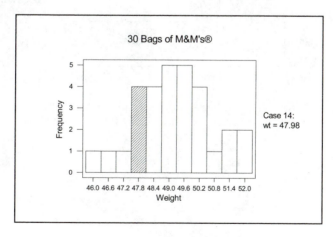

g. continued

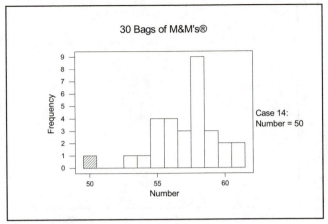

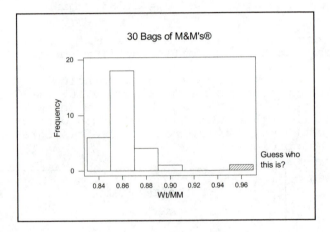

h. After bag 14 was weighed and before the M&M's were counted, someone ate a few of them, approximately 5 of them.

Draw a diagram of a normal curve with its corresponding percentages for standard deviations away from the mean (see Figure 2.32 in ES10-p107). Add the percentages from the left to the right, until the desired z-value is reached. The sum equals the percentile.

2.199 a. $0.50-0.20 = 0.30$; therefore, $z \approx -0.8$ or -0.9

b. $0.95-0.50 = 0.45$; therefore, $z \approx +1.6$ or $+1.7$

c.

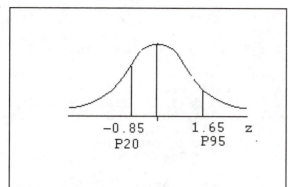

2.201 z-scores must be changed to percentiles in order to make the comparisons. Percentages are obtained from the empirical rule.

$z = 2$ corresponds to P_{97}

$z = 1$ corresponds to P_{84}

$z = -1$ corresponds to P_{16}

$z = 0$ corresponds to P_{50}

Therefore, Joan has the higher relative score for fitness, agility, and flexibility. Jean scored highest in posture, while they scored the same in strength.

2.203 Data summary: $n = 8$, $\sum x = 31,825$, $\sum x^2 = 126,894,839$

a. $\overline{x} = \sum x/n = 31,825/8 = \underline{3978.1}$

b. $SS(x) = \sum x^2 - ((\sum x)^2/n)$

$= 126,894,839 - (31,825^2/8) = 291,010.88$

$s^2 = SS(x)/(n-1) = 291,010.88/7 = 41,572.982$

$s = \sqrt{s^2} = \sqrt{41572.98} = \underline{203.9}$

c. $\bar{x} \pm 2s = 3978.1 \pm 2(203.9)$

$\quad = 3978.1 \pm 407.8 \quad$ or $\quad \underline{3570.3}$ to $\underline{4385.9}$

2.205 a.

Variable	N	Mean	StDev	Variance	Sum	Sum of Squares
Calories	40	**111.88**	**44.92**	2017.55	4475.00	579325.00
Sodium(mg)	40	**566.3**	**238.4**	56824.0	22650.0	15041700.0

b. Calories: $111.88 \pm 2(44.92)$ or between 22.04 and 201.72

Sodium: $566.3 \pm 2(238.4)$ or between 89.5 and 1043.1 mg

Only 2 of the brands of soups fall outside this calorie interval, so 95% are included within the interval. Only 1 brand falls outside the sodium content interval, so 98% are included within the interval.

Yes, Chebyshev's theorem is satisfied, both percentages are at least 75%.

c. Sodium: 566.3 ± 238.4 or between 327.9 and 804.7 mg

The empirical rule predicts that 68% of the brands of soups' sodium content will fall between 327.9 and 804.7, provided the distribution is normally distributed. In fact, these limits include 27 of the brands out of 40 or 67.5%. Therefore, these results suggest the sodium content of the soups does satisfy that part of the empirical rule.

2.207 a.

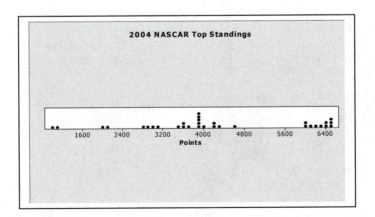

b.

Variable	N	Mean	StDev	Sum	Sum of Squares
Points Top 32	32	**4248**	**1624**	135948	659330342

c.

Variable	Minimum	Q1	Median	Q3	Maximum
Points Top 32	986	3170	3902	6058	6506

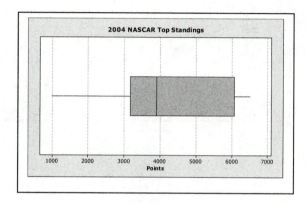

d. 4248 ± 2(1624) or between 1000 and 7496. 31 of the 32, or 96.95%, fall within this interval. This satisfies Chebyshes's requirement of at least 75%.

e. 4248 ± 1624 or between 2624 and 5872.   18 of the 32, or 56.3%, fall within this interval. This does not satify the empirical rules claim of 68%.

f. The percentages found, 56.3% and 96.95%, are not consistent with the empirical rule. The points distribution for the top 32 is part of a skewed left distribution. There is a ''wide'' cluster near the top and then the distribution tails out to the left. By using the top 32, much of the left tail has been chopped off. The distribution is not normal.

g.

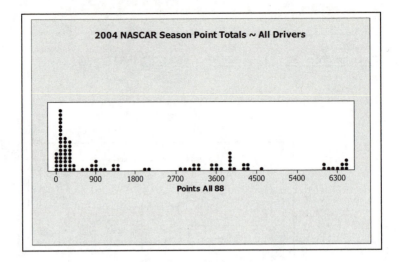

g(b).

					Sum of
Variable	N	Mean	StDev	Sum	Squares
Points All 88	88	1946	2145	171209	733484657

g(c).

Variable	Minimum	Q1	Median	Q3	Maximum
Points All 88	34.0	176	835	3679	6506

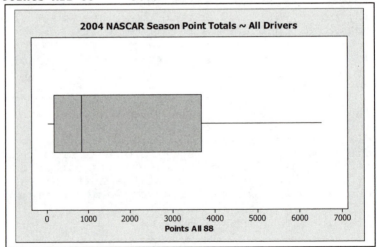

g(d). 1946 ± 2(2145) or between -2344 and 6236. Yes, 82 of the 88, or 93.2%, fall within this interval. This satisfies Chebyshes's requirement of at least 75%.

g(e). 1946 ± 2145 or between -199 and 4091. No, 73 of the 88, or 83.0%, fall within this interval. This does not satify the empirical rules claim of 68%.

g(f). The distribution is at least trimodal, there are at least three clusters of points n the dotpot that are distinctly separate from the others - there is a cluster below 400, one between 2700 and 4500, and one above 6000 points. This distribution is definitely different than the distribution of the top 32, although the 'top 32' are visable on the dotplot. The distribution is definitely not a normal or an approximately normal distribution.

2.209 There are many possible answers for this question; only one of those possibilities is shown.
  a. 70, 77.5, 77.5, 77.5, 85 yields s = 5.30, which is the smallest standard deviation for a sample of 5 data with 70 and 85.
  b. 70, 76, 85, 89, 95 yields s = 10.02.
  c. 70, 85, 90, 99, 110 yields s = 15.02.
  d. In order to increase the standard deviation the data had to become more dispersed.

2.211
  a.

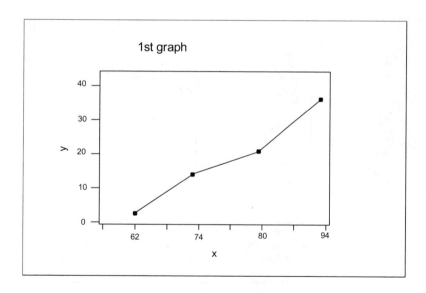

b.

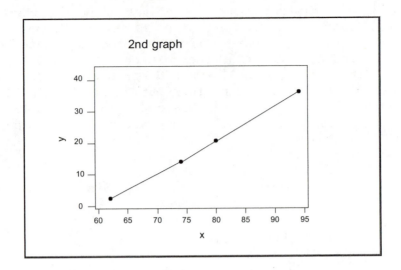

c. The line graph in (a) suggests an accelerated rate of increase from 1980 to 1995, while the line graph in (b) suggests that the rate of increase has been constant from 1962 to 1995.

2.213 a.

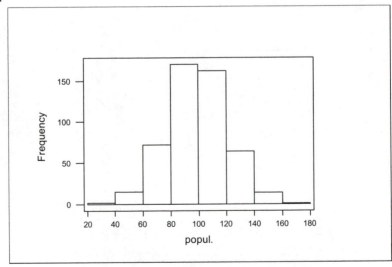

b. & C.

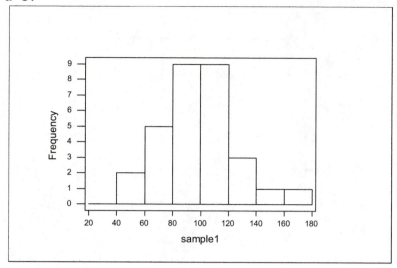

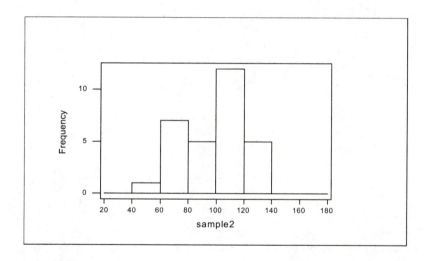

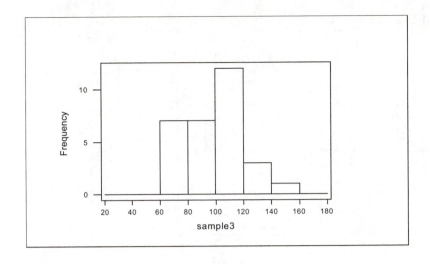

d.
Variable	N	Mean	Median	StDev
popul.	500	98.932	99.190	20.915
sample1	30	98.35	96.19	25.53
sample2	30	96.84	101.26	21.12
sample3	30	99.25	100.75	20.01
sample4	30	97.45	95.28	18.83

Variable	Minimum	Maximum	Q1	Q3
popul.	31.792	162.786	84.358	113.016
sample1	53.65	162.79	83.17	114.72
sample2	53.18	128.83	77.83	112.62
sample3	65.26	141.59	80.76	115.00
sample4	57.90	151.98	88.52	107.45

e. Yes, the sample statistics calculated closely resemble the population parameters.

2.215 Samples of size 30 usually demonstrated some of the properties of the population. As the sample size was increased, more of the properties of the population were shown. The suggested distributions in this exercise seem to require sample sizes greater than 30 for a closer match to the population.

2.217

CredHrs	Freq	xf	x^2f
3	75	225	675
6	150	900	5400
8	30	240	1920
9	50	450	4050
12	70	840	10080
14	300	4200	58800
15	400	6000	90000
16	1050	16800	268800
17	750	12750	216750
18	515	9270	166860
19	120	2280	43320
20	60	1200	24000
Σ	3570	55155	890655

Summary: n = 3570; Σxf = 55,155, Σx^2f = 890,655

a.

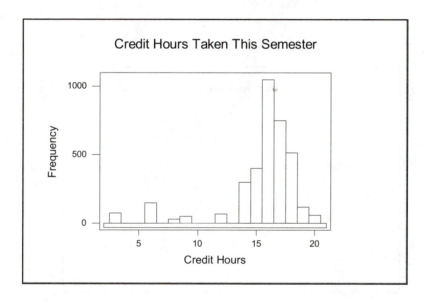

b. mean: $\overline{x}$ = $\Sigma xf / \Sigma f$ = 55,155/3570= 15.449 = <u>15.4</u>
   median: $d(\tilde{x})$ = $(\Sigma f+1)/2$ = (3570+1)/2 = 1785.5[th]; $\tilde{x}$ = <u>16</u>
   mode: mode = <u>16</u>
   midrange: midrange = (H+L)/2 = (3+20)/2 = <u>11.5</u>
   midquartile: midquartile = (Q1+Q3)/2 = (15+17)/2 = <u>16</u>

c. $nk/100 = (3570)(25)/100 = 892.5$; $d(Q_1) = 893^{rd}$;
   $Q_1 = \underline{15}$ and $Q_3 = \underline{17}$

d. $nk/100 = (3570)(15)/100 = 535.5$; $d(P_{15}) = 536th$
   $P_{15} = \underline{14}$
   $nk/100 = (3570)(12)/100 = 428.4$; $d(P_{12}) = 429th$
   $P_{12} = \underline{14}$

e. range: range = $H - L = 20 - 3 = \underline{17}$
   variance: $SS(x) = \Sigma x^2 - ((\Sigma x)^2/n)$
   $$= 890655 - (55155^2/3570) = 38533.42437$$

   $s^2 = SS(x)/(n-1) = 38533.42437 /3569 = \underline{10.7967}$

   standard deviation:
   $s = \sqrt{s^2} = \sqrt{10.7967} = 3.2858 = \underline{3.3}$

2.219 a.

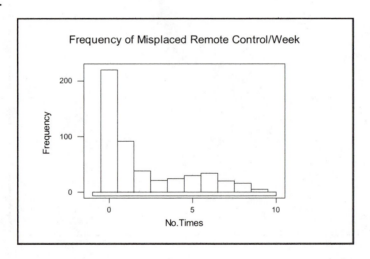

b. $n = \Sigma f = 500$, $\Sigma xf = 994$, $\sum x^2 f = 5200$
   mean = 1.988, median = 1, mode = 0, midrange = 4.5

c. $SS(x) = \sum x^2 f - ((\sum xf)^2/n)$
   $= 5200 - (994^2/500) = 3223.928$

   $s^2 = SS(x)/(\sum f-1) = 3223.928/499 = 6.46078 = \underline{6.46}$

   $s = \sqrt{s^2} = \sqrt{6.46078} = 2.5418 = \underline{2.5}$

d. $Q_1 = 0$, $Q_3 = 4$, $P_{90} = 6$

e. midquartile = 2

f. 5-number summary: 0, 0, 1, 4, 9

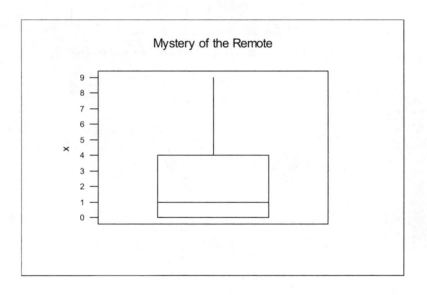

2.221   Summary: n = 220; Σxf = 219,100, Σx^2f = 224,470,000

a.

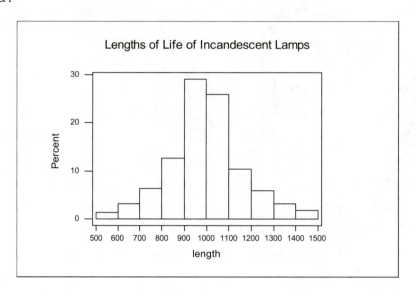

b. $\overline{x}$ = $\Sigma xf/\Sigma f$ = 219,100/220 = 995.90909 = <u>995.9</u>

c. SS(x) = Σx^2f - $((\Sigma xf)^2/\Sigma f)$
              = 224,470,000 - $(219,100^2/220)$ = 6,266,318.182

   s^2 = SS(x)/$(\Sigma f-1)$ = 6,266,318.182/219 = 28,613.32503

   s = $\sqrt{s^2}$ = $\sqrt{28613.32503}$ = 169.15473 = <u>169.2</u>

2.223 a.

Class limits	f
-1.00-0.00	1
0.00-1.00	6
1.00-2.00	10
2.00-3.00	7
3.00-4.00	6
4.00-5.00	3
5.00-6.00	3
6.00-7.00	1
7.00-8.00	2
8.00-9.00	0
9.00-10.0	1
Σ	40

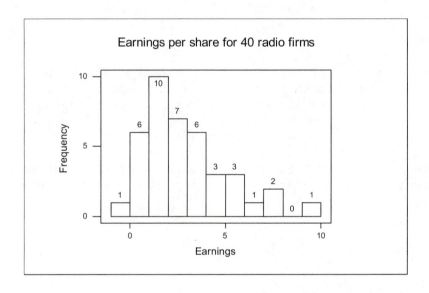

b. $d(\tilde{x}) = (n+1)/2 = 20.5$th;
median is in the class \$2.00-\$3.00.

CHAPTER 3 ▽   DESCRIPTIVE ANALYSIS AND PRESENTATION OF BIVARIATE DATA

Chapter Preview

Chapter 3 deals with the presentation and analysis of bivariate (two variables) data. There are three main categories of bivariate data.

1.   <u>Two Qualitative Variables</u>

This type of data is best presented in a contingency table and/or bar graph. Variations of the contingency table are given in Section 1 of Chapter 3.

2.   <u>One Qualitative Variable and One Quantitative Variable</u>

This type of data can be presented and/or summarized in table form or graphically. More statistical techniques are available because of the one quantitative variable. Dot plots, box plots, and stem-and-leaf diagrams can represent the data for each different value of the qualitative variable.

3.   <u>Two Quantitative Variables</u>

Initially, this type of data is best presented in a scatter diagram. If a relationship seems to exist, based on the scatter plot, then linear correlation and regression techniques will be performed.

An article reported by ESPN on Kevin Garnett of the Minnesota Timberwolves basketball team is presented in this chapter's opening section 3.1.

SECTION 3.1 EXERCISES

3.1   a.   Yes. A relationship seems to exist, the higher number of personal fouls go with the higher values of points scored per game.

       b.   Somewhat. Explanations will vary.

Exercise 3.3,presents two qualitative variables in the form of contingency tables and bar graphs. A contingency table is made up of rows and columns. Rows are horizontal and columns are vertical. Adding across the rows gives marginal row totals. Adding down the columns gives marginal column totals. The sum of the marginal row totals should be equal to the sum of the marginal column totals, which in turn, should be equal to the sample size.

Computer and/or calculator commands to construct a cross-tabulation table can be found in ES10-p150.

3.3 a.

	On Airplane	Hotel Room	All Other	Marginal total
Business	35.5%	9.5%	5.0%	50%
Leisure	25.0%	16.5%	8.5%	50%
Marginal total	60.5%	26.0%	13.5%	100%

b.

	On Airplane	Hotel Room	All Other	Marginal total
Business	71.0%	19.0%	10.0%	100%
Leisure	50.0%	33.0%	17.0%	100%
Marginal total	60.5%	26.0%	13.5%	100%

The table shows the distribution of ratings for business and leisure separately. For example, 71% of business travelers would like more space on the airplanes while 50% of leisure travelers would like more space on the airplanes.

c.

	On Airplane	Hotel Room	All Other	Marginal total
Business	58.7%	36.5%	37.0%	50%
Leisure	41.3%	63.5%	63.0%	50%
Marginal total	100%	100%	100%	100%

The table shows the distribution of business travelers and leisure travelers for each of the categories. For example, for more space in the hotel room, 36.5% of the responses were from business travelers and 63.5% from leisure travelers.

3.5 a. Population: Adults
   Variables: Gender; Age would like to remain rest of life

b.

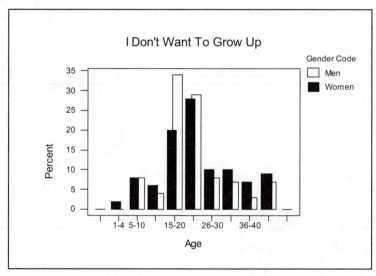

c. There does not seem to be a great difference, however it appears that the women tended to pick ages above 25 more often than did men.

3.7   a. 3350
       b. Two variables, political affiliation and television network preferred, are paired together. Both variables are qualitative.
       c. 880
       d. 46.9% [1570/3350]
       e. 19.2% [203/1060]
       f. 5.9% [197/3350]

3.9   Eastern: $\overline{x} = \Sigma x/n = 35.5/8 = 4.438$
                 $d(\tilde{x}) = (n+1)/2 = (8+1)/2 = 4.5^{th}$; $\tilde{x} = 4.55$
       Western: $\overline{x} = \Sigma x/n = 38.7/8 = 4.838$
                 $d(\tilde{x}) = (n+1)/2 = (8+1)/2 = 4.5^{th}$; $\tilde{x} = 4.6$

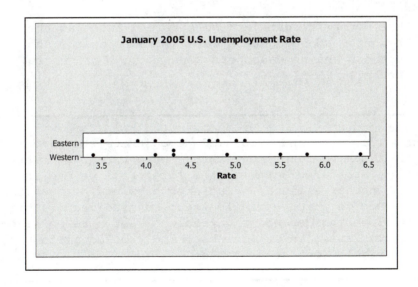

Exercise 3.11 demonstrates the statistical methods that can be used on "one qualitative, one quantitative" type data. Be sure to split the data based on the qualitative variable. The effect is a side-by-side comparison of the quantitative variable for each different value of the qualitative variable.

Computer and/or calculator commands to construct multiple dotplots or boxplots can be found in ES10-p49.

3.11 a.

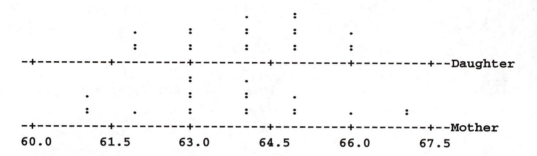

b. The mother heights are more spread out than the daughter heights. No daughters were as short as the shortest mothers and no daughters were as tall as the tallest mothers.

Exercises 3.11c, 3.15 and 3.18-3.22 demonstrate the numerical approach that can be taken now that we have two quantitative variables. A scatter diagram is the first tool we use in determining whether a linear relationship exists between the two variables. Decide which variable is to be predicted. This variable will be the dependent variable.* Let x be equal to the independent variable (input variable) and y be equal to the dependent variable (output variable).

                                                                    ...

c.

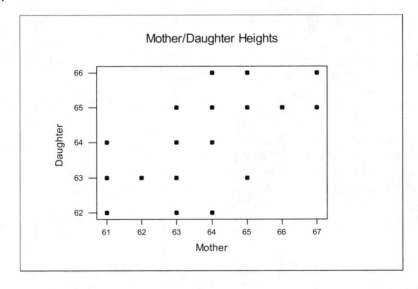

d. As mothers' heights increased, the daughters' heights
   also tended to increase.

3.13 The input variable most likely would be height. Based on height, weight is often predicted or given in a range of acceptable values depending on the size of a person's frame.

3.15 a.

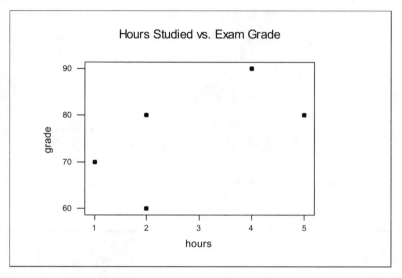

b. As hours studied increased, there seems to be a trend for the exam grades to also increase.

3.17 a. age, height,
b. Age = 3 yrs., height = 87 cm.
c. Answers will vary – whether a child's growth is above or below normal, etc.

3.19

Hours Studied vs. Exam Grade

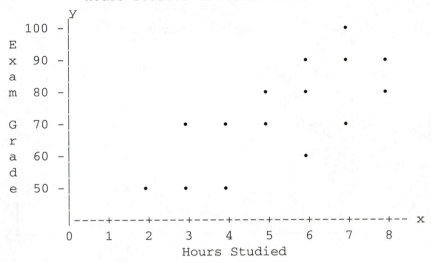

3.21

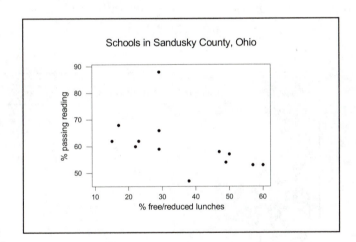

3.23 a. Answers will vary but one might expect that a relationship would exist between a stadium's distance from home plate to center field fence and the number of seats.

b. Answers will vary. One might expect that a larger distance would make for a bigger stadium which in turn would mean more seating.

c.

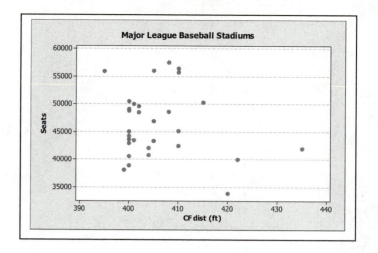

d. Scatter diagram indicates that there is no relationship
   between the center field distance and the number of seats.

3.25 a.

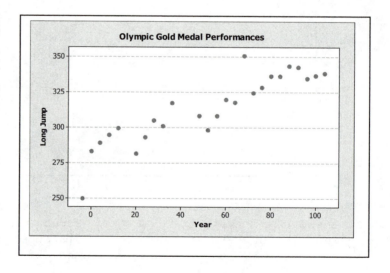

-- 106 --

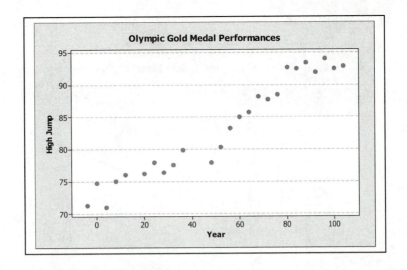

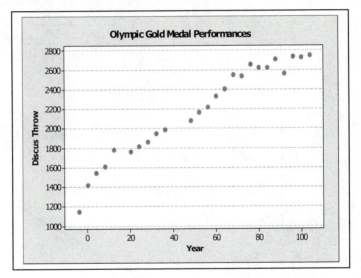

b. There is a very strong increasing pattern in all three, and even though it is not a perfectly straight line, they all seem to follow a fairly elongated pattern.

c. These three scatter diagrams indicate that our Olympic athletes, at least the gold medal winning ones, jump higher and longer and throw the discus further. All of these are strength skills and indicate that these athletes are stronger today. It may or may not be reasonable to anticipate that the general population will follow a similar pattern.

d. Prediction for 2008: 350 inches (108,350)

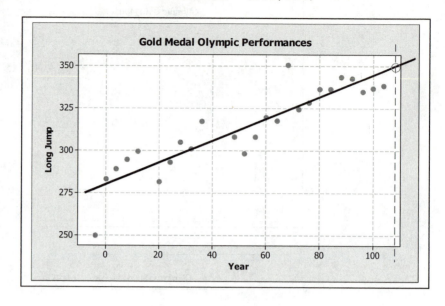

Prediction for 2008: 96 inches (108,96)

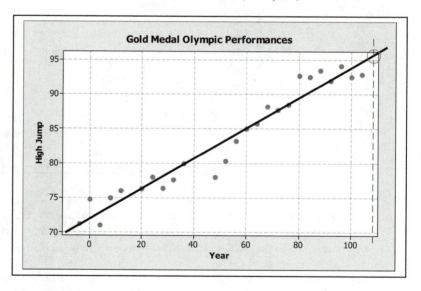

Prediction for 2008: 2900 inches (108,2900)

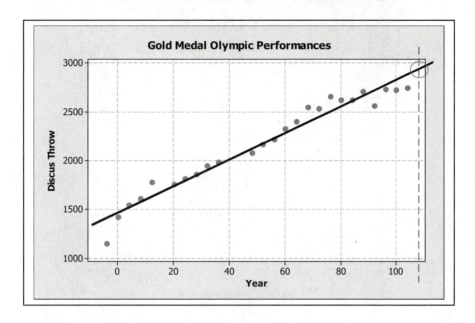

e.

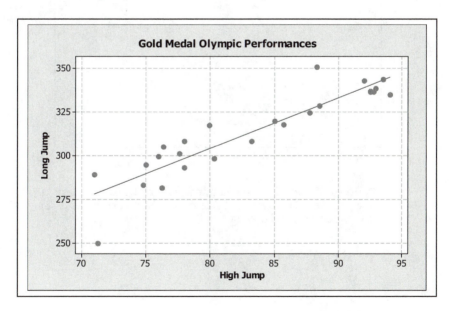

They seem to display a linear relationship. This
should not be surprising.

3.27 a.

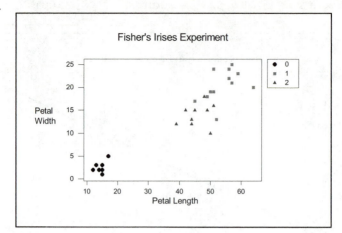

b.

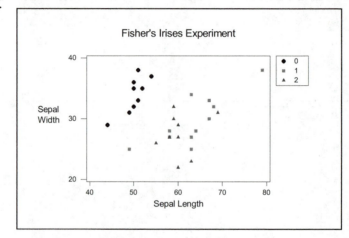

c. Type 0 displays a different pattern than do types 1 and 2, while types 1 and 2 seem to blend together on the scatter diagram.

d.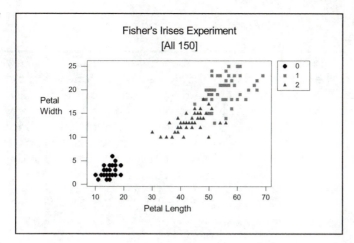

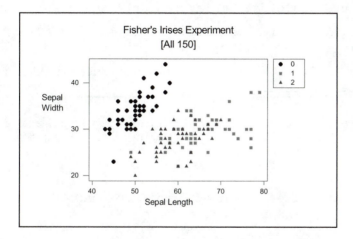

e. Aside from the quantity of points on the scatter diagrams the same patterns are apparent. The random sample of 30 yields the same relationship ideas as the full set of 150. Clearly the 150 are more "impressive" in demonstrating the relationship of the variables as well as the types.

Based on the scatter diagram, we might suspect whether or not a linear relationship exists. If the y's increase linearly as the x's increase, there exists a relationship called <u>positive correlation</u>. If the y's decrease linearly as the x's increase, there exists a relationship called <u>negative correlation</u>. The measure of the strength of this linear relationship is denoted by r, the coefficient of linear correlation.

<u>r = linear correlation coefficient</u>

1. r has a value between -1 and +1, i.e. $-1 \leq r \leq +1$

2. r = -1 specifies perfect negative correlation. All of the data points would fall on a straight line slanted downward.

3. r = +1 specifies perfect positive correlation. All of the data points would fall on a straight line slanted upward.

4. r ≈ 0 indicates little or no consistent linear pattern or a horizontal pattern.

3.29 a. Become closer to a straight line with a positive slope;
 b. Become closer to a straight line with a negative slope

3.31 Coefficient values near zero indicate that there is very little or no linear correlation.

<u>Calculating r - the linear correlation coefficient</u>

Preliminary Calculations:

1. Set up a table with the column headings: x, y, x^2, xy and y^2.

2. Insert the bivariate data into corresponding x and y columns. Perform the various algebraic functions to fill in the remaining columns.

3. Sum all columns, that is, find Σx, Σy, Σx^2, Σxy, Σy^2.

4. Double check calculations and summations.   . . .

---

5. Calculate: SS(x) - the sum of squares of x

             SS(y) - the sum of squares of y

             SS(xy) - the sum of squares of xy

  where:

$$SS(x) = \Sigma x^2 - ((\Sigma x)^2/n)$$
$$SS(y) = \Sigma y^2 - ((\Sigma y)^2/n)$$
$$SS(xy) = \Sigma xy - ((\Sigma x \cdot \Sigma y)/n)$$

Final Calculation:

6. Calculate r:

$$r = \frac{SS(xy)}{\sqrt{SS(x)\,SS(y)}}$$
          (round to the nearest hundredth)

7. Retain the <u>summations</u> and the <u>sums of squares</u>, as they will be needed for later calculations.

NOTE: Remember $SS(x) \neq \Sigma x^2$, $SS(y) \neq \Sigma y^2$ and $SS(xy) \neq \Sigma xy$.

The computer and/or calculator command to calculate the correlation coefficient can be found in ES10-p166.

3.33 a.

x	y	x^2	xy	y^2
2	80	4	160	6400
5	80	25	400	6400
1	70	1	70	4900
4	90	16	360	8100
2	60	4	120	3600
14	380	50	1110	29,400

$SS(x) = \Sigma x^2 - ((\Sigma x)^2/n) = 50 - (14^2/5) = \underline{10.8}$

$SS(y) = \Sigma y^2 - ((\Sigma y)^2/n) = 29,400 - (380^2/5) = \underline{520}$

$SS(xy) = \Sigma xy - ((\Sigma x \cdot \Sigma y)/n) = 1110 - (14 \cdot 380/5) = \underline{46}$

b. $r = SS(xy)/\sqrt{SS(x) \cdot SS(y)} = 46/\sqrt{10.8 \cdot 520} = 0.6138 = \underline{0.61}$

3.35   a.

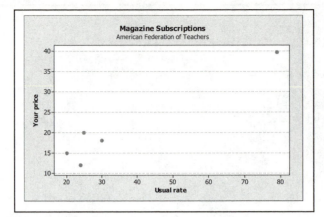

n = 5, $\sum x$ = 177.83, $\sum y$ = 104.69, $\sum x^2$ = 8730.09,
$\sum xy$ = 4763.48, $\sum y^2$ = 2671.44

b.   SS(x) = 8730.09 − (177.83^2/5) = 2405.38822
c.   SS(y) = 2671.44 − (104.69^2/5) = 479.44078
d.   SS(xy) = 4763.48 − (177.83·104.69/5) = 1040.07546

e.   r = SS(xy)/$\sqrt{SS(x)\cdot SS(y)}$

     = 1040.07546/$\sqrt{2405.38822 \cdot 479.44078}$

     = 0.9685 = 0.97

3.37   a. Manatees, powerboats
        b. Number of registrations, manatee deaths
        c. As one increases other does also
        d. Answers will vary. Possible: Restrict the number of boat
           registrations.

         Estimating r - the linear correlation coefficient

1. Draw as small a rectangle as possible that encompasses all of
   the data on the scatter diagram. (Diagram should cover a "square
   window" - same length and width)
2. Measure the width.
3. Let k = the number of times the width fits along the length
              or in other words: length/width.   . . .

3.39 a. Estimate r to be near -0.75

b.

Data	x	y	x²	xy	y²
1	2	12	4	24	144
2	4	13	16	52	169
3	5	9	25	45	81
4	6	7	36	42	49
5	6	12	36	72	144
6	7	8	49	56	64
7	9	6	81	54	36
8	9	9	81	81	81
9	10	7	100	70	49
10	12	5	144	60	25
Σ	70	88	572	556	842

$SS(x) = \Sigma x^2 - ((\Sigma x)^2/n) = 572 - (70^2/10) = 82.0$
$SS(y) = \Sigma y^2 - ((\Sigma y)^2/n) = 842 - (88^2/10) = 67.6$
$SS(xy) = \Sigma xy - ((\Sigma x \cdot \Sigma y)/n) = 556 - (70 \cdot 88/10) = -60.0$

$r = SS(xy)/\sqrt{SS(x) \cdot SS(y)} = -60.0/\sqrt{82.0 \cdot 67.6} = \underline{-0.806}$
$= \underline{-0.81}$

* What is the r value in exercise 3.39 telling you? (See the bottom
of the next page for answer.)

3.41 Answers will vary: positive vs. negative; nearness to
straight line, etc.

3.43   a.

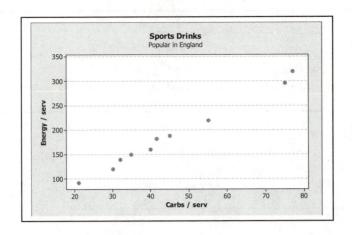

b. Yes, a straight line relationship appears. As carb/serving
   increase, so does the energy/serving.

c. Summations from extensions tables: n = 10, $\sum x$ = 451.7,
      $\sum y$ = 1868, $\sum x^2$ = 23528.8, $\sum xy$ = 96642.4, $\sum y^2$ = 397448

   SS(x) = $\sum x^2$ - (($\sum x)^2/n$) = 23528.8 - ($451.7^2/10$) =   3125.511
   SS(y) = $\sum y^2$ - (($\sum y)^2/n$) = 397448 - ($1868^2/10$) = 48505.6
   SS(xy) = $\sum xy$ - (($\sum x \cdot \sum y)/n$) = 96642.4 - ($451.7 \cdot 1868/10$)
         = 12264.84

   r = SS(xy)/$\sqrt{SS(x) \cdot SS(y)}$ = 12264.84/$\sqrt{3125.511 \cdot 48505.6}$
     = 0.996

d. There is a strong, almost perfect, positive correlation
   between the carbs/serving and the energy/serving for sports
   drinks.

---

*(3.39) As age increases, the number of irrelevant answers given
during a controlled experiment decreases. Can you make a
prediction or generalization based on this information?

e.

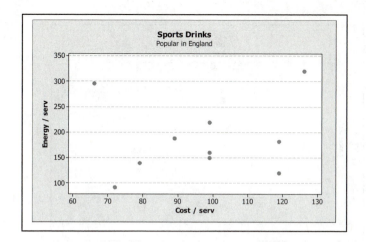

Appears to have no linear relationship between the cost/serving of a sports drink and the energy/serving.

Summations from extensions tables: n = 10, $\sum x$ = 967, $\sum y$ = 1868, $\sum x^2$ = 97303, $\sum xy$ = 182680, $\sum y^2$ = 397448

SS(x) = $\sum x^2$ - (($\sum x$)^2/n) = 97303 - (967^2/10) = 3794.1
SS(y) = $\sum y^2$ - (($\sum y$)^2/n) = 397448 - (1868^2/10) = 48505.6
SS(xy) = $\sum xy$ - (($\sum x \cdot \sum y$)/n) = 182680 - (967·1868/10)
        = 2044.4

r = SS(xy)/$\sqrt{SS(x) \cdot SS(y)}$ = 2044.4/$\sqrt{3794.1 \cdot 48505.6}$
  = 0.1507 = 0.15

There is little or no correlation between the cost/serving of a sports drink and the energy/serving.

3.45 a.

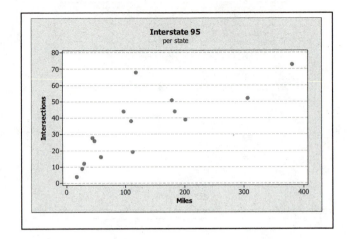

b. There is a somewhat linear pattern. As the number of
   miles per state increases, so does the number of
   intersections in that state. This relationship would
   make sense, the more miles would naturally have more
   intersections, but not an exact fit.
c. r = 0.792 = 0.79
d. Yes, the correlation coefficient fits the scatter diagram.
   There is a positive relationship but it is not that
   strong, there is some variation in the data.
e. Connecticut has more intersections than it has miles of
   interstate.
f.

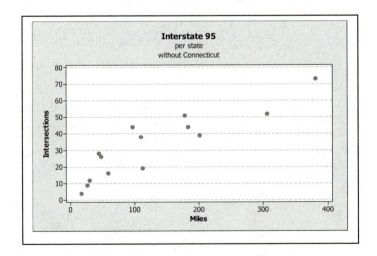

Pattern more closely follows a straight line.

g. r = 0.892

h. Correlation coefficient is a stronger positive value
    indicating a stronger relationship, which corresponds to
    the scatter diagram.

3.47 a.

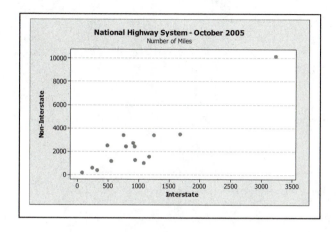

b. A linear pattern exists. There is one ordered pair that
   is quite different than all of the rest, much larger mile
   numbers - Texas.

c. r = 0.908 = 0.91

d.

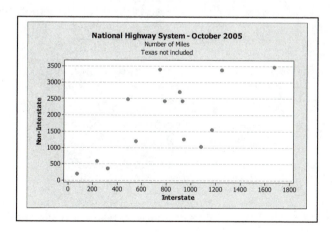

r = 0.665 = 0.67

e. Without Texas, there appears to be more variation; still a positive correlation.

3.49 a. Summations from extensions tables: $n = 7$, $\sum x = 315$, $\sum y = 219.04$, $\sum x^2 = 14875$, $\sum xy = 11261.2$, $\sum y^2 = 10140.8$

$SS(x) = \sum x^2 - ((\sum x)^2/n) = 14875 - (315^2/7) = 700$
$SS(y) = \sum y^2 - ((\sum y)^2/n) = 10140.8 - (219.04^2/7) = 3286.7254$
$SS(xy) = \sum xy - ((\sum x \cdot \sum y)/n) = 11261.2 - (315 \cdot 219.04/7)$
$\qquad = 1404.4$

$r = SS(xy)/\sqrt{SS(x) \cdot SS(y)} = 1404.4/\sqrt{700 \cdot 3286.7254}$
$\quad = 0.926 = \underline{0.93}$

b.

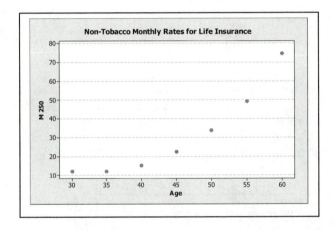

c. The relationship between age and monthly rate is not linear, it does however show a very definite and well-defined increasing pattern.

d. It is this elongated pattern that causes the calculated r to be so large.

e. The pattern shown on the scatter diagrams showing age versus monthly rate should not surprise us, after all we have always been told the rate for insurance increases (accelerates) as the insured person's issue age increases, thus the "upward-bending" pattern.

f. Answers will vary depending on choice of data.

3.51 Typically bigger fires require more fire trucks and result in more damage, thus a positive correlation would be expected. It certainly does not mean that, if fewer fire engines were dispatched that there would be less damage.

SECTION 3.4 EXERCISES

If a linear relationship exists between two variables, that is,

1. its scatter diagram suggests a linear relationship

2. its calculated r value is not near zero

the techniques of linear regression will take the study of bivariate data one step further. Linear regression will calculate an equation of a straight line based on the data. This line, also known as the line of best fit, will fit through the data with the smallest possible amount of error between it and the actual data points. The regression line can be used for generalizing and predicting over the sampled range of x.

FORM OF A LINEAR REGRESSION LINE

$$\hat{y} = b_0 + b_1 x$$

where $\hat{y}$ (y hat) = predicted y
        b_0 (b sub zero) = y intercept
        b_1 (b sub one) = slope of the line
        x = independent data value.

3.53

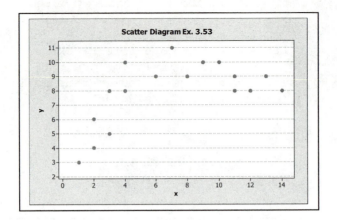

No. The pattern shown is definitely not that of a straight line, thus the results from using any of the linear techniques would be meaningless.

3.55 a. $\sum x^2 = 13,717$, SS(x) = 1396.9
$\sum y^2 = 15,298$, SS(y) = 858.0
$\sum xy = 14,257$, SS(xy) = 919.0

b. Summations (Σ's) are sums of data values. Sum of squares (SS( )) are parts of complex formulas; they are calculated separately as preliminary values.

3.57 a. $\hat{y} = 14.9 + 0.66(20) = \underline{28.1}$

$\hat{y} = 14.9 + 0.66(50) = \underline{47.9}$

b. Yes, the line of best fit is made up of all points that satisfy its equation.

CALCULATING $\hat{y} = b_0 + b_1 x$ - THE EQUATION OF THE LINE OF BEST FIT

1. Retrieve preliminary calculations from previous r calculations.

2. Calculate b_1 where $b_1 = \dfrac{SS(xy)}{SS(x)}$

3. Calculate b_0 where $b_0 = \dfrac{1}{n}(\sum y - b_1 \sum x)$ . . .

$\hat{y}$ = predicted value of y (based on the regression line)

NOTE: See Review Lessons for additional information about the concepts of slope and intercept of a straight line.

<u>DRAWING THE LINE OF BEST FIT ON THE SCATTER DIAGRAM</u>

1. Pick two *x*-values that are <u>within</u> the interval of the data *x*-values. (one value near <u>either</u> end of the domain)

2. Substitute these values into the calculated $\hat{y} = b_0 + b_1 x$ equation and find the corresponding $\hat{y}$ values.

3. Plot these points on the scatter diagram in such a manner that they are distinguishable from the actual data points.

4. Draw a straight line connecting these two points. This line is a graph of the line of best fit.

5. Plot a third point, the ordered pair $(\overline{x}, \overline{y})$ as an additional check. It should be a point on the line of best fit.

OR:

Computer and/or calculator commands to find the equation of the line of best fit and also draw it on a scatter diagram can be found in ES10-pp180-181.

3.59 a.

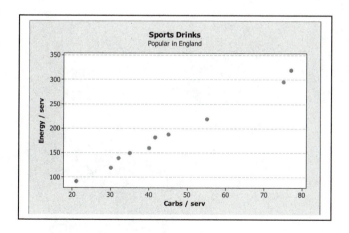

A linear relationship appears between carbs/serving and energy/serving for the sports drinks in the sample. The ordered pairs follow very closely to a straight line.

b. Summations from extensions tables: $n = 10$, $\sum x = 451.7$,
   $\sum y = 1868$, $\sum x^2 = 23528.8$, $\sum xy = 96642.4$

$SS(x) = \sum x^2 - ((\sum x)^2/n) = 23528.8 - (451.7^2/10) = 3125.511$
$SS(xy) = \sum xy - ((\sum x \cdot \sum y)/n) = 96642.4 - (451.7 \cdot 1868/10)$
$\qquad = 12264.84$

$b_1 = SS(xy)/SS(x) = 12264.84/3125.511 = 3.924$

$b_0 = [\sum y - b_1 \cdot \sum x]/n = [1868 - (3.924 \cdot 451.7)]/10 = 9.553$

$\qquad \hat{y} = 9.55 + 3.924x$

c. $\hat{y} = 9.55 + 3.924(40) = 166.51$

d. $\hat{y} = 9.55 + 3.924(65) = 264.61$

3.61 a. The slope of 4.71 indicates that for each increase in
height of one inch, college women's weight increased by
4.71 pounds.

b. The scale for the y-axis starts at $y = 95$ and the scale
for the x-axis starts at $x = 60$. The y-intercept of -186.5
occurs when $x = 0$, so the x-axis would have to include $x = 0$ and the the y-axis would have to be extended down.

3.63 a. $\hat{y} = 185.7 - 21.52x$, when $x = 3$
   $\hat{y} = 185.7 - 21.52(3) = \underline{121.14}$ or $\underline{\$12,114.}$
b. $\hat{y} = 185.7 - 21.52x$, when $x = 6$
   $\hat{y} = 185.7 - 21.52(6) = \underline{56.58}$ or $\underline{\$5,658.}$
c. $21.52(\$100) = \underline{\$2,152}$ [the "slope" in dollars]

3.65 a. $\hat{y} = -5359 + 0.9956x$, when $x = 500,000$
   $\hat{y} = -5359 + 0.9956(500,000) = \underline{492,411}$ or $\underline{\$492,411,000}$
b. $\hat{y} = -5359 + 0.9956x$, when $x = 1,000,000$
   $\hat{y} = -5359 + 0.9956(1,000,000) = \underline{990,241}$ or $\underline{\$990,241,000}$
c. $\hat{y} = -5359 + 0.9956x$, when $x = 1,500,000$
   $\hat{y} = -5359 + 0.9956(1,500,000) = \underline{1,488,041}$ or
   $\underline{\$1,488,041,000}$

3.67 The vertical scale shown on figure 3.25 is located at x = 58 and therefore is not the y-axis; the y = 80 occurs at x = 58. Remember, the x-axis is the vertical line located at x = 0.

3.69 (61, 95)   (67, 130)

$$b_1 \approx \frac{Y_2 - Y_1}{x_2 - x_1} = \frac{130 - 95}{67 - 61} = \frac{35}{6} = 5.83$$

$$b_0 \approx y - b_1 x = 130 - 5.83(67) = -260.61$$

3.71 a.

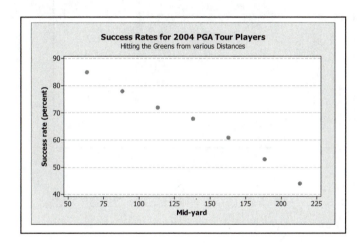

b. Yes, the ordered pairs follow very closely to a straight line pattern.

c. n = 7, $\sum x$ = 966, $\sum y$ = 461, $\sum x^2$ = 150808, $\sum xy$ = 59018, $\sum y^2$ = 31583

SS(x) = $\sum x^2$ - $((\sum x)^2/n)$ = 150808 - $(966^2/7)$ = 17500
SS(y) = $\sum y^2$ - $((\sum y)^2/n)$ = 31583 - $(461^2/7)$ = 1222.8571
SS(xy) = $\sum xy$ - $((\sum x \cdot \sum y)/n)$ = 59018 - (966·461/7)= -4600

$$r = SS(xy)/\sqrt{SS(x) \cdot SS(y)} = -4600/\sqrt{17500 \cdot 1222.8571}$$
$$= \underline{-0.994}$$

d. Very strong negative correlation - strong relationship between yards to green and success rate. The longer the yards to the green, the lower the success rate of hitting the green.

e. Yes, as noted in part b, the ordered pairs follow along a straight line pattern. A straight line would be the best fit for this data.

f. b_1 = SS(xy)/SS(x) = -4600/17500 = -0.26286 = -0.263
b_0 = [$\sum y$ - $b_1 \cdot \sum x$]/n = [461 - (-0.263·966)]/7 = 102.15
$\hat{y}$ = 102.15 - 0.263x

g.

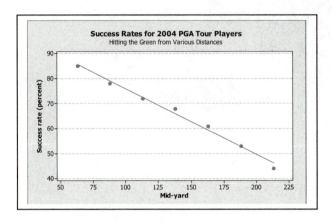

h. $\hat{y}$ = 102.15 - 0.263x, when x = 90
$\hat{y}$ = 102.15 - 0.263(90) = 78.48 = 78.5%

3.73 a. Yes
b. The ordered pairs should start in the upper left and proceed diagonally to the lower right.
c.

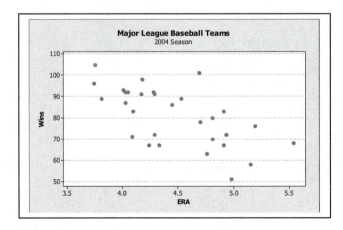

d. Yes, as the ERA increases, the number of wins decrease or as the ERA decreases, the number of wins increases.

e. Wins = 163 - 18.5 ERA   or $\hat{y} = 163 - 18.5x$

f. Wins decrease by 18.5 for every one increase in ERA. This determined by the slope, -18.5.

g. Yes

3.75 a.

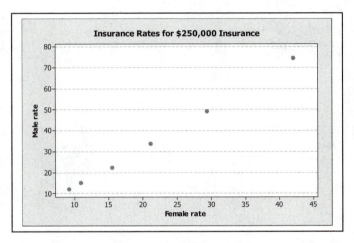

Yes, the scatter diagram shows a linear relationship. The ordered pairs follow very closely a straight line from lower left to upper right.

Summations from extensions table: n = 7, $\sum x$ = 137.11, $\sum y$ = 219.04, $\sum x^2$ = 3599.71, $\sum xy$ = 6022.95, $\sum y^2$ = 10140.8

b. SS(x) = $\sum x^2 - ((\sum x)^2/n)$ = 3599.71 - (137.11²/7) = 914.117
    SS(y) = $\sum y^2 - ((\sum y)^2/n)$ = 10140.8 - (219.04²/7)
        = 3286.725
    SS(xy) = $\sum xy - ((\sum x \cdot \sum y)/n)$ = 6022.95 - (137.11·219.04/7)
        = 1732.582

   r = SS(xy)/$\sqrt{SS(x) \cdot SS(y)}$ = 1732.582/$\sqrt{914.117 \cdot 3286.725}$

   = 0.9996

There is a strong linear relationship because the
points all lay very close to a straight line and the
r-value is just about equal to 1.0, perfect positive
correlation.

c. b_1 = SS(xy)/SS(x) = 1732.582/914.117 = 1.895
    b_0 = [Σy - $b_1 \cdot \Sigma$x]/n = [219.04 - (1.895·137.11)]/7
    b_0 = -5.826

    $\hat{y}$ = -5.83 + 1.90x

d. male monthly rate = - 5.83 + (1.90·15.00) = -5.83 +
    28.50 = $22.67

e. Males paid a higher rate. The slope of 1.90 means
    that males pay $1.90 for every $1.00 that females
    pay.

3.77 a.

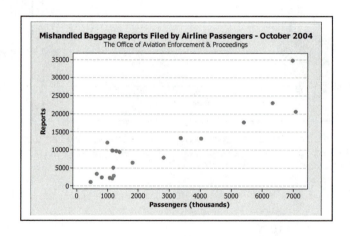

b. Scatter diagram suggests that linear regression could be
   useful. The ordered pairs follow in a straight line
   format.

c. Reports = 1427 + 3.47 Passengers   OR $\hat{y}$ = 1427 + 3.47x

d.

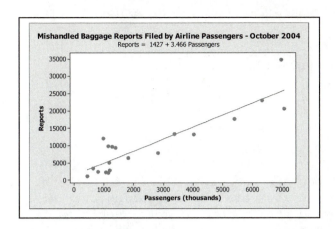

3.79 a. With population and drivers parallel it means that both
      are increasing at the same rate. Not all of the
      population is in the category of drivers. The number of
      non-drivers appears to remain constant. If population and
      drivers were not parallel, then one would be increasing at
      a faster rate than the other.

  b. When the lines cross for drivers and motor vehicles, it
      shows that there were more drivers than vehicles before
      1971, then more vehicles than drivers after 1973. From
      1971 to 1973, the number of drivers and vehicles was about
      the same.

  c. Numbers for both motor vehicles and drivers are increasing
      as time increases, but before 1973, the number of drivers
      were increasing at a faster rate than the number of cars.

  d. Numbers for both motor vehicles and drivers are increasing
      as time increases, but after 1973, the number of motor
      vehicles is increasing at a faster rate than the number of
      drivers.

  e. Answers will vary. One possibility is that drivers will
      not surpass motor vehicles after 2003. People are too
      used to having at least one car.

f. Motor vehicles: (1982, 165) & (2000, 215)

$$m = \frac{215 - 165}{2000 - 1982} = \frac{50}{18} = 2.777 = 2.8$$

Drivers: (1982, 155) & (2000, 180)

$$m = \frac{180 - 155}{2000 - 1982} = \frac{25}{18} = 1.388 = 1.4$$

Both slopes are positive, therefore both the number of motor vehicles and drivers are increasing over time. During this time period, the number of motor vehicles is increasing at twice the rate of the number of drivers. The motor vehicles are increasing at a rate of 2.8 million cars per year whereas the number of drivers is increasing at the rate of 1.4 million per year.

CHAPTER EXERCISES

Exercises 3.81-3.85

To find percentages based on the grand total
  • divide each count by the grand total

To find percentages based on the row totals
  • divide each count by its corresponding row total
  • each row should add up to 100%

To find percentages based on the column totals
  • divide each count by its corresponding column total
  • each column should add up to 100%

3.81 a.

	Elem.	Jr.H.	Sr.H.	Coll.	Adult	Total
Fear	37	28	25	27	21	138
Do not	63	72	75	73	79	362
Total	100	100	100	100	100	500

b.

	Elem.	Jr.H.	Sr.H.	Coll.	Adult	Total
Fear	7.4%	5.6%	5.0%	5.4%	4.2%	27.6%
Do not	12.6%	14.4%	15.0%	14.6%	15.8%	72.4%
Total	20.0%	20.0%	20.0%	20.0%	20.0%	100.0%

c.

	Elem.	Jr.H.	Sr.H.	Coll.	Adult	Total
Fear	37%	28%	25%	27%	21%	27.6%
Do not	63%	72%	75%	73%	79%	72.4%
Total	100%	100%	100%	100%	100%	100%

d.

	Elem.	Jr.H.	Sr.H.	Coll.	Adult	Total
Fear	26.8%	20.3%	18.1%	19.6%	15.2%	100%
Do not	17.4%	19.9%	20.7%	20.2%	21.8%	100%
Total	20%	20%	20%	20%	20%	100%

e.

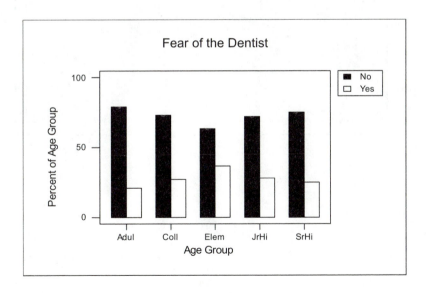

3.83 a.

Yr	Lab. Retr.	Gold Retr.	Grmn Shep	Beag	York Terr	Dach shun	Row Totals
03	144,896	52,520	43,938	45,021	38,246	39,468	364,089
04	146,692	52,550	46,046	44,555	43,522	40,770	374,135
Tot	291,588	105,070	89,984	89,576	81,768	80,238	738,224

b.

Yr	Lab. Retr.	Gold Retr.	Grmn Shep	Beag	York Terr	Dach shun	Row Totals
03	19.63%	7.11%	5.95%	6.10%	5.18%	5.35%	49.32%
04	19.87%	7.11%	6.24%	6.04%	5.90%	5.52%	50.68%
Tot	39.50%	14.22%	12.19%	12.14%	11.08%	10.87%	100.00%

c.

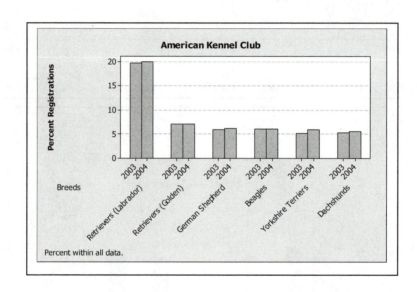

d.

Yr	Lab. Retr.	Gold Retr.	Grmn Shep	Beag	York Terr	Dach shun	Row Totals
03	39.80%	14.43%	12.07%	12.37%	10.50%	10.84%	100.00%
04	39.21%	14.05%	12.31%	11.91%	11.63%	10.90%	100.00%
Tot	39.50%	14.22%	12.19%	12.14%	11.08%	10.87%	100.00%

e.

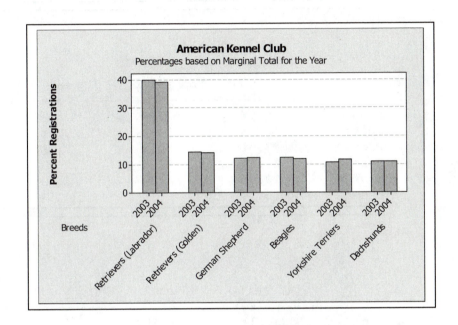

3.85 a.

	1st shift	2nd shift	3rd shift	Total
Sand	87	110	72	269
Shift	16	17	4	37
Drop	12	17	16	45
Corebrk.	18	16	33	67
Broken	17	12	20	49
Other	8	18	22	48
Total	158	190	167	515

b.	1st shift	2nd shift	3rd shift	Total
Sand	16.9%	21.4%	14.0%	52.2%
Shift	3.1%	3.3%	0.8%	7.2%
Drop	2.3%	3.3%	3.1%	8.7%
Corebrk.	3.5%	3.1%	6.4%	13.0%
Broken	3.3%	2.3%	3.9%	9.5%
Other	1.6%	3.5%	4.3%	9.3%
Total	30.7%	36.9%	32.4%	100%

c.	1st shift	2nd shift	3rd shift	Total
Sand	55.1%	57.9%	43.1%	52.2%
Shift	10.1%	8.9%	2.4%	7.2%
Drop	7.6%	8.9%	9.6%	8.7%
Corebrk.	11.4%	8.4%	19.8%	13.0%
Broken	10.8%	6.3%	12.0%	9.5%
Other	5.1%	9.5%	13.2%	9.3%
Total	100%	100%	100%	100%

d.	1st shift	2nd shift	3rd shift	Total
Sand	32.3%	40.9%	26.8%	100%
Shift	43.2%	45.9%	10.8%	100%
Drop	26.7%	37.8%	35.6%	100%
Corebrk.	26.9%	23.9%	49.3%	100%
Broken	34.7%	24.5%	40.8%	100%
Other	16.7%	37.5%	45.8%	100%
Total	30.7%	36.9%	32.4%	100%

e.

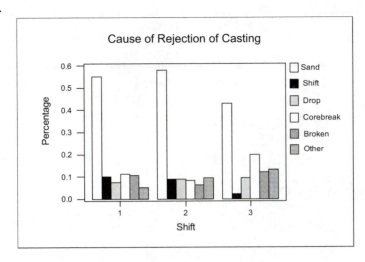

Cause of Rejection of Casting

3.87 a. The purpose of a correlation analysis is to determine whether two variables are linearly related or not. The result of correlation analysis is the numerical value of the linear correlation coefficient, r.

b. The purpose of a regression analysis is to determine the equation of the line of best fit that describes the linear relationship of the two variables. The result is the equation.

3.89 a.

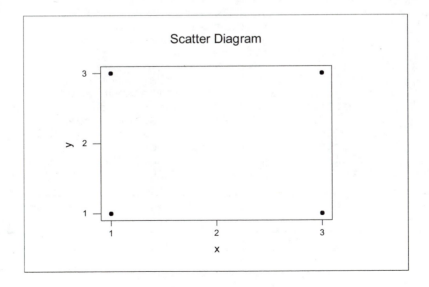

Scatter Diagram

**Looking at the scatter diagram in exercise 3.89a, what would you expect for the correlation coefficient? (See the bottom of this page for answer.)

b. Summations from extensions table: $n = 4$, $\Sigma x = 8$, $\Sigma y = 8$, $\Sigma x^2 = 20$, $\Sigma xy = 16$, $\Sigma y^2 = 20$

$$SS(x) = \Sigma x^2 - ((\Sigma x)^2/n) = 20 - (8^2/4) = 4.0$$
$$SS(y) = \Sigma y^2 - ((\Sigma y)^2/n) = 20 - (8^2/4) = 4.0$$
$$SS(xy) = \Sigma xy - ((\Sigma x \cdot \Sigma y)/n) = 16 - (8 \cdot 8/4) = 0.0$$

$$r = SS(xy)/\sqrt{SS(x) \cdot SS(y)} = 0.0/\sqrt{4.0 \cdot 4.0} = \underline{0.00}$$

At this point, is there much sense in calculating the line of best fit?
No, the r value shows that no linear relationship exists. Regression analysis would be useless.
NOTE: The line of best fit is a horizontal line (indicating a lack of a linear relationship).

c. $b_1 = SS(xy)/SS(x) = 0.0/4.0 = 0.0$

$b_0 = [\Sigma y - b_1 \cdot \Sigma x]/n = [8 - (0.0 \cdot 8)]/4 = 2.0$

$\underline{\hat{y} = 2.0 + 0.0x}$

3.91 a. The points will be scatter across both the x and y intervals leaving a "scatter-gun" appearance.
b. The points will be scatter across both the x and y intervals leaving a pattern that stretches from lower left to upper right and is somewhat elongated.
c. The points will be scatter across both the x and y intervals leaving a pattern that stretches from lower left to upper right and is quite elongated.
d. The points will be scatter across both the x and y intervals leaving a pattern that stretches from upper left to lower right and is somewhat elongated.
e. The points will be scatter across both the x and y intervals leaving a pattern that stretches from upper left to lower right and is quite elongated.

*(3.89) There is no linear trend or pattern to the points in any direction. Therefore, the r value will probably be close to 0.00.

3.93 a. Answers will vary.

b. Answers will vary but should include that the ordered pairs would be scattered in no definite pattern.

c. Answers will vary.

d.

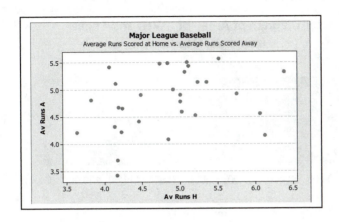

e. Answers will vary.

3.95 a.

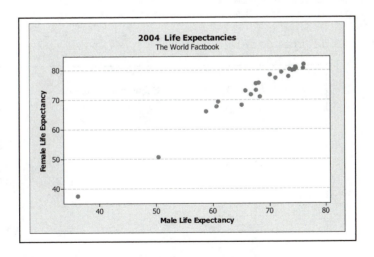

b. Yes, as the male life expectancy increases, so does the female life expectancy and in a linear format.

c. Female Life Expectancy = - 1.40 + 1.11 Male Life Expectancy or $\hat{y} = -1.40 + 1.11x$

d. For every one additional male life expectancy year, the female life expectancy increases by 1.11 years.

3.97 a. Scatter diagram:

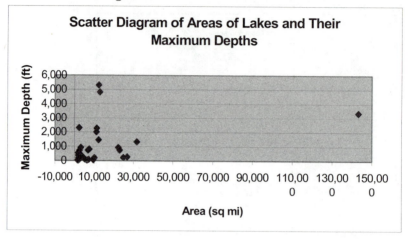

b. r = 0.3709   There is one data point that is located away from all the other ordered pairs.   The separation between this one point and the others greatly effects the value of the coefficient.

3.99 a.   Code the data.
b.

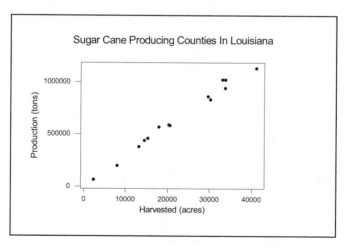

c. The pattern of the data is very elongated and shows an increasing relationship. This overall pattern is linear in nature.

d. Summations from extensions table: $n = 14$, $\sum x = 3131$, $\sum y = 9090$, $\sum x^2 = 867915$, $\sum xy = 2510500$

$$SS(x) = \sum x^2 - ((\sum x)^2/n) = 867915 - (3131^2/14)$$
$$= 167689.2143$$

$$SS(xy) = \sum xy - \sum x \cdot \sum y)/n) = 2510500 - (3131 \cdot 9090/14)$$
$$= 477586.4286$$

$$b_1 = SS(xy)/SS(x) = 477586/167689 = 2.84805$$

$$b_0 = [\sum y - b_1 \cdot \sum x]/n = [9090 - 2.84805 \cdot 3131)]/14 = 12.34$$

$$\hat{y} = 12.3 + 2.85x$$

e. The slope, 2.85, means that for every 100 acres there is, on the average, 2,850 tons of sugar cane produced.

3.101 a.

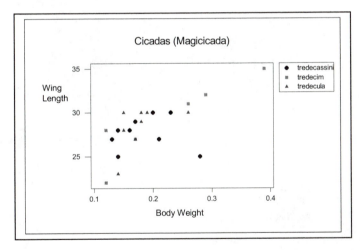

b. The overall pattern is linear, moderately strong and shows an increasing relationship. All three species are intermingled to a certain extent. The exceptions are: all the points in the upper right portion of the diagram are tredecim, and the tredecassini are all in the central part of the pattern.

c. Summations from extensions table: $n = 24$, $\Sigma x = 4.6$,
$\Sigma y = 680$, $\Sigma x^2 = 0.978400$, $\Sigma xy = 133.050$, $\Sigma y^2 = 19448.0$

$SS(x) = \Sigma x^2 - ((\Sigma x)^2/n) = 0.978400 - (4.6^2/24)$
$= 0.0967333$
$SS(y) = \Sigma y^2 - ((\Sigma y)^2/n) = 19448.0 - (680^2/24) = 181.333$
$SS(xy) = \Sigma xy - ((\Sigma x \cdot \Sigma y)/n) = 133.050 - (4.6 \cdot 680/24)$
$= 2.71667$

$r = SS(xy)/\sqrt{SS(x) \cdot SS(y)}$
$= 2.71667/\sqrt{0.0967333 \cdot 181.333} = \underline{0.649}$

d. $b_1 = SS(xy)/SS(x) = 2.71667/0.0967333 = 28.084$

$b_0 = [\Sigma y - b_1 \cdot \Sigma x]/n = [680 - (28.084 \cdot 4.6)]/24 = 22.95$

$\hat{y} = \underline{23.0 + 28.1x}$

e. Wing length = 23.0 + 28.1(body weight) = 23.0 +
28.1(0.2) = 28.62 mm.   Probably the tredecassini
species.

3.103 a. Numerator of formula (3-1):

Numerator = $\Sigma(x-\overline{x})(y-\overline{y})$

$= \Sigma[xy - \overline{x}y - x\overline{y} + \overline{x}\,\overline{y}]$

$= \Sigma xy - \overline{x} \cdot \Sigma y - \overline{y} \cdot \Sigma x + n\overline{x}\,\overline{y}$

$= \Sigma xy - [(\Sigma x/n) \cdot \Sigma y] - [(\Sigma y/n) \cdot \Sigma x] + [n \cdot (\Sigma x/n)(\Sigma y/n)]$

$= \Sigma xy - [(\Sigma x \cdot \Sigma y/n) - (\Sigma x \cdot \Sigma y/n) + (\Sigma x \cdot \Sigma y/n)]$

$= \Sigma xy - [(\Sigma x \cdot \Sigma y)/n]$

$= SS(xy)$

Denominator of formula (3-1):
Denominator = $(n-1)s_x s_y$

$= (n-1) \cdot \sqrt{SS(x)/(n-1)} \cdot \sqrt{SS(y)/(n-1)}$

$= \sqrt{SS(x) \cdot SS(y)}$

Therefore, formula (3-1) is equivalent to formula (3-2).

b. The numerators of formula (3-5) and (3-7) were shown to be equal in (a) above. (See numerator.)

The denominators are equal by definition (formula 2-9).

CHAPTER 4 ∇ PROBABILITY

Chapter Preview

Chapter 4 deals with the basic theory and concepts of probability. Probability, in combination with the descriptive techniques in the previous chapters, allows us to proceed into inferential statistics in later chapters.

Some history of M&M's Milk Chocolate Candies is presented in this chapter's Section 4.1, Sweet Statistics.

SECTION 4.1 EXERCISES

4.1   a. Most: yellow, blue and orange
           Least: brown, red, and green
      b. Not exactly, but similar

4.3

Color	Count	(multiply each Table 4.2 percent by 40)
Brown	5	
Yellow	6	
Red	6	
Blue	9	
Orange	8	
Green	6	
	40	

Computer and/or calculator commands to generate random integers and tally the findings can be found in ES10-pp101.

Variations in the quantity of random integers and the interval are necessary for exercises 4.5 and 4.6.

4.5 Note: Each will get different results. MINITAB results on
one run were:
a. Relative frequency for: 1 - 0.22, 2 - 0.16, 3 - 0.14,
   4 - 0.22, 5 - 0.16, 6 - 0.10
b. Relative frequency for: H - 0.58, T - 0.42

SECTION 4.2 EXERCISES

4.7 P'(5) = 9/40 = 0.225

4.9 a. 25.0/(25.0+14.2+10.8) = 0.50 = 50%
b. 0.50
c. They are both asking the same question but in two
   different formats, one is asking for a percentage and
   the other is asking for a probability.

4.11 a. (23.90 + 9.80 + 1.90) = 35.6% = 0.356
b. (20.90 + 13.60 + 4.40) = 38.9% = 0.389
c. (23.90 + 25.50 + 20.90) = 70.30% = 0.703
d. (13.60 + 4.40) = 18% = 0.18

4.13 a. Seattle: 37/390 = 0.09; San Diego: 371/390 = 0.95
b. Seattle: 188/390 = 0.48; San Diego: 104/390 = 0.27
c. San Diego would be the better choice, warm 95% of the time
   versus 9% for Seattle. It is sunny 65% of the time versus
   29% for Seattle.

4.15 {0, 1, 2, 3, 4, 5, 6, 7, 8, 9}

See the possible outcomes for rolling a pair of dice in ES10-p209.
(Also shown in numerical form in ES10-p239)

4.17 P(5) = 4/36; P(6) = 5/36; P(7) = 6/36; P(8) = 5/36;
P(9) = 4/36; P(10) = 3/36; P(11) = 2/36; P(12) = 1/36

4.19 Each student will get different results. These are the
results I obtained: [Note: 12 is an ordered pair (1,2)]

```
12  65  15  32  54     12  52  63  64  62
66  44  42  45  42     35  54  66  54  32
31  12  23  33  26     33  32  23  46  64
63  63  35  54  52     55  56  26  11  44
11  61  46  11  45     55  15  33  43  11
```

a. P'(white die is odd) = 27/50 = <u>0.54</u>

b. P'(sum is 6) = 7/50 = <u>0.14</u>

c. P'(both dice show odd number) = 14/50 = <u>0.28</u>

d. P'(number on color die is larger) = 16/50 = <u>0.32</u>

e. The answers in Exercise 4.18 were determined by using a
sample space and are therefore the theoretical
probabilities. The answers above were obtained
experimentally and are expected to be similar in value,
but not exactly the same.

<u>A REGULAR (BRIDGE) DECK OF PLAYING CARDS</u>

52 cards 26 red, 26 black 4 suits

(diamonds, hearts, clubs, spades)

each suit - 13 cards - 2,3,4,5,6,7,8,9,10,Jack,Queen,King,Ace

face cards = Jack, Queen, and King

4.21 Let J = jack, Q = queen, K = king, H = heart, C = club,
D = diamond, S = spade.
S = {JH, JC, JD, JS, QH, QC, QD, QS, KH, KC, KD, KS}

NOTE: See additional information about tree diagrams in Review Lessons, Tree Diagrams, at the end of this manual.

4.23 a. S = {$1, $5, $10, $20}

b.

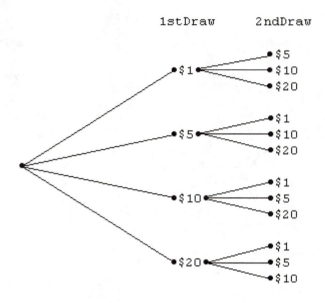

In exercise 4.25, add the rows and columns first, to find marginal totals.

4.25 a.

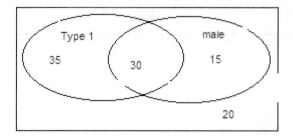

Both representations show that 30 patients are both male and have Type 1 diabetes. There are 65 Type 1 diabetics and 45 males. The 20 female Type 2 diabetics are neither Type 1 nor

b. (35+20)/100 = <u>0.55</u>

c. (15+20)/100 = <u>0.35</u>

4.27 Let U = used part and D = defective part

Given info: P(U) = 0.60, P(U or D) = 0.61, P(D) = 0.05

P(U and D) = <u>0.04</u>; 4% are both used and defective

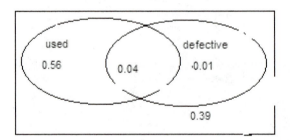

<u>PROPERTIES OF PROBABILITY</u>

1. $0 \leq P(A) \leq 1$ The probability of an event must be a value between 0 and 1, inclusive.

2. $\sum P(A) = 1$ The sum of all the probabilities for each event in the sample space equals 1.

4.29   The three success ratings (highly successful, successful, and not successful) appear to be non intersecting, and their union appears to be the entire sample space. If this is true, none of the three sets of probabilities are appropriate.

Judge A has a total probability of 1.2. The total must be exactly 1.0.

Judge B has a negative probability of -0.1 for one of the events. All probability numbers are between 0.0 and 1.0.

Judge C has a total probability of 0.9. The total must be exactly 1.0.

4.31   a. You can expect a 1 to occur approximately 1/6th of the time when you roll a die repeatedly.

b. When one coin is tossed one time, there are two possible outcomes, heads or tails. Each is as likely as the other, therefore each has a probability of occurring of 1/2. 50% of the tosses are expected to be heads, the other 50% tails.

4.33   a.    b.

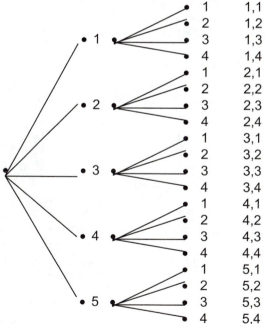

```
1    1,1
2    1,2
3    1,3
4    1,4
1    2,1
2    2,2
3    2,3
4    2,4
1    3,1
2    3,2
3    3,3
4    3,4
1    4,1
2    4,2
3    4,3
4    4,4
1    5,1
2    5,2
3    5,3
4    5,4
```

4.35 a.

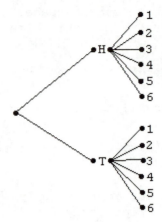

b. {(H,1),(H,2),(H,3),(H,4),(H,5),(H,6),
    (T,1),(T,2),(T,3),(T,4),(T,5),(T,6)}

NOTE: P' is the notation for an experimental or empirical
probability.

4.37 Each student will get different results. These are the
results I obtained:

n(heads)/10	P'(head)/set of 10	Cum.P'(head)
6	0.6	6/10 = 0.60
3	0.3	9/20 = 0.45
5	0.5	14/30 = 0.47
5	0.5	19/40 = 0.48
7	0.7	26/50 = 0.52
4	0.4	30/60 = 0.50
6	0.6	36/70 = 0.51
6	0.6	42/80 = 0.52
6	0.6	48/90 = 0.53
5	0.5	53/100 = 0.53

3	0.3	56/110 = 0.51
4	0.4	60/120 = 0.50
7	0.7	67/130 = 0.52
3	0.3	70/140 = 0.50
6	0.6	76/150 = 0.51
3	0.3	79/160 = 0.49
7	0.7	86/170 = 0.51
7	0.7	93/180 = 0.52
4	0.4	97/190 = 0.51
6	0.6	103/200 = 0.52

p'   Observed Probability of Heads in Sets of Ten

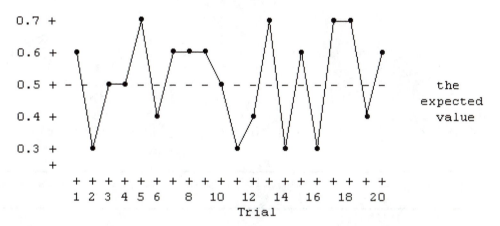

The observed probability varies above and below 0.5, but
seems to average approximately 0.5.

Cumulative Observed Probability of Heads from Sets of Ten

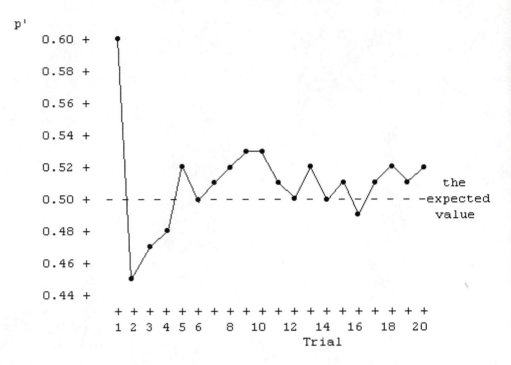

4.39 P(red) = P(green) = P(blue) = P(yellow) = P(purple) = 1/5

The sum of all the probabilities over a sample space is equal to one.   ∑P(x) = 1

4.41 P(P) = 4 · P(F)
        P(P) + P(F) = 1
        4·P(F) + P(F) = 1
        5·P(F) = 1
        P(F) = 1/5
        P(P) = 4·P(F) = 4·(1/5) = 4/5

4.43 a. 1/13   b. 12:1

4.45   a. P(all same suit) = 1/(158,753,389,899 + 1)
              = 0.00000000000629908

b. P(Royal Flush) = 1/(649,739 + 1) = 0.00000153908

c. $0.00000000000629909 = 6.29908 \times 10^{-12}$
    $0.00000153908 = 1.53908 \times 10^{-6}$

4.47   a. S = {HH, HT, TH, TT} and equally likely.

b. S = {HH, HT, TH, TT} and not equally likely; the
       possibilities are the same however.

SECTION 4.3 EXERCISES

4.49   a. (80+55)/300 = 135/300 = 0.45
b. 80/(80+120) = 80/200 = 0.40
c. 55/(55+45) = 55/100 = 0.55

4.51   a. 5128/(5128+9729) = 5128/14857 = 0.345 = 0.35
b. (1375+2383+160+1718)/14857 = 5636/14857 = 0.379 = 0.38
c. (1718+7843)/14857 = 9561/14857 = 0.6435 = 0.64
d. (221+188)/14857 = 409/14857 = 0.0275 = 0.03
e. (188+221+1375)/(188+221+1375+1+7+160)
        = 1784/1952 = 0.9139 = 0.91
f. (1718+7843)/(1718+7843+2383+961) = 9561/12905 = 0.7408
                                            = 0.74
g. (1718+7843)/9729 = 9561/9729 = 0.9827 = 0.98

4.53   a. 18471/31262 = 0.5908 = 0.59
b. 12791/31262 = 0.4091 = 0.41
c. 10986/31262 = 0.3514 = 0.35
d. 5030/18471 = 0.2723 = 0.27
e. 209/694 = 0.3011 = 0.30
f. 7676/12791 = 0.6001 = 0.60
g. 7676/12791 = 0.6001 = 0.60
h. Same answers, different ways of asking same question.

4.55 a. A column total would not work - some of the modes of
     transportation give a total and then are sub-divided into
     subsets. Some categories would be counted twice.
   b. 9036/120191 = 0.0751 = 0.08
   c. 9036/11644 = 0.7760 = 0.78
   d. (5627+847+3408+1049)/120191 = 10931/120191 = 0.0909 = 0.09
   e. 5627/(5627+133+847+3408+1049+3401)
$$= 5627/14465 = 0.3890 = 0.39$$

SECTION 4.4 EXERCISES

COMPLEMENT - Probability of A complement = $P(\overline{A})$

$$P(\overline{A}) = P(\text{not A}) = 1 - P(A)$$

4.57 a. 1 - 0.7 = 0.3
     b. 1 - 0.78 = 0.22

4.59 1 - 0.66 = 0.34

4.61 P(A or B) = 0.4 + 0.5 - 0.1 = <u>0.8</u>

4.63 P(A or B) = P(A) + P(B) - P(A and B)
     0.7 = 0.4 + 0.5 - P(A and B)
     0.7 = 0.9 - P(A and B)
     -0.2 = -P(A and B)
     0.2 = P(A and B)

4.65 Let A = work part-time; $\overline{A}$ = work fulltime;
     B = earn more than $20,540
     P(A) = 0.37; $P(\overline{A})$ = 1 - 0.37 = 0.63; P(B) = 0.50;
     $P(\overline{A}$ and B) = 0.32
     $P(\overline{A}$ or B) = 0.63 + 0.50 - 0.32 = 0.81

4.67   Let U = used part and D = defective part

Given info: P(U) = 0.60,   P(U or D) = 0.61,   P(D) = 0.05

P(U or D) = P(U) + P(D) - P(U and D)
    0.61 = 0.60 + 0.05 - P(U and D)

P(U and D) = 0.60 + 0.05 - 0.61

P(U and D) = <u>0.04</u>; 4% are both used and defective

Same answer as in exercise 4.27, using a Venn diagram.

<u>PROBABILITY - THE MULTIPLICATION RULE & THE CONDITIONAL</u>

The <u>probability of event A given event B</u> has occurred is a conditional probability, written as P(A|B).

For any two events, the <u>probability of events A and B</u> occurring simultaneously is equal to:

1. the probability of event A times the probability of event B, given event A has already occurred: that is:
$$P(A \text{ and } B) = P(A) \cdot P(B|A)$$

OR

2. the probability of event B times the probability of event A, given event B has already occurred: that is:
$$P(A \text{ and } B) = P(B) \cdot P(A|B)$$

4.69   P(A and B) = P(A) · P(B|A) = 0.7 · 0.4 = 0.28

4.71   P(A and B) = P(A) · P(B|A)
       0.3 = 0.6 · P(B|A)   (divide both sides by 0.6)
       0.5 = P(B|A)

4.73   Given: P(accurate reading) = 0.98; P(clean (not user)) = 0.90
       therefore P(user) = 0.10.

P(user and accurate reading (fails test)) = 0.10 · 0.98
                                          = 0.098

4.75 Given: P(extra deductions) = P(E) = 0.10;
          P(extra deductions and deny it) = P(E and D) = 0.09

P(D|E) = P(D and E)/P(E)
P(D|E) = 0.09/0.10 = 0.90 or 90%

4.77 tree diagram

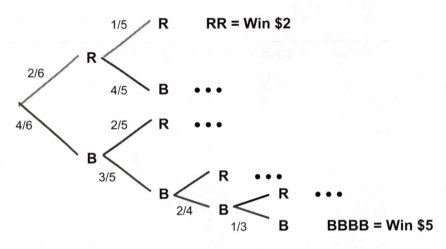

1ST Draw   2nd Draw   3rd Draw   4th Draw

b. 2/5 or 1/5, depending on whether or not the first pick was
   red or blue

c. $\dfrac{4}{6}\cdot\dfrac{3}{5}\cdot\dfrac{2}{4}\cdot\dfrac{1}{3}=\dfrac{1}{15}=0.067$

d. P($2) = $\dfrac{2}{6}\cdot\dfrac{1}{5}$ = 0.067 Winning $2 and winning $5 have the
   same probability.

4.79 1st: P(A and B) = P(B) · P(A|B) = 0.4 · 0.2 = 0.08

P(A or B) = P(A) + P(B) - P(A and B)

= 0.30 + 0.40 - 0.08 = <u>0.62</u>

4.81 1st: P(A or B) = P(A) + P(B) - P(A and B)

$$0.66 = 0.4 + 0.3 - P(A \text{ and } B)$$
$$0.66 = 0.7 - P(A \text{ and } B)$$
$$-0.04 = -P(A \text{ and } B)$$
$$0.04 = P(A \text{ and } B)$$

P(A and B) = P(B and A) = P(B) · P(A|B)

$$0.04 = 0.3 \cdot P(A|B) \quad \text{(divide both sides by 0.3)}$$
$$0.133 = P(A|B)$$

4.83 a. P(B) = 1 - P($\overline{B}$) = 1 - 0.4 = 0. 6

b. P(A or B) = P(A) + P(B) - P(A and B)

$$1.0 = P(A) + 0.6 - 0.3$$
$$1.0 = P(A) + 0.3$$
$$0.7 = P(A)$$

c. P(B and A) = P(B) · P(A|B)

$$0.3 = 0.6 \cdot P(A|B) \quad \text{(divide both sides by 0.6)}$$
$$0.5 = P(A|B)$$

4.85 Given: P(A) = 0.5; P(A|B) = 0.25; P(B|A) = 0.2

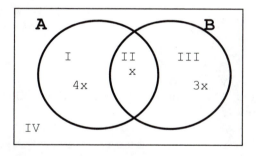

P(B|A) = P(II)/P(A) = x/5x = 0.2
→ P(A) = 0.5 = 5x   → x = 0.1

a. P(B) = 4x = 0.4

b. P($\overline{A}$) = 1 - P(A) = 1 - 0.5 = 0.5
   P(IV) = 1 - P(I + II + III) = 1 - 0.8 = 0.2
   P($\overline{B}$|$\overline{A}$) = P(IV)/P($\overline{A}$) = 0.2/0.5 = 0.4

<u>Mutually Exclusive Events</u> - events that cannot occur at the same time (they have no sample points in common).

<u>Not Mutually Exclusive Events</u> - events that can occur at the same time (they have sample points in common).

4.87 a. <u>Not</u> <u>mutually</u> <u>exclusive</u>. *One head* belongs to both events, therefore the two event intersect.

   b. <u>Not</u> <u>mutually</u> <u>exclusive</u>. All sales that *exceed $1000* also *exceed $100*, therefore $1200 belongs to both events and the two events have an intersection.

   c. <u>Not</u> <u>mutually</u> <u>exclusive</u>. The student selected could be both *male* and *over 21*, therefore the two events have an intersection.

   d. <u>Mutually</u> <u>exclusive</u>. The total cannot be both *less than* 7 and *more than 9* at the same time, therefore there is no intersection between these two events.

PROBABILITY - THE ADDITION RULE

P(A or B) = P(A) + P(B) - P(A and B) if A and B are <u>not</u> mutually exclusive

The <u>probability of event A or event B</u> is equal to the probability of event A plus the probability of event B, minus the probability of events A and B occurring at the same time(otherwise that common probability is counted twice).

P(A or B) = P(A) + P(B) if A and B are mutually exclusive

The <u>probability of event A or event B</u> is equal to the probability of event A plus the probability of event B, if A and B have nothing in common (i.e., they cannot occur at the same time).

P(A and B) = 0 if A and B are mutually exclusive

The <u>probability of events A and B</u> occurring at the same time is impossible if A and B are mutually exclusive.

4.89 If two events are mutually exclusive, then there is no intersection. The event, *A and B*, is the intersection. If no intersection, then P(A and B) = 0.0.

4.91 a. $P(\overline{A})$ = 1 - 0.3 = 0.7

   b. $P(\overline{B})$ = 1 - 0.4 = 0.6

   c. P(A or B) = 0.3 + 0.4 = 0.7

   d. P(A and B) = 0.0   (Mutually exclusive events have no intersection.)

4.93 a. Yes, they can not occur at the same time; i.e., a student can not be both male and female.
   b. No, they can occur at the same time; i.e., a student can be both male and registered for statistics.
   c. No, they can occur at the same time; i.e., a student can be both female and registered for statistics.
   d. Yes, the probability of being female at this college plus the probability of being male at this college equals one.
   e. No, the two events do not include all of the students.
   f. Yes, in both situations there are no common elements shared by the two events.
   g. No, two complementary events comprise the sample sample; two mutually exclusive events do not necessarily make up the whole sample space.

4.95 a. A & C and A & E are mutually exclusive because they cannot occur at the same time.

   b. P(A or C) = P(A) + P(C) = 6/36 + 6/36 = 12/36
      P(A or E) = P(A) + P(E) = 6/36 + 5/36 = 11/36
      P(C or E) = P(C) + P(E) - P(C and E)
                = 6/36 + 5/36 - 1/36 = 10/36

4.97 a. 'daytime' and 'evening' are mutually exclusive because
       they cannot occur at the same time.
   b. 'preschool' and 'levels' are mutually exclusive because
       they cannot occur at the same time.
   c. 'daytime' and 'preschool' are not mutually exclusive, they
       can occur at the same time.
   d. P(preschool) = (26 + 29)/(105 + 74) = 55/179 = 0.307
   e. P(daytime) = 105/179 = 0.5865 = 0.587
   f. P(not levels) = 1 - P(levels) = 1 - (75 + 39)/179
$$= 1 - 114/179$$
$$= 1 - 0.6368$$
$$= 0.3632 = 0.363$$
   g. P(preschool or evening) = 55/179 + 74/179 - 29/179
$$= 100/179$$
$$= 0.5586 = 0.559$$
   h. P(preschool and daytime) = 26/179 = 0.1452 = 0.145
   i. P(daytime|levels) = 75/(75 + 39) = 75/114 = 0.6578 = 0.658
   j. P(adult & diving|evening) = 6/74 = 0.0810 = 0.081

4.99 (0.62 · 0.58) + (0.43 · 0.42) = 0.3596 + 0.1806 = 0.5402
$$= 0.54$$

SECTION 4.6 EXERCISES

Independent Events - when there are independent events, the
occurrence of one event **has no effect** on the probability of the
other event.

4.101 a. independent   b. not independent

   c. independent   d. independent

   e. not independent (If you do not own a car, how can your car
       have a flat tire?)

   f. not independent

```
┌─────────────────────────────────────────────────────────────────────┐
│              PROBABILITY - THE MULTIPLICATION RULE                    │
│                                                                       │
│  For two independent events:                                          │
│                                                                       │
│  1. the probability of events A and B occurring simultaneously        │
│     is equal to the probability of event A times the probability      │
│     of event B                                                        │
│                     P(A and B) = P(A) · P(B)                          │
│                                                                       │
│  2. the conditional probabilities are equal to the single event       │
│     probabilities                                                     │
│                  P(A|B) = P(A)    AND    P(B|A) = P(B)                 │
│                                                                       │
│             FORMULAS FOR CONDITIONAL PROBABILITIES                     │
│                                                                       │
│  P(A|B) = P(A and B)     and     P(B|A) = P(A and B)                  │
│              P(B)                            P(A)                      │
│                                                                       │
│  As in Section 4.3, conditionals can also be computed without the     │
│  formulas above.  Suppose P(A|B) is desired.  The word given,(|), in  │
│  the conditional tells what the newly reduced sample space is.  The   │
│  number of elements in the reduced sample space, n(B), becomes the    │
│  denominator in the probability fraction.  The numerator is the       │
│  number of elements in the reduced sample space that satisfy the      │
│  first event, n(A and B).  Therefore:  P(A|B) = n(A and B)            │
│                                                      n(B)              │
└─────────────────────────────────────────────────────────────────────┘
```

4.103 $P(A \text{ and } B) = P(A) \cdot P(B) = 0.7 \cdot 0.4 = \underline{0.28}$

4.105 $P(A \text{ and } B) = P(A) \cdot P(B)$
$0.3 = 0.6 \cdot P(B)$ (divide both sides by 0.6)
$0.5 = P(B)$

4.107 a. $P(A \text{ and } B) = P(A) \cdot P(B) = 0.3 \cdot 0.4 = \underline{0.12}$

b. $P(B|A) = P(B) = \underline{0.4}$

c. $P(A|B) = P(A) = \underline{0.3}$

4.109 a. P(A and B) = P(B) · P(A|B)

          0.20   = 0.4 · P(A|B);   therefore, P(A|B) = <u>0.5</u>

    b. P(A and B) = P(A) · P(B|A)

          0.20 = 0.3 · P(B|A);   therefore, P(B|A) = <u>0.667</u>

    c. No, A and B are not independent events.

Note: A independent of B and A independent of C <u>does not</u> imply that B and C are independent.

4.111 a. P(A) = 12/52 = 3/13 and P(A|B) = 6/26 = 3/13

      Therefore, A and B are <u>independent</u> events.

    b. P(A) = 12/52 = 3/13 and P(A|C) = 3/13

      Therefore, A and C are <u>independent</u> events.

    c. P(B) = 26/52 = 1/2 and P(B|C) = 13/13 = 1

      Therefore, B and C are <u>dependent</u> events.

4.113 a. P(no life insurance) = 1 - 0.49 = <u>0.51</u>

    b. P(18-24 purchase life insurance) = <u>0.15</u>

    c. P(no life insurance and 25-34 will purchase life insurance)

        = (0.51)(0.26) = <u>0.1326</u>

4.115 a. P(2 households have 3 or more vehicles)
      = (0.17)(0.17) = 0.0289
    b. P(neither household has 3 or more vehicles)
      = (0.83)(0.83) = 0.6889
    c. P(4 households have 3 or more vehicles)
      = (0.17)(0.17)(0.17)(0.17) = 0.0008

4.117 P(3 grandparents are primary caregivers)

      = (0.42)(0.42)(0.42) = 0.0741

4.119 Let C = correct decision, I = incorrect decision

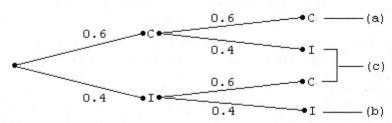

1st person   2nd person

a. P(right decision) = P(C1 and C2) = 0.6·0.6 = <u>0.36</u>
b. P(wrong decision) = P(I1 and I2) = 0.4·0.4 = <u>0.16</u>

c. P(delay) = P[(C1 and I2) or (I1 and C2)]
          = 0.6 · 0.4 + 0.4 · 0.6 = 0.24 + 0.24 = <u>0.48</u>

4.121 a. P(odd) = <u>3/5</u>

b. P(neither odd) = (2/5)·(2/5) = 4/25 = <u>0.16</u>

P(exactly one odd) = [(2/5)·(3/5)] + [(3/5)·(2/5)]
                   = 12/25 = <u>0.48</u>

P(both odd) = (3/5)·(3/5) = 9/25 = <u>0.36</u>

4.123 a. Whether or not the events part-time and graduate
        within five years are independent.
     b. No. Whether a student is part-time or full-time will make
        a difference in how soon he/she will graduate.
     c. P(PT and G) = P(PT)·P(G|PR) + P(PT)·P(G|PU)
                    = (0.42)(0.419) + (0.42)(0.551)
                    = 0.17598 + 0.23142 = <u>0.4074</u>

4.125 a. Two events are mutually exclusive if they cannot occur at the same time or they have no elements in common.

b. Two events are independent if the occurrence of one has no effect on the probability of the other.

c. Mutually exclusive has to do with whether or not the events share common elements; while independence has to do with the effect one event has on the other event's probability.

4.127 a. $P(G|H) = P(G \text{ and } H)/P(H) = 0.1/0.4 = \underline{0.25}$

b. $P(H|G) = P(G \text{ and } H)/P(G) = 0.1/0.5 = \underline{0.2}$

c. $P(\overline{H}) = 1 - P(H) = 1 - 0.4 = \underline{0.6}$

d. $P(G \text{ or } H) = P(G) + P(H) - P(G \text{ and } H)$
$= 0.5 + 0.4 - 0.1 = \underline{0.8}$

e. $P(G \text{ or } \overline{H}) = P(G) + P(\overline{H}) - P(G \text{ and } \overline{H})$
$= 0.5 + 0.6 - 0.4 = \underline{0.7}$

f. No. G and H have an intersection, $P(G \text{ and } H) = 0.1$; therefore, they are not mutually exclusive.

g. No. $P(G|H)$ does not equal $P(G)$

4.129 a. $P(M \text{ and } N) = \underline{0.0}$ (they are mutually exclusive)

b. $P(M \text{ or } N) = P(M) + P(N) = 0.3 + 0.4 = \underline{0.7}$

c. $P(M \text{ or } \overline{N}) = P(\overline{N}) = 1 - P(N) = 1 - 0.4 = \underline{0.6}$ (M is a subset of $\overline{N}$ since M and N are mutually exclusive.)

d. $P(M|N) = \underline{0.0}$ (they are mutually exclusive)

e. $P(M|\overline{N}) = P(M \text{ and } \overline{N})/P(\overline{N}) = 0.3/0.6 = \underline{0.5}$

f. No. Mutually exclusive events are disjoint, therefore
   they must be dependent.

4.131 a. P(satisfied|unskilled) = (150+100)/(250+150) = <u>0.625</u>

b. P(satisfied|skilled female) = 25/100 = <u>0.25</u>

c. Compare P(satisfied|skilled female) to
   P(satisfied|unskilled female)

P(satisfied|skilled female) = 25/100 = <u>0.25</u>

P(satisfied|unskilled female) = 100/150 = <u>0.667</u>

Since these two probabilities are not equal, therefore the
events are <u>not</u> <u>independent</u>.

CHAPTER EXERCISES

4.133 a. 324/1897 = 0.1708 = 0.17
b. 323/1897 = 0.1703 = 0.17
c. (289 + 485)/1897 = 774/1897 = 0.4080 = 0.41
d. Without replacement – need two different violations. If
   the first violation was returned before second selection,
   it would be possible to draw the same one again.

4.135 P[(med or sh) and (mod or sev)]
                 = (90 + 121 + 35 + 54)/1000 = <u>0.300</u>

4.137 a. S = {GGG, GGR, GRG, GRR, RGG, RGR, RRG, RRR}

b. P(exactly one R) = <u>3/8</u>

c. P(at least one R) = <u>7/8</u>

4.139 P(boy) is approximately equal to 7/8

4.141 a. 133/332 = 0.4006 = 0.40
b. 164/332 = 0.4939 = 0.49
c. 21/332 = 0.0632 = 0.06
d. (164+199−90)/332 = 273/322 = 0.8222 = 0.82
e. 38/95 = 0.40
f. 90/199 = 0.4522 = 0.45

4.143 a. Answers will vary, may include Brazil, Spain, India, etc.
b. The condition is, 'based on countries included.'

c. Total number = 49.8+27.4+42.3+3.8+22.1+13.8+9.8+132.4
= 301.4
132.4/301.4 = 0.4393 = 0.44 = 44%

d. 132.4/301.4 = 0.4393 = 0.44

e. Both are asking the same question but require the answer
in a different format, percentage versus decimal form.

4.145 If event A is a subset of event B, then P(A and B) = P(A)

P(A or B) = P(A) + P(B) − P(A and B)
= P(A) + P(B) − P(A) = P(B)

Note the wording: pink seedless denotes pink <u>and</u> seedless. Use
formulas accordingly.

4.147 a. P(seedless) = (10+20)/100 = <u>0.30</u>

b. P(white) = (20+40)/100 = <u>0.60</u>

c. P(pink and seedless) = 10/100 = <u>0.10</u>

d. P(pink or seedless) = (10+20+30)/100 = <u>0.60</u>

e. P(pink|seedless) = 0.10/0.30 = <u>0.333</u>

f. P(seedless|pink) = 0.10/0.40 = <u>0.25</u>

> Rearrange probability formulas in order needed. Remember
> (A and B) = P(A) · P(B|A) is the same as P(B and A) = P(B)· P(A|B)
> since A and B is the same as B and A.

4.149 a. P(A and B) = P(B)·P(A|B) = (0.36)·(0.88) = <u>0.3168</u>

b. P(B|A) = P(A and B)/P(A) = 0.3168/0.68 = <u>0.4659</u>

c. <u>No.</u> P(A) does not equal P(A|B)

d. <u>No.</u> P(A and B) does not equal 0.0

e. It would mean that the two events "candidate wants job"
and "RJB wants candidate" could not both happen.

4.151 a. P(both damage free) = (10/15)·(9/14) = <u>0.429</u>

b. P(exactly one) = (10/15)·(5/14) + (5/15)·(10/14)
$$= \underline{0.476}$$

c. P(at least one)= 0.429 + 0.476 = <u>0.905</u>

4.153 P(satisfactory) = P(all good) = p^6

a. P(satisfactory|p=0.9) = 0.9^6 = <u>0.531</u>

b. P(satisfactory|p=0.8) = 0.8^6 = <u>0.262</u>

c. P(satisfactory|p=0.6) = 0.6^6 = <u>0.047</u>

4.155 a. 149070013/293027571 = 0.5087 = 0.51
b. 60836722/293027571 = 0.2076 = 0.21
c. 97756380/293027571 = 0.3336 = 0.33
d. (149070013+36251160-21172956)/293027571
   = 164148217/293027571 = 0.56018 = 0.56
e. 29713748/149070013 = 0.1993 = 0.20
f. 97756380/195939689 = 0.4989 = 0.50
g. P(M) is not = P(M|F) since P(M and F) = 0
   The events of Male and female are mutually exclusive.

4.157 a. <u>False.</u> If mutually exclusive, P(R or S) is found by
adding 0.2 and 0.5.

b. <u>True.</u> 0.2 + 0.5 - (0.2·0.5) = 0.6

c. <u>False.</u> If mutually exclusive, P(R and S) must be equal to
zero; there is no intersection.

d. <u>False.</u> 0.2 + 0.5 = 0.7, not 0.6

4.159

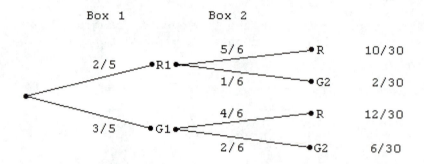

Let G2 = green ball is selected from Box 2 and R1 = red ball
is selected from box 1.

P(G2) = P[(R1 and G2) or (G1 and G2)]
     = P(R1)·P(G2|R1) + P(G1)·P(G2|G1)
     = (2/5)·(1/6) + (3/5)·(2/6) = 2/30 + 6/30 = <u>8/30</u>

4.161 a.

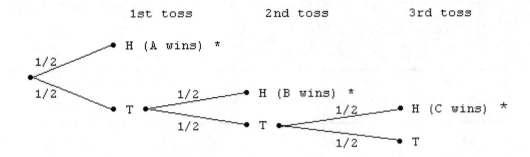

P(A wins on 1st turn) = <u>1/2</u>

P(B wins on 1st turn) = P(A does not)·P(B wins)
= (1/2)·(1/2) = <u>1/4</u>

P(C wins on 1st turn) = P(A does not)·P(B does not)
·P(C wins)
= (1/2)·(1/2)·(1/2) = <u>1/8</u>

b.

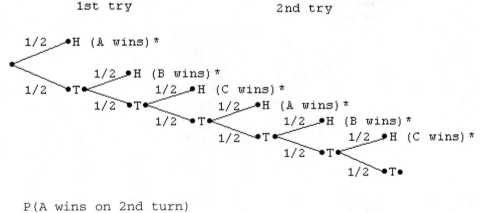

P(A wins on 2nd turn)
= P(A not on 1st)·P(B not)·P(C not)·P(A wins on 2nd)
= (1/2)·(1/2)·(1/2)·(1/2) = 1/16

P(A wins on 1st try or 2nd try) = 1/2 + 1/16 = <u>9/16</u>

P(B wins on 1st try or 2nd try) = 1/4 + 1/32 = <u>9/32</u>

P(C wins on 1st try or 2nd try) = 1/8 + 1/64 = <u>9/64</u>

Exercise 4.163 involves many possibilities and given conditionals.
These are clues that a tree diagram should be used. Assign
probabilities to the branches.

4.163

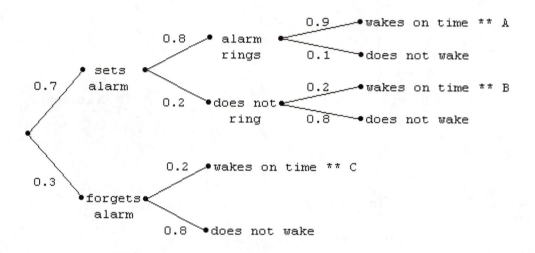

P(wakes on time) = P(A or B or C) = P(A) + P(B) + P(C)
                 = (0.7)(0.8)(0.9) + (0.7)(0.2)(0.2)
                     + (0.3)(0.2)

                 = 0.504 + 0.028 + 0.060 = <u>0.592</u>

4.165 P(2-wk and S) = 3/13 => 12 weeks of the 52 are 2-wk sculptor's, and that is 6 S showings - leaving 14 1-wk S showings - using 26 of the weeks.
Therefore the 22 painters must have 18 1-wk and 4 2-wk showings.

a. (18x1 + 4x2)/52 = 26/52 = 0.50
b. (14x1 + 6x2)/52 = 26/52 = 0.50
c. (18+14)/52 = 32/52 = 0.615
d. (4x2 + 6x2)/52 = 20/52 = 0.385

4.167 Let AW = Team A wins game, BW = Team B wins game

a. P(AW) = 0.60; P(A wins a one game series) = $\underline{0.60}$

b. P(A wins best of 3 game series) =

$$= P[(AW1,AW2) \text{ or } (AW1,BW2,AW3) \text{ or } (BW1,AW2,AW3)]$$

$$= (0.6)(0.6) + (0.6)(0.4)(0.6) + (0.4)(0.6)(0.6)$$

$$= \underline{0.648}$$

c. P(A wins best of 7 game series)
$$= P(\text{A wins in 4 games}) + P(\text{A wins in 5 games}) +$$
$$P(\text{A wins in 6 games}) + P(\text{A wins in 7 games})$$

$$= 1 \cdot (0.6)^4 + 4 \cdot (0.6)^4 (0.4)^1 + 10 \cdot (0.6)^4 (0.4)^2$$
$$+ 20 \cdot (0.6)^4 (0.4)^3$$

$$= 0.1296 + 0.20736 + 0.20736 + 0.16589 = \underline{0.710}$$

d. (a) 0.70   (b) 0.784   (c) 0.874

e. (a) 0.90   (b) 0.972   (c) 0.997

f. The larger the number of games in the series, the greater the chance that the ''best'' team will win.
The greater the difference between the two teams individual chances, the more likely the ''best'' team wins.

4.169 a. The sample space for tossing 3 coins with no conditions is:
{HHH, HHT, HTH, THH, HTT, TTH, THT, TTT}
The condition, at least one is a head, removes TTT from consideration and leaves the sample space being:
{HHH, HHT, HTH, THH, HTT, TTH, THT}
P(3 heads|at least one of the coins shows a head) = 1/7

b. P(3 heads|at least one of the coins shows a head)
$$= P(3H \text{ and at least one } H)/P(\text{at least one } H)$$
$$= P(3H \text{ and at least one } H)/(1 - P(0H))$$
$$= (1/8)/(1 - 1/8) = (1/8)/(7/8) = 1/7$$

PROBABILITY

BASIC PROPERTIES: $0 \leq \text{each probability} \leq 1$ (1)

$$\sum_{\text{over } S} P(A) = 1 \qquad (2)$$

Finding probabilities	From an equally likely sample space	By formula, given certain probabilities
P(A), any event A	$P(A) = \dfrac{n(A)}{n(S)}$ (3)	-does not apply-
P($\overline{A}$), complementary event	$P(\overline{A}) = \dfrac{n(\overline{A})}{n(S)}$ (4)	$P(\overline{A}) = 1.0 - P(A)$ (11)
Any 2 events, no special conditions or relations known: P(A|B), conditional event P(A or B), union of 2 events P(A and B), intersection of 2 events	$P(A\|B) = \dfrac{n(A \text{ and } B)}{n(B)}$ (5) $P(A \text{ or } B) = \dfrac{n(A \text{ or } B)}{n(S)}$ (6) $P(A \text{ and } B) = \dfrac{n(A \text{ and } B)}{n(S)}$ (7)	$P(A\|B) = \dfrac{P(A \text{ and } B)}{P(B)}$ (12) $P(A \text{ or } B) = P(A) + P(B) - P(A \text{ and } B)$ (13) $P(A \text{ and } B) = P(A) \cdot P(B\|A)$ (14)
2 events, known to be mutually exclus. P(A or B) P(A and B) P(A|B)	$P(A \text{ or } B) = \dfrac{n(A) + n(B)}{n(S)}$ (8)	$P(A \text{ or } B) = P(A) + P(B)$ (15) $P(A \text{ and } B) = 0$ (16) $P(A\|B) = 0$ (17)
2 events, known to be independent P(A and B) P(A|B)	$P(A \text{ and } B) = \dfrac{n(A \text{ and } B)}{n(S)}$ (9) $P(A\|B) = \dfrac{n(A \text{ and } B)}{n(B)} = \dfrac{n(A)}{n(S)}$ (10)	$P(A \text{ and } B) = P(A) \cdot P(B)$ (18) $P(A\|B) = P(A)$ (19)

Resulting Properties:

(20) If $P(A)+P(B) = P(A \text{ or } B)$; then A and B are mutually exclusive.

(21) If $P(A) \cdot P(B) = P(A \text{ and } B)$; then A and B are independent.

(22) If $P(A|B) = P(A)$, then A and B are independent.

(23) If $P(A \text{ and } B) = 0$, then A and B are mutually exclusive.

(24) If $P(A \text{ and } B) \neq 0$, then A and B are not mutually exclusive.

The Relationship between Independence and Mutually Exclusive

(25) If events are independent, then they are NOT mutually exclusive.

(26) If events are mutually exclusive, then they are NOT independent.

CHAPTER 5 ▽ PROBABILITY DISTRIBUTIONS (DISCRETE VARIABLES)

Chapter Preview

Chapter 5 combines the "ideas" of a frequency distribution from Chapter 2 with probability from Chapter 4. This combination results in a discrete probability distribution. The main elements of this type distribution will be covered in this chapter. The elements include:

1. discrete random variables
2. discrete probability distributions
3. the mean and standard deviation of a discrete probability distribution
4. binomial probability distribution
5. the mean and standard deviation of a binomial distribution.

An article, published by the National Sleep Foundarion, showing the number of caffeinated beverages adult American say they drink daily, is presented in the opening section, 5.1 Caffeine Drinking.

SECTION 5.1 EXERCISES

5.1   a. 22%
  b. 4+ cups or cans
  c. Number of daily cups or cans of caffeinated beverage.
  d. Yes. The events listed (none, one, two, three, four+) are non-overlapping.

SECTION 5.2 EXERCISES

5.3   One of the random variables is the number of siblings that a classmate has. The possible values for the random variable are
x = 0, 1, 2, 3, ..., n.
The other random variable is the length of the last conversation a class mate had with their mother. The random variable will be a numerical value between 0 and 60 minutes for most classmates.

Random Variable - a numerical quantity whose value depends on the conditions and probabilities associated with an experiment.

Discrete Random Variable - a numerical quantity taking on or having a finite or countably infinite number of values.

Use x to denote a discrete random variable. (x is often a count of something; ex. the number of home runs in a baseball game)

Continuous Random Variable - a numerical quantity taking on or having an infinite number of values. (often a measurement of something; ex. a person's height)

5.5 a. The variable ''number of siblings'' is discrete because it is a count.
The variable ''length of conversation'' is continuous because time is a measureable value.
b. The variable "number of dinner guests" is discrete because it is a count of people.
c. The variable "number of miles" is continuous because distance is a measureable value.

5.7 a. The random variable is the *number of new jobs at a given company*.
b. The random variable is discrete because it represents a count.

5.9 The random variable is the *distance from center to arrow*.
x = 0 to n, where n = radius of the target, measured in inches, including all possible fractions. The variable is continuous.

5.11 a. The random variable is the *average amount of time spent on various activities*.
b. The random variable is continuous because time is a measureable value

SECTION 5.3 EXERCISES

Probability distributions look very much like frequency
distributions. The probability P(x) takes the place of frequency.
The frequency *f* column contains integers (counts), whereas the
probability *P(x)* column contains fractions or decimals between 0
and 1. The probability P(x) relates to relative frequency, the
frequency divided by the size of the data set.

The two main properties of a probability distribution are:

1. $0 \leq$ each P(x) ≤ 1, each probability is a number between 0
    and 1 inclusive.

2. $\sum$ P(x) = 1, the sum of all the probabilities should be
    equal to 1.

Remember to always:

1. make sure each entry in the P(x) column is between 0
   and 1, and

2. sum your P(x) column and check that it is equal
   to 1.

Both properties <u>must</u> exist.

5.13

x	0	1
P(x)	1/2	1/2

<u>Function Notation</u>

P(x) $\Rightarrow$ an equation with x as its variable, which assigns
probabilities to the corresponding or given values.

P(0) $\Rightarrow$ replace x on the right side of the equation with 0
and evaluate.

P(3) $\Rightarrow$ replace x on the right side of the equation with 3
and evaluate.   . . .

```
ex.:  P(x) =   x + 1         P(0) = 0 + 1 =   1      P(3) = 3 + 1 =   4
                26                    26      26              26      26

NOTE: Only evaluate P(x) for the x values in its domain, otherwise
      P(x) = 0.  The domain of a variable is the specified set of
      replacements (x-values).
```

5.15 a. The values of x in a probability distribution form a set of
        mutually exclusive events because they can never overlap.
        Each possible outcome is assigned a unique numerical
        value.
     b. All possible outcomes (values of the random variable) are
        accounted for.

5.17 a.

x	P(x)
1	0.12
2	0.18
3	0.28
4	0.42
Σ	1.00

P(x) is a probability function:

1. Each P(x) is a value between
   0 and 1.

2. The sum of the P(x)'s is 1.

b.

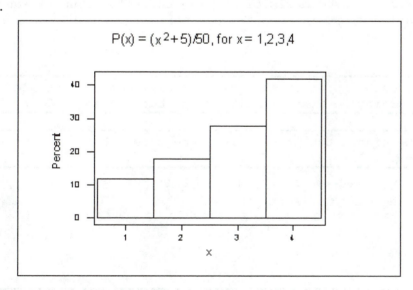

$$P(x) = (x^2+5)/50, \text{ for } x = 1,2,3,4$$

** What shape distribution does the histogram in exercise 5.17b
depict?   (See the bottom of the next page for the answer.)

5.19

x	0	1	2	3
P(x)	0.20	0.30	0.40	0.10

Notice that each P(x) is a value between 0.0 and 1.0, and the sum of all P(x) values is exactly 1.0.

5.21 a.

Number	Proportion
0	0.196
1	0.131
2	0.162
3	0.168
4	0.121
5	0.082
6	0.054
7 to 10	0.072
11 or more	0.014

b. Student now chooses among multiple acceptances

5.23 The percentages do not sum to 1.00, a requirement for a probability distribution. The variable is not a random variable, it is an attribute variable. Random variables are numerical.

For more information on the computer commands to generate discrete data according to a probability distribution, see ES10-p278.

Note: The higher the probability, the more often the number will be generated.

5.25 a. Everyone's generated values will be different. Listed here is one such sample.

```
2  2  3  2  2  5  3  3  2  2  2  1  4  3  3  1
5  1  3  3  3  4  3  3  5
```

**(5.17b) A J-shaped distribution.

---

b. Sample obtained

x	rel.freq.
1	0.12
2	0.28
3	0.40
4	0.08
5	0.12
ALL	1.00

c. Given Distribution

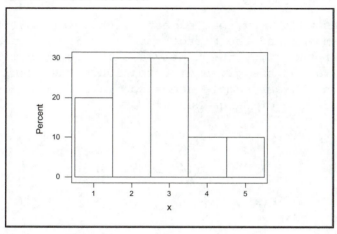

Sample Results

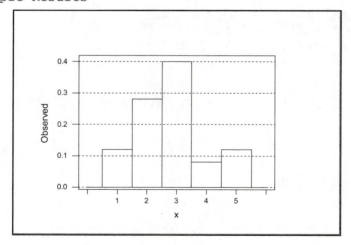

d. The distribution of the sample is somewhat similar to that of the given distribution. The two highest probabilities in the random data occurred at x = 2 and 3, matching the two highest probabilities for the given distribution.

e. Results will vary, but expect: occasionally a sample will have no 4 or no 5 in it, 2 or 3 is almost always the most frequent number, 4 or 5 is almost always the least frequent number, the histograms seem to vary but yet almost always look somewhat like the histogram of the given distribution.

f. Results will vary, but expect: little variability among the samples and the histograms, 4 or 5 occur nearly 10% most of the time, the histograms are quite similar to the histogram of the given distribution. The results indicate that the larger sample seems to stabilize the overall results (Law of Large Numbers).

SECTION 5.4 EXERCISES

5.27 $\sigma^2 = \sum[(x - \mu)^2 \cdot P(x)]$

$= \sum[x^2 - 2x\mu + \mu^2) \cdot P(x)]$

$= \sum[x^2 \cdot P(x) - 2x\mu \cdot P(x) + \mu^2 \cdot P(x)]$

$= \sum[x^2 \cdot P(x)] - 2\mu \cdot \sum[x \cdot P(x)] + \mu^2 \cdot [\sum P(x)]$

$= \sum[x^2 \cdot P(x)] - 2\mu \cdot [\mu] + \mu^2 \cdot [1]$

$= \sum[x^2 \cdot P(x)] - 2\mu^2 + \mu^2$

$= \sum[x^2 \cdot P(x)] - \mu^2$ or $\sum[x^2 \cdot P(x)] - \{\sum[x \cdot P(x)]\}^2$

5.29 The sum of the number values, once each. Nothing of any meaning.

The mean and standard deviation of a probability distribution are μ and σ respectively. They are parameters since we are using theoretical probabilities.

$$\mu = \sum[xP(x)] \qquad \sigma^2 = \sum[x^2P(x)] - \{\sum[xP(x)]\}^2 \Rightarrow \sigma = \sqrt{\sigma^2}$$

$$\text{OR}$$

$$\sigma^2 = \sum[x^2P(x)] - \mu^2 \qquad\qquad \Rightarrow \sigma = \sqrt{\sigma^2} \text{ (easier formula)}$$

5.31

x	R(x)	xR(x)	$x^2R(x)$
0	0.2	0.0	0.0
1	0.2	0.2	0.2
2	0.2	0.4	0.8
3	0.2	0.6	1.8
4	0.2	0.8	3.2
$\sum$	1.0 ck	2.0	6.0

$\mu = \sum[xR(x)] = 2.0$

$\sigma^2 = \sum[x^2R(x)] - \{\sum[xR(x)]\}^2 = 6.0 - \{2.0\}^2 = 2.0$

$\sigma = \sqrt{\sigma^2} = \sqrt{2.0} = 1.4$

5.33 a.

x	P(x)	xP(x)	$x^2P(x)$
0	0.60	0.00	0.00
1	0.30	0.30	0.30
2	0.07	0.14	0.28
3	0.02	0.06	0.18
4	0.01	0.04	0.16
$\sum$	1.0 ck	0.54	0.92

$\mu = \sum[xP(x)] = 0.54$

$\sigma^2 = \sum[x^2P(x)] - \{\sum[xP(x)]\}^2 = 0.92 - \{0.54\}^2 = 0.6284$

$\sigma = \sqrt{\sigma^2} = \sqrt{0.6284} = 0.7927 = 0.79$

b.

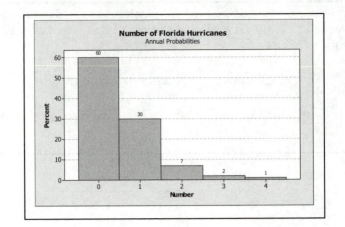

5.35

x	P(x)	xP(x)	$x^2P(x)$
10	0.4	4.0	40.0
11	0.2	2.2	24.2
12	0.2	2.4	28.8
13	0.1	1.3	16.9
14	0.1	1.4	19.6
Σ	1.0 ck	11.3	129.5

$\mu = \Sigma[xP(x)] = 11.3$

$\sigma^2 = \Sigma[x^2P(x)] - \{\Sigma[xP(x)]\}^2 = 129.5 - \{11.3\}^2 = 1.81$

$\sigma = \sqrt{\sigma^2} = \sqrt{1.81} = 1.34536 = 1.35$

5.37   a.

x	P(x)	xP(x)	$x^2P(x)$
1	0.6	0.6	0.6
2	0.1	0.2	0.4
3	0.1	0.3	0.9
4	0.1	0.4	1.6
5	0.1	0.5	2.5
Σ	1.0 ck	2.0	6.0

$\mu = \Sigma[xP(x)] = 2.0$

$\sigma^2 = \Sigma[x^2P(x)] - \{\Sigma[xP(x)]\}^2 = 6.0 - \{2.0\}^2 = 2.0$

$\sigma = \sqrt{\sigma^2} = \sqrt{2.0} = 1.4$

b. $\mu - 2\sigma = 2.0 - 2(1.4) = -0.8$

$\mu + 2\sigma = 2.0 + 2(1.4) = 4.8$

The interval from -0.8 to 4.8 encompasses the numbers 1, 2, 3 and 4.

c. The total probability associated with these values of x is 0.9.

---

* How does this value of 0.9 in exercise 5.37 compare with Chebyshev's theorem?   (See the bottom of this page for answer.)

5.39   a. vary, close to -$0.20
b. -$0.20;
c. close, no, need mean = 0 for a fair game

---

*(5.37) Chebyshev's theorem states that for any shape distribution, at least 75% of the data is within 2 standard deviations of the mean.   90% is well over the minimum limit of 75%.

BINOMIAL EXPERIMENTS must have:

1. n independent repeated trials
   a) n - the number of times the trial is repeated
   b) independent - the probabilities of the outcomes
       remain the same throughout the entire experiment

2. two possible outcomes for each trial
   a) success - the outcome or group of outcomes that is
               the focus of the experiment
   b) failure - the outcome or group of outcomes not
               included in success

3. p = probability of success on any one trial
   q = probability of failure on any one trial (q = 1 - p)

4. x = number of successes when the experiment of all n
       trials is completed. x can range in value from
       0 through n. However, when the experiment is
       completed, x will have exactly one value, that is, the
       number of successes that occurred.

5.41 a. Each question is in itself a separate trial with its own
          outcome having no effect on the outcomes of the other
          questions.

   b. There are four different ways that one correct and three
      wrong answers can be obtained in four questions, each with
      the same probability. The sum of the 4 probabilities is
      the same value as 4 times one of them.

   c. The 1/3 is the probability of success for each question,
      i.e., the probability of choosing the right answer from the
      3 choices. The 4 is the number of independent trials, i.e.,
      the number of questions.
      The expected average would be the sample size times the
      probability of success; if the probability of guessing one
      answer correctly is 1/3, then it seems reasonable that on
      the average one should be able to guess 1/3 of all questions
      correctly.

5.43 The number of defective items should be fairly small and therefore easier to count.

FACTORIALS

$n! = n(n-1)(n-2)\ldots1$ ex.: $3! = 3\cdot2\cdot1 = 6$

$0! = 1$ (this is defined this way so that the algebra of factorials
                      will work)

$$\binom{n}{x} = \binom{n(\text{trials})}{n(\text{successes}) \quad n(\text{failures})} = \frac{n!}{x!(n-x)!}$$

ex.:

$$\binom{8}{3} = \binom{8\text{trials}}{3\text{successes} \quad 5\text{failures}} = \frac{8!}{3!5!} = \frac{8\cdot7\cdot6\cdot5\cdot4\cdot3\cdot2\cdot1}{3\cdot2\cdot1\cdot5\cdot4\cdot3\cdot2\cdot1} \text{ or}$$

$$= \frac{8\cdot7\cdot6\cdot5!}{3\cdot2\cdot1\cdot5!} = 56$$

EXPONENTS

$b^n = b\cdot b\cdot b\cdots$ (n times)   ex.: $.2^3 = (.2)(.2)(.2) = .008$

NOTE: See additional information about factorial notation in Review Lessons.

5.45   a. $4! = 4\cdot3\cdot2\cdot1 = 24$

b. $7! = 7\cdot6\cdot5\cdot4\cdot3\cdot2\cdot1 = 5,040$

c. $0! = 1$ (by definition)

d. $\dfrac{6!}{2!} = \dfrac{6\cdot5\cdot4\cdot3\cdot2\cdot1}{2\cdot1} = 6\cdot5\cdot4\cdot3 = 360$

e. $\dfrac{5!}{3!\cdot2!} = \dfrac{5\cdot4\cdot3\cdot2\cdot1}{3\cdot2\cdot1\cdot2\cdot1} = 10$

f. $\dfrac{6\cdot5\cdot4\cdot3\cdot2\cdot1}{4\cdot3\cdot2\cdot1\cdot2\cdot1} = 15$

g. $(0.3)^4 = (0.3)(0.3)(0.3)(0.3) = 0.0081$

h. $\dfrac{7\cdot6\cdot5\cdot4\cdot3\cdot2\cdot1}{3\cdot2\cdot1\cdot4\cdot3\cdot2\cdot1} = 35$

i. $\dfrac{5!}{2!\cdot3!} = \dfrac{5\cdot4\cdot3\cdot2\cdot1}{2\cdot1\cdot3\cdot2\cdot1} = 10$

j. $\dfrac{3!}{0!\cdot3!} = \dfrac{3\cdot2\cdot1}{1\cdot3\cdot2\cdot1} = 1$

$\dbinom{4}{1}(0.2)^1(0.8)^3 = \dbinom{4}{1}\cdot(0.2)^1\cdot(0.8)^3$ The use of the multiplication dot is optional. They are sometimes used to emphasize that each of the three parts to a binomial must be evaluated separately first, then multiplication can take place.

---

k. $4\cdot(0.2)(0.8)(0.8)(0.8) = 0.4096$

l. $1\cdot1\cdot(0.7)^5 = 0.16807$

5.47 Binomial properties:

n = 100 trials (shirts),
two outcomes (first quality or irregular),
p = P(irregular),
x = n(irregular); any integer value from 0 to 100.

5.49 a. x is not a binomial random variable because the trials are
not independent. The probability of success (get an ace)
changes from trial to trial. On the first trial it is
4/52. The probability of an ace on the second trial
depends on the outcome of the first trial; it is 4/51 if
an ace is not selected, and it is 3/51 if an ace was
selected. The probability of an ace on any given trial
continues to change when the experiment is completed
without replacement.

b. x is a binomial random variable because the trials are
independent. n = 4, the number of independent trials; two
outcomes, success = ace and failure = not ace; p = P(ace)
= 4/52 and q = P(not ace) = 48/52; x = n(aces drawn in 4
trials) and could be any number 0, 1, 2, 3 or 4. Further,
the probability of success (get an ace) remains 4/52 for
each trial throughout the experiment, as long as the card
drawn on each trial is replaced before the next trial
occurs.

5.51 a.

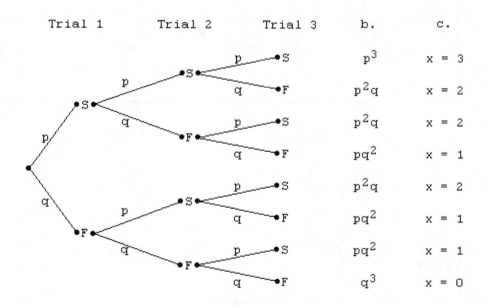

e. $P(x) = \binom{3}{x} p^x q^{3-x}$, for x = 0, 1, 2, 3

$$P(x) = \binom{n}{x} p^x q^{n-x} \text{ for } x = 0,1,2,\ldots n$$

where: $P(x)$ = probability of x successes

n = the number of independent trials

$$\binom{n}{x} = \frac{n!}{x!\,(n-x)!} = \text{binomial coefficient} = \text{the}$$

number of combinations of
successes and failures that
result in exactly x successes in
n trials.   . . .

p^x = probability of x successes, that is, $p \cdot p \cdot p \cdots$,
x times. Remember x is the number of
successes, therefore every time a success
occurs, the probability p is multiplied in.

q^{n-x} = probability of (n-x) failures, that is,
$q \cdot q \cdot q \cdots$, (n-x) times. This is the probability
for "all of the rest of the trials."

Check: The sum of the exponents should equal n.

5.53 $P(x) = \binom{3}{x}(0.5)^x (0.5)^{3-x}$

$P(0) = \binom{3}{0}(0.5)^0 (0.5)^3 = 1(1)(0.125) = 0.125$

$P(2) = \binom{3}{2}(0.5)^2 (0.5)^1 = 3(0.25)(0.5) = 0.375$

$P(3) = \binom{3}{3}(0.5)^3 (0.5)^0 = 1(0.125)(1) = 0.125$

5.55 a. 0.3585
  b. 0.0159
  c. 0.9245

Exercise 5.57, parts a and c show more detailed solutions. Review
factorials on page 183 of this manual, if necessary.

5.57 a. $\binom{4}{1}(0.3)^1(0.7)^3 = \dfrac{4!}{1!3!}(0.3)^1(0.7)^3 = 4(0.3)(0.343) = 0.4116$

  b. $\binom{3}{2}(0.8)^2(0.2)^1 = 0.384$

  c. $\binom{2}{0}(1/4)^0(3/4)^2 = \dfrac{2!}{0!2!}(1/4)^0(3/4)^2 = 1(1)(9/16) = 0.5625$

  d. $\binom{5}{2}(1/3)^2(2/3)^3 = 0.329218$

  e. $\binom{4}{2}(0.5)^2(0.5)^2 = 0.375$

  f. $\binom{3}{3}(1/6)^3(5/6)^0 = 0.0046296$

5.59 By inspecting the function we see the binomial properties:

1. n = 5,

2. p = 1/2 and q = 1/2 (p + q = 1),

3. The two exponents x and 5-x add up to n = 5, and

4. x can take on any integer value from zero to n = 5;

  therefore it is binomial.

By inspecting the probability distribution:

x	T(x)
0	1/32
1	5/32
2	10/32
3	10/32
4	5/32
5	1/32
Σ	32/32 = 1.0

It is a probability distribution.

1. Each T(x) is between 0 and 1.

2. ΣT(x) = 1.0

$$T(x) = \binom{5}{x}(1/2)^x(1/2)^{5-x} \text{ for } x = 0, 1, \ldots, 5$$

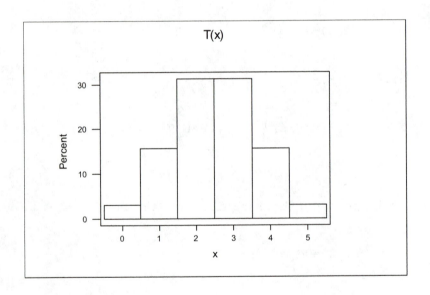

Use Table 2, the Binomial Probability Table, to find the needed
probabilities. Locate n and p, then the particular x or x's. If
more than one x is needed, add the probabilities.

Notation for a Binomial Probability Distribution

$P(x|B(n,p))$ where   x =   whatever values (# of successes) are
                                needed for the problem (can be more
                                than one, just list possibilities)
                         B =   binomial distribution
                         n =   number of trials
                         p =   probability of success

Use Table 2, (Appendix B, ES10-pp807-809), the Binomial
Probabilities table, to find needed probabilities.

5.61 $P(x = 8, 9, 10|B(n = 10, p = 0.90)) = P(8) + P(9) + P(10)$

$$= 0.194 + 0.387 + 0.349 = 0.930$$

5.63   a. $P(x = 2|B(n = 3, p = 0.66)) = \binom{3}{2}(0.66)^2(0.34)^1$

$$= 0.4443 = 0.444$$

b. $P(x = 8|B(n = 12, p = 0.66)) = \binom{12}{8}(0.66)^8(0.34)^4$

$$= 0.2382 = 0.238$$

c. $P(x = 20|B(n = 30, p = 0.66)) = \binom{30}{20}(0.66)^{20}(0.34)^{10}$

$$= 0.1526 = 0.153$$

Terminology in probability problems
For a binomial problem with **n = 10**:

    at least 5 successes $\Rightarrow$ x = 5,6,7,8,9 or 10
    at most 5 successes $\Rightarrow$ x = 0,1,2,3,4 or 5

    at most 9 successes $\Rightarrow$ x = 0,1,2,3,4,5,6,7,8 or 9;
        Since **P(0) + P(1) + ... + P(9)** + P(10) = 1, then
            P(0) + P(1) + ... + P(9) = 1 - P(10).
      Therefore use: P(at most 9) = 1 - P(10).

    at least 2 successes $\Rightarrow$ x = 2,3,4,5,6,7,8,9,10
        Since P(0) + P(1) + **P(2) + P(3) +...+ P(9) + P(10)** = 1,
        then
            P(2) + P(3) +...+ P(9) + P(10) = 1 - [P(0) + P(1)].
      Therefore use: P(at least 2) = 1 - [P(0) + P(1)].

5.65   a. $P(x = 5 | B(n = 5, p = 0.90)) = 0.590$

     b. $P(x = 4,5 | B(n = 5, p = 0.90)) = 0.328 + 0.590 = 0.918$

5.67   P(shut down) = $P(x \geq 2)$, where x represents the number
      defective in the sample of n = 10.

     $P(x \geq 2) = 1.0 - [P(x = 0) + P(x = 1)]$

        $P(x = 0) = \binom{10}{0} (0.005)^0 (0.995)^{10} = 0.9511$

        $P(x = 1) = \binom{10}{1} (0.005)^1 (0.995)^9 = 0.0478$

     $P(x \geq 2) = 1.0 - [0.9511 + 0.0478] = 0.0011$

5.69 $P(x = 1, 2, 3, 4, 5, 6 | B(n = 6, p = 0.5))$
        $= 1 - P(x = 0) = 1 - \binom{6}{0} (0.5)^0 (0.5)^6 = 0.984$

5.71 a. $P(x < 2 | B(n = 5, p = 0.334)) = \binom{5}{0} (0.334)^0 (0.666)^5 +$

$\binom{5}{1} (0.334)^1 (0.666)^4 = [0.13103 + 0.32856] = 0.4596$

b. $P(x > 3 | B(n = 5, p = 0.334)) = [(0.04144 + 0.00422)]$
$= 0.0457$

c. $P(x = 5 | B(n = 5, p = 0.334)) = 0.0042$

5.73 $P(\text{replacement}) = \text{risk} = 1 - (0.886 + 0.107) = 0.007 = 0.7\%$

Computer and/or calculator commands found in ES10-pp292-293 will
save time in answering Exercise 5.75

5.75 a. $P(x \leq 5 | B(n = 20, p = 0.48)) = 0.03132$
b. $P(x \geq 3 | B(n = 20, p = 0.48)) = 1.0000 - 0.00038 = 0.99962$

Additional information on computer and/or calculator commands for
binomial probabilities can be found in ES10-p298.

5.77

x	P(x)	x	P(x)	x	P(x)
1*	0.0000	8	0.1009	15	0.0351
2	0.0003	9	0.1328	16	0.0177
3	0.0015	10	0.1502	17	0.0079
4	0.0056	11	0.1471	18	0.0031
5	0.0157	12	0.1254	19	0.0010
6	0.0353	13	0.0935	20	0.0003
7	0.0652	14	0.0611	21*	0.0001

* any other probabilities are each less than 0.00005

5.79 a. $P(10 \cdot x \cdot 20 | B(n = 50, p = 0.22)) = 0.9988 - 0.3130$
$= 0.6858$

b. $P(30 \cdot x \cdot 40 | B(n = 50, p = 0.58)) = 0.9997 - 0.5539$
$= 0.4458$

c. $P(30 \cdot x \cdot 40 | B(n = 50, p = 0.75)) = 0.8363 - 0.0063$
$= 0.8300$

d. $P(30 \bullet x \bullet 40 | B(n = 50, p = 0.77)) = 0.7436 - 0.0022$
$$= 0.7414$$

e. The values of p are almost complements: p = 0.22 and p = 0.77; the values of x are complements: 10 to 20 and 30 to 40. The roles of x and p have been completely interchanged, thus very similar answers.

f. In parts b, c and d, as p increased, the probability of the interval from 30 to 40 increased

5.81 $P(10 \bullet x \bullet 20 | B(n = 35, p = 0.40)) = 0.9867 - 0.0575 = 0.9292$

5.83 a. Using p = 0.555.

Row	xTwoShots	ShaqP2	ShaqCumP2
1	0	0.198025	0.19802
2	1	0.493950	0.69197
3	2	0.308025	1.00000

Row	x15Shots	ShaqP15	ShaqCumP15
1	0	0.000005	0.00001
2	1	0.000099	0.00010
3	2	0.000868	0.00097
4	3	0.004690	0.00566
5	4	0.017549	0.02321
6	5	0.048153	0.07137
7	6	0.100093	0.17146
8	7	0.160502	0.33196
9	8	0.200176	0.53214
10	9	0.194178	0.72631
11	10	0.145307	0.87162
12	11	0.082375	0.95400
13	12	0.034246	0.98824
14	13	0.009856	0.99810
15	14	0.001756	0.99985
16	15	0.000146	1.00000

Shaq had already made 9 of 13 free throws and he had two more
attempts. That would make 15 free throw attempts for the game.
The chances of him making 2 more seem to be against him. The
binomial probability of him making 2 of 2 is only 0.308, while
the probability that he makes more than 9 in a set of 15 is
0.274. Neither are strong probabilities.

b. Using p = 0.829.

Row	xTwoShots	KobeP2	KobeCmP2
1	0	0.029241	0.02924
2	1	0.283518	0.31276
3	2	0.687241	1.00000

Row	x10Shots	KobeP10	KobeCumP10
1	0	0.000000	0.00000
2	1	0.000001	0.00000
3	2	0.000023	0.00002
4	3	0.000292	0.00032
5	4	0.002480	0.00280
6	5	0.014426	0.01722
7	6	0.058282	0.07550
8	7	0.161455	0.23696
9	8	0.293522	0.53048
10	9	0.316218	0.84670
11	10	0.153301	1.00000

Kobe had made 6 of 8 previous attempts and these 2 would make a
total of 10 for the game. The probability of him making 2
shots in a set of 2 is 0.687 and the probability that he makes
8 or more in a set of 10 is 0.763 - both are strong
probabilities.

SECTION 5.6 EXERCISES

> For a <u>binomial distribution</u> the <u>mean</u> can be calculated using μ = **np**
> and the <u>standard deviation</u> by using σ = $\sqrt{\textbf{npq}}$.
>
> Remember for <u>any type</u> discrete probability distribution:
> $\mu = \Sigma[xP(x)]$ and $\sigma = \sqrt{\sigma^2}$ where $\sigma^2 = \Sigma[x^2P(x)] - [\Sigma[xP(x)]]^2$
>
> Both formulas work for the binomial, however np and $\sqrt{npq}$ are
> quicker and less prone to computational error.

5.85 μ = np = 30·0.6 = 18

$\sigma = \sqrt{npq} = \sqrt{30 \cdot 0.6 \cdot 0.4} = \sqrt{7.2} = 2.68 = 2.7$

5.87   a. Use extension table in exercise 5.86

$\mu = \Sigma[xP(x)] = 0.549 = 0.55$

$\sigma^2 = \Sigma[x^2P(x)] - \{\Sigma[xP(x)]\}^2 = 0.819 - \{0.549\}^2 = 0.5176$

$\sigma = \sqrt{\sigma^2} = \sqrt{0.5176} = 0.7194 = 0.72$

   b. The mean and standard deviation of the probability
      distribution round to exactly the mean and standard
      deviation of the given binomial distribution.

5.89   a. μ = np = 50·(0.5) = 25.0

$\sigma = \sqrt{npq} = \sqrt{50 \cdot (0.5) \cdot (0.5)} = 3.5355 = 3.5$

   b. μ = np = 40·(0.11) = 4.4

$\sigma = \sqrt{npq} = \sqrt{40 \cdot 0.11 \cdot 0.89} = 1.9789 = 1.98$

   c. μ = np = 400·0.06 = 24.0

$\sigma = \sqrt{npq} = \sqrt{400 \cdot 0.06 \cdot 0.94} = 4.7497 = 4.7$

d. $\mu = np = 50 \cdot 0.88 = 44.0$

$\sigma = \sqrt{npq} = \sqrt{50 \cdot 0.88 \cdot 0.12} = 2.298 = 2.3$

5.91 a. x might be approximated by using a binomial random variable. There are only two possible outcomes for each trial of the experiment, there are n = 12 repeated trials. But are the trials independent? [This might depend on whether you are traveling alone or with someone.] Probably not, but the binomial distribution might be a reasonable estimator. x is the count of people and can take on values from zero to 12.

b. $P(x = 4,5 \mid B(n = 12, p = 0.41)) = [(0.2054 + 0.2284)]$
$\qquad\qquad\qquad\qquad\qquad\qquad\qquad = 0.4338$

c. $\mu = np = 12 \cdot 0.41 = 4.92$

$\sigma = \sqrt{npq} = \sqrt{12 \cdot 0.41 \cdot 0.59} = 1.70376 = 1.7$

d.

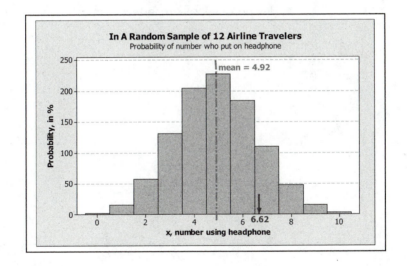

5.93 $\mu = np = 200$ and $\sigma = \sqrt{npq} = 10$, therefore:
$\qquad\qquad npq = 100$
$\qquad\qquad 200q = 100$
$\qquad\qquad q = 100/200 = 0.5$
$\qquad\qquad p = 1 - q = 0.5$
$\qquad\qquad n = 200/0.5 = 400$

5.95 Given that n = 15 and p = 0.4
μ = np = 15(0.4) = 6

$\sigma = \sqrt{npq} = \sqrt{15 \cdot 0.4 \cdot 0.6} = 1.897 = 1.9$

μ + 2σ = 6.0 + 2(1.9) = 6.0 + 3.8 = 9.8
greater than 9.8 gives x = 10,11,12,13,14,15

P(x ≥ 10│B(n = 15,p = 0.4)) = 0.03383

5.97 a.
x	P(x)
0	0.886385
1	0.107441
2	0.005969
3	0.000201
4*	0.000005

*the rest are each less than 0.0000005

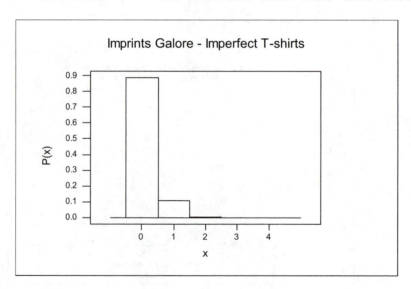

b. P(x = 0│B(n = 12,p = 0.01)) = 0.886385

c.
x	cumP(x)
0	0.88638
1	0.99383
2	0.99979
3*	1.00000

P(x = 0,1│B(n = 12,p = 0.01)) = 0.99383

*the rest are each 1.00000

d. $\mu = np = 12 \cdot (0.01) = 0.12$

$\sigma = \sqrt{npq} = \sqrt{12 \cdot (0.01) \cdot (0.99)} = 0.344674 = 0.345$

e. $\mu - \sigma = 0.12 - 0.345 = -0.225$ and
$\mu + \sigma = 0.12 + 0.345 = 0.465$ includes only x = 0,
therefore the proportion of the distribution is 0.88638

f. $\mu - 2\sigma = 0.12 - 2(0.345) = -0.57$ and
$\mu + 2\sigma = 0.12 + 2(0.345) = 0.81$ includes only x = 0,
therefore the proportion of the distribution is 0.88638

g. The two percentages do not agree with the empirical rule;
the shape is skewed right, not a normal shape. Part (f)
does agree with Chebyshev's proportion of at least 0.75.
Chebyshev's works for any shape distribution.

h.

C4	Count	Simulation	Expected
0	177	177/200 = 0.885	0.886385
1	21	21/200 = 0.105	0.107441
2	2	2/200 = 0.01	0.005969
3	0	0/200 = 0.00	0.000201
4	0	0/200 = 0.00	0.000005
N=	200		

	Simulation	Expected
Mean	0.125	0.12
St. Dev.	0.36059	0.345

i. All simulations were very close to the expected
probabilities.

CHAPTER EXERCISES

5.99 1. Each probability, P(x), is a value between zero and one
inclusive.

2. The sum of all the P(x) is exactly one.

5.101 a.

x	f(x)	
0	(3/4)/[(0!)(3!)] = 0.125	
1	(3/4)/[(1!)(2!)] = 0.375	
2	(3/4)/[(2!)(1!)] = 0.375	
3	(3/4)/[(3!)(0!)] = 0.125	
	Σ 1.000	

f(x) is a probability function since:

i) $0 \leq$ each f(x) ≤ 1
ii) $\sum f(x) = 1.0$

b.

x	f(x)
9	0.25
10	0.25
11	0.25
12	0.25
Σ	1.00

f(x) is a probability function since:

i) $0 \leq$ each f(x) ≤ 1
ii) $\sum f(x) = 1.0$

c.

x	f(x)
1	1.00
2	0.50
3	0.00
4	-0.50
Σ	1.00

f(x) is NOT a probability function since:

f(x=4) is not between 0 and 1

d.

x	f(x)
0	1/25
1	3/25
2	7/25
3	13/25
Σ	24/25

f(x) is NOT a probability function since:

$\sum f(x) = 24/25$, not 1.0

5.103 a. P(exactly 14 arrive) = P(x = 14) = 0.1

b. P(at least 12 arrive) = P(x = 12, 13 or 14)
   = P(12) + P(13) + P(14)
   = 0.2 + 0.1 + 0.1 = 0.4

c. P(at most 11 arrive) = P(x = 10 or 11)
   = P(10) + P(11)
   = 0.4 + 0.2 = 0.6

5.105

x	P(x)	xP(x)	x²P(x)
1	0.10	0.10	0.10
2	0.15	0.30	0.60
3	0.25	0.75	2.25
4	0.35	1.40	5.60
5	0.15	0.75	3.75
$\sum$	1.0	3.30	12.30

$\mu = \sum[xP(x)] = 3.30 = 3.3$

$\sigma^2 = \sum[x^2P(x)] - \{\sum[xP(x)]\}^2 = 12.30 - \{3.30\}^2 = 1.41$

$\sigma = \sqrt{\sigma^2} = \sqrt{1.41} = 1.1874 = 1.187$

5.107 a. $P(x = 0, 1, 2|B(n = 10, p = 0.10)) = P(0) + P(1) + P(2)$

$= 0.349 + 0.387 + 0.194 = 0.930$

b. $P(x = 2, 3, 4, \ldots, 10|B(n = 10, p = 0.10))$

$= 1 - P(x = 0, 1|B(n = 10, p = 0.10))$

$= 1.000 - [0.349 + 0.387] = 0.264$

5.109 $P(x = 5|B(n = 10, p = 0.70)) = 0.103$

5.111 a. $P(x > 4|B(n = 15, p = 0.70)) = 1 - P(x \bullet 4)$

$= 1 - (0+ + 0.001) = 1 - 0.001 = 0.999$

b. $P(x = 10|B(n = 15, p = 0.7)) = 0.206$

c. $P(x < 10|B(n = 15, p = 0.70)) = (0+ + 0.001 + 0.003 + 0.012 + 0.035 + 0.081 + 0.147) = 0.279$

5.113 No, the variable is attribute, not numerical.

5.115 2/3(9) = 6 or more votes for a proposal to be accepted.

P(x = 6, 7|B(n = 7, p = 0.5)) = 0.055 + 0.008 = 0.063

5.117 If x is the random variable n(defective), then *success* is *defective*. On the first selection, P(defective) = 3/10. However, the P(defective) changes for the next selection: it is either 3/9 or 2/9 depending on whether or not the first selection resulted in a defective or not. Since the probability of defective changes, the trials are not independent. Thus, the experiment is not binomial.

5.119 a. P(accepted) = P(x = 0, 1|B(n = 10, p = 0.05))
= P(0) + P(1) = 0.599 + 0.315 = 0.914

b. P(not accepted) = P(x = 2, 3, ... ,10|B(n = 10, p = 0.20))
= 1 - P(x = 0, 1|B(n = 10, p = 0.20))
= 1 - [0.107 + 0.268] = 0.625

c. Even though the P(defective) changes from trial to trial, if the population is very large, the probabilities are very similar. For example, suppose the population has 10 thousand items and 50 are defective. P(defective) on the first trial is 50/10,000 = 0.0050; if after 10 trials 45 defectives have been selected, p(defective) will be 45/9990 = 0.0045.

5.121 $\sigma^2 = \sum x^2 P(x) - \mu^2$ (Formula 5.3b)
$100 = \sum x^2 P(x) - 50^2$ or $\sum x^2 P(x) = 2600$

5.123 a.

x	f	xf
39	1	39
40	2	80
41	3	123
42	4	168
43	6	258
44	7	308
45	8	360
46	4	184
47	3	141
48	1	48
49	1	49
	40	1758

Mean   43.95

Estimated probability of germination = 43.95/50
= 0.879 or 0.88

b.   B(50, 0.88) partly listed below

x	B(50,0.88)	Cum B(50,0.88)
31	0.000002	0.00000
32	0.000008	0.00001
33	0.000032	0.00004
34	0.000118	0.00016
35	0.000395	0.00056
36	0.001208	0.00176
37	0.003352	0.00512
39	0.018974	0.03250
40	0.038264	0.07076
41	0.068439	0.13920
42	0.107547	0.24675
43	0.146731	0.39348
44	0.171186	0.56466
45	0.167382	0.73205
46	0.133420	0.86547
49	0.011424	0.99832
50	0.001675	1.00000

c.

x	B(50,0.88)	40(B(50,0.88))	Expected	Actual
39	0.018974	0.75895	1	1
40	0.038264	1.53055	2	2
41	0.068439	2.73756	3	3
42	0.107547	4.30188	4	4
43	0.146731	5.86923	6	6
44	0.171186	6.84744	7	7
45	0.167382	6.69527	7	8
46	0.133420	5.33681	5	4
47	0.083269	3.33078	3	3
48	0.038165	1.52661	2	1
49	0.011424	0.45694	0	1

d. The distribution for the expected results and the distribution for the actual occurrences are very similar.

5.125 Tool Shop: x = profit

x	P(x)	x·P(x)
100,000	0.10	10,000.0
50,000	0.30	15,000.0
20,000	0.30	6,000.0
-80,000	0.30	-24,000.0
Σ	1.00ck	7,000.0

mean profit = $\mu = \Sigma[x \cdot P(x)] = 7,000.0$

Book Store: x = profit

x	P(x)	x·P(x)
400,000	0.20	80,000.0
90,000	0.10	9,000.0
-20,000	0.40	-8,000.0
-250,000	0.30	-75,000.0
Σ	1.00ck	6,000.0

mean profit = $\mu = \sum[x \cdot P(x)] = 6{,}000.0$

The Tool Shop has a slightly higher mean profit.

5.127 $\mu = \sum[x \cdot P(x)]$
    $= (1) \cdot (1/n) + (2) \cdot (1/n) + \ldots + (n) \cdot (1/n)$
    $= (1/n) \cdot [1 + 2 + 3 + \ldots + n]$
    $= (1/n) \cdot [(n)(n+1)/2]$
    $= (n + 1)/2$

CHAPTER 6 ∇ NORMAL PROBABILITY DISTRIBUTIONS

Chapter Preview

Chapter 6 continues the presentation of probability distributions started in Chapter 5. In this chapter, the random variable is a continuous random variable (versus a discrete random variable in Chapter 5); therefore, the probability distribution is a continuous probability distribution. There are many types of continuous distributions, but this chapter will limit itself to the most common, namely, the normal distribution. The main elements of a normal probability distribution to be covered are:
1. how probabilities are found
2. how they are represented
3. how they are used.

Intelligence and aptitude test score measuring are presented in this chapter's opening section, Intelligence Scores.

SECTION 6.1 EXERCISES

6.1   a. It's a quotient defined by
        [100 x (Mental Age/Chronological Age)]
       b. I.Q.: 100, 16; SAT: 500, 100; Standard score: 0, 1;
       c. z = (I.Q. − 100)/16;   z = (SAT − 500)/100
       d. 2, 132, 700
       e. The percentages are the same (other than for round-off)

SECTION 6.2 EXERCISES

6.3   a. Proportion
       b. Percentage
       c. Probability

Continuous Random Variable
   - a numerical quantity that can take on values on a certain
     interval

STANDARD NORMAL DISTRIBUTION

- bell shaped, symmetric curve
- $\mu = 0$, $\sigma = 1$
- distribution for the standard normal score z

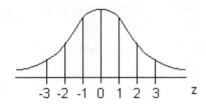

$$z = 0 \implies \mu$$

$z = 1 \implies \mu + 1\sigma$	$z = -1 \implies \mu - 1\sigma$
$z = 2 \implies \mu + 2\sigma$	$z = -2 \implies \mu - 2\sigma$
$z = 3 \implies \mu + 3\sigma$	$z = -3 \implies \mu - 3\sigma$

- area under the curve = 1

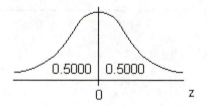

- symmetric, therefore

The Empirical Rule and Table 3

  ≈ 68% of the data lies within 1 standard deviation of the
    mean.
    - Note z = 1, gives .3413. Since symmetric, from z = -1
      to z = 1 would be 2(.3413) = .6828 ≈ 68%.

  ≈ 95% of the data lies within 2 standard deviations of the
    mean.
    - Note z = 2, gives .4772. Since symmetric, from z = -2
      to z = 2 would be 2(.4772) = .9544 ≈ 95%.

  ≈ 99.7% of the data lies within 3 standard deviations of
    the mean.
    - Note z = 3, gives .4987. Since symmetric, from z = -3
      to z = 3 would be 2(.4987) = .9974 ≈ 99.7%.

6.5 a. A bell-shaped distribution with a mean of 0 and a standard
    deviation of 1.

   b. The variable is z; the standard score and this
    distribution is the *standard* or reference used to
    determine the probabilities for all other normal
    distributions.

6.7   a. <u>0.4032</u>   b. <u>0.3997</u>

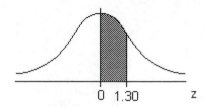   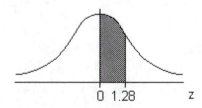

c. <u>0.4993</u>   d. <u>0.4761</u>

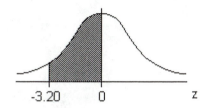   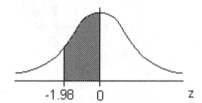

6.9 P(z > 2.03) =
0.5000 − 0.4788 = <u>0.0212</u>

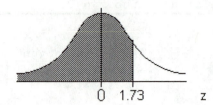

6.11 P(−1.39 < z < 0.00) = <u>0.4177</u>

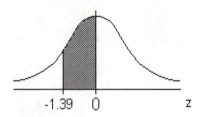

6.13 P(−1.83 < z < 1.23) =
0.4664 + 0.3907 = <u>0.8571</u>

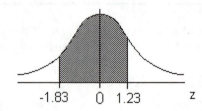

6.15 a. <u>0.4394</u>   b. 0.5000 − 0.4394 = <u>0.0606</u>

        c. 0.5000 + 0.4394 = <u>0.9394</u>   d. 2(0.4394) = <u>0.8788</u>

6.17 a. <u>0.5000</u>   b. 0.5000 - 0.3531 = <u>.1469</u>

c. 0.5000 + 0.4893 = <u>0.9893</u>   d. 0.4452 + 0.5000 = <u>0.9452</u>

e. 0.5000 - 0.4452 = <u>0.0548</u>

6.19 a. <u>0.4906</u>   b. 0.4821 + 0.4904 = <u>0.9725</u>

c. 0.5000 - 0.0517 = <u>0.4483</u>   d. 0.5000 + 0.4306 = <u>0.9306</u>

6.21 a. <u>0.2704</u>   b. 0.3790 + 0.4738 = <u>0.8528</u>

c. 0.5000 - 0.3944 = <u>0.1056</u>   d. 0.5000 + 0.4599 = <u>0.9599</u>

6.23 P(0.75 < z < 2.25) =
        0.4878 - 0.2734 = <u>0.2144</u>

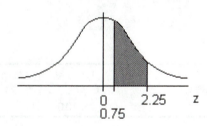

6.25 a. 0.3849 + 0.3888 = <u>0.7737</u>   b. 0.4599 + 0.4382 = <u>0.8981</u>

c. 0.4032 + 0.4951 = <u>0.8983</u>   d. 0.4998 - 0.1368 = <u>0.3630</u>

Look at the *inside* of Table 3 (Appendix B, ES10-p810), and get as close as possible to the probability desired. Locate the position (row and column) on the outside of the table. This will be the corresponding z value. Remember the negative sign if the z value is to the left of μ.

NOTE: NORMAL CURVE

TABLE 3
- inside ⇒ probabilities
- outside ⇒z-values

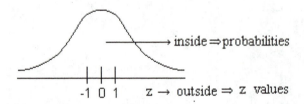

inside ⇒ probabilities

-1 0 1   z → outside ⇒ z values

6.27   a. <u>1.14</u>   b. <u>0.47</u>   c. <u>1.66</u>

　　　d. <u>0.86</u>   e. <u>1.74</u>   f. <u>2.23</u>

Subtract the area desired from 0.5000 first. Table 3 (Appendix B, ES10-p810) works from 0 to a *z* value.
Remember that one half of the distribution is .5000.

6.29   a. <u>1.65</u>   b. <u>1.96</u>   c. <u>2.33</u>

Draw a picture of a normal distribution with the desired area shaded in.
Fill in the needed probabilities.
Locate the appropriate probability in Table 3 (Appendix B, ES10-p810). Locate the corresponding z value.

6.31   -1.28 or +1.28

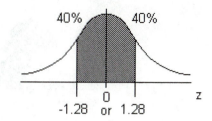

40%   40%

0
-1.28 or 1.28   z

6.33   0.2500   0.2500

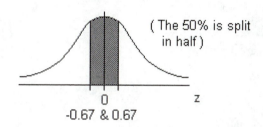

(The 50% is split
in half)

0
-0.67 & 0.67   z

6.35 a.   z = <u>0.84</u>

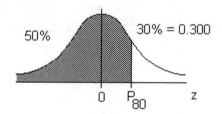

50%   30% = 0.300

0   P₈₀   z

b.   0.75/2 = 0.375   -1.15 and +1.15

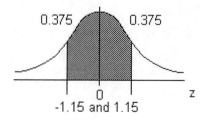

0.375   0.375

0
-1.15 and 1.15   z

6.37

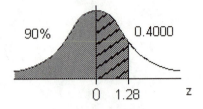

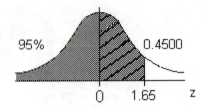

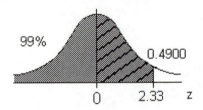

SECTION 6.4 EXERCISES

$\underline{x}$ is still used to denote a <u>continuous random variable</u>, but x is now referred to on an interval, not a single value.(ex.: a < x < b) As before, any letter may be used; x is the most common.

6.39 a. 0.7606
       b. 0.0386
       c. 0.2689

6.41 $z = (x - \mu)/\sigma = (58 - 43)/5.2 = \underline{2.88}$

<u>Applications of the Normal Distribution</u>

1. Draw a sketch of the desired area, noting given μ and σ.

2. Write the desired probability question in terms of the given variable - usually x (ex.: P(x > 10)).

3. Transform the given variable x into a z value using
   $z = (x - \mu)/\sigma$.

6.43 Use formula $z = (x - \mu)/\sigma$:

a. $P[x > 60] = P[z > (60 - 60)/10]$

   $= P[z > 0.00] = \underline{0.5000}$

b. $P[60 < x < 72] = P[(60 - 60)/10 < z < (72 - 60)/10]$
   $= P[0.00 < z < 1.20] = \underline{0.3849}$

c. $P[57 < x < 83] = P[(57 - 60)/10 < z < (83 - 60)/10]$
   $= P[-0.30 < z < 2.30]$
   $= 0.1179 + 0.4893 = \underline{0.6072}$

d. $P[65 < x < 82] = P[(65 - 60)/10 < z < (82 - 60)/10]$
   $= P[0.50 < z < 2.20]$
   $= 0.4861 - 0.1915 = \underline{0.2946}$

e. $P[38 < x < 78] = P[(38 - 60)/10 < z < (78 - 60)/10]$
   $= P[-2.20 < z < 1.80]$
   $= 0.4861 + 0.4641 = \underline{0.9502}$

f. P[x < 38] = P[z < (38 - 60)/10]
            = P[z < -2.20]
            = 0.5000 - 0.4861 = 0.0139

6.45 a. P(100 < x < 120) = P[(100 - 100)/16 < z < (120 - 100)/16]
                         = P[0.00 < z < 1.25] = 0.3944

      b. P(x > 80) = P[z > (80 - 100)/16]
               = P[z > - 1.25]
               = 0.3944 + 0.5000 = 0.8944

6.47 Use formula z = (x - μ)/σ:

      a. P[7200 < x < 10800]
               = P[(7200 - 9000)/1800 < z < (10800 - 9000)/1800]
               = P[-1.00 < z < 1.00]
               = 0.3413 + 0.3413 = 0.6826 = 68.26%

      b. P[5400 < x < 12600]
               = P[(5400 - 9000)/1800 < z < (12600 - 9000)/1800]
               = P[-2.00 < z < 2.00]
               = 0.4772 + 0.4772 = 0.9544 = 95.44%

      c. P[3600 < x < 14400]
               = P[(3600 - 9000)/1800 < z < (14400 - 9000)/1800]
               = P[-3.00 < z < 3.00]
               = 0.4987 + 0.4987 = 0.9974 = 99.74%

      d. 0.6826 ≈ 68%;   0.9544 ≈ 95%;   0.9974 ≈ 99.7%

6.49 a. P[17 < x < 22]
               = P[(17 - 44.5)/17.1 < z < (22 - 44.5)/17.1]
               = P[-1.61 < z < -1.32]
               = 0.4463 - 0.4066 = 0.0397 = 4.0%

      b. P(x < 25) = P[z < (25 - 44.5)/17.1]
                   = P[z < -1.14]
                   = 0.5000 - 0.3729 = 0.1271 = 12.7%

c. P(x > 21) = P[z > (21 - 44.5)/17.1]
                 = P[z > -1.37]
                 = 0.5000 + 0.4147 = <u>0.9147</u> = <u>91.5%</u>

d. P[45 < x < 65]
         = P[(45 - 44.5)/17.1 < z < (65 - 44.5)/17.1]
         = P[0.03 < z < 1.20]
         = 0.3849 - 0.0120 = <u>0.3729</u> = <u>37.3%</u>

e. P(x > 75) = P[z > (75 - 44.5)/17.1]
                 = P[z > 1.78]
                 = 0.5000 - 0.4625 = <u>0.0375</u> = <u>3.8%</u>

6.51 a. P(x > 8) = P[z > (8 - 4.6)/1.4]
                 = P[z > 2.43]
                 = 0.5000 - 0.4925 = <u>0.0075</u> = <u>0.75%</u>

b. P(x < 4) = P[z < (4 - 4.6)/1.4]
                 = P[z < -0.43]
                 = 0.5000 - 0.1664 = <u>0.3336</u> = <u>33.4%</u>

6.53 a.

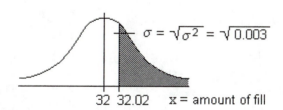

$\sigma = \sqrt{\sigma^2} = \sqrt{0.003}$

32 32.02   x = amount of fill

P[x > 32.02] = P[z > (32.02 - 32.00)/$\sqrt{0.003}$]
                 = P[z > 0.37]
                 = 0.5000 - 0.1443 = <u>0.3557</u>

b. 100(0.3557) = 35.57 ≈ <u>36 bottles</u>

6.55   a. from (24 - 0.4) to (24 + 0.6) or <u>from 23.6 mm to 24.6 mm</u>

b. P[23.6 < x < 24.6]
  = P[(23.6 - 24.0)/0.13 < z < (24.6 - 24.0)/0.13]
  = P[-3.08 < z < 4.62]
  = 0.4990 + 0.499997 = <u>0.998997</u> = <u>99.9%</u>

c. P[x > 24.5] = P[z > (24.5 - 24.0)/0.13]
      = P[z > 3.85]
      = 0.5000 - 0.4999 = <u>0.0001</u> = <u>0.01%</u>

d. P[23.65 < x < 24.35]
  = P[(23.65 - 24.0)/0.13 < z < (24.35 - 24.0)/0.13]
  = P[-2.69 < z < 2.69]
  = 0.4964 + 0.4964 = <u>0.9928</u> = <u>99.3%</u>

Draw a picture first, filling in the given and calculated
probabilities.
Based on the probability, determine the z value.
Use z = (x - μ)/σ   to find x.

6.57   a.

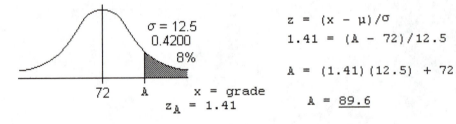

$$\sigma = 12.5$$
0.4200
8%

72   A   x = grade
$z_A = 1.41$

$$z = (x - \mu)/\sigma$$
$$1.41 = (A - 72)/12.5$$

$$A = (1.41)(12.5) + 72$$

$$A = \underline{89.6}$$

b.

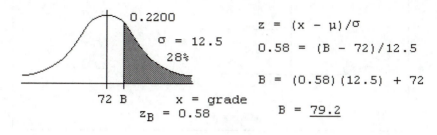

$$z = (x - \mu)/\sigma$$

$$0.58 = (B - 72)/12.5$$

$$B = (0.58)(12.5) + 72$$

$$B = \underline{79.2}$$

c.

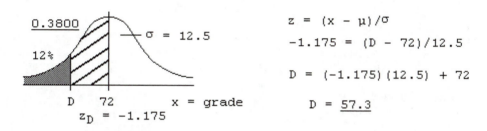

$$z = (x - \mu)/\sigma$$

$$-1.175 = (D - 72)/12.5$$

$$D = (-1.175)(12.5) + 72$$

$$D = \underline{57.3}$$

6.59

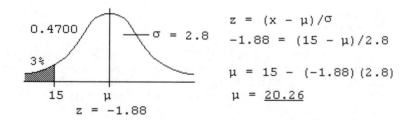

$$z = (x - \mu)/\sigma$$

$$-1.88 = (15 - \mu)/2.8$$

$$\mu = 15 - (-1.88)(2.8)$$

$$\mu = \underline{20.26}$$

6.61

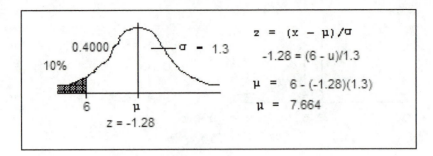

6.63 a. $6/30 = 0.20 = 20\%$

b. $P(x < 47.9) = P[z < (47.9 - 47.9)/\text{std dev}]$
$= P[z < 0.00] = 0.5000 = 50\%$

c. $z = (x - \mu)/\sigma$
$-1.65 = (47.9 - \mu)/1.5$
$-2.475 = 47.9 - \mu$
$\mu = 50.375$

d. $z = (x - \mu)/\sigma$
$-1.65 = (47.9 - \mu)/1.0$
$-1.65 = 47.9 - \mu$
$\mu = 49.55$

e. $z = (x - \mu)/\sigma$
$-2.33 = (47.9 - \mu)/1.5$
$-3.495 = 47.9 - \mu$
$\mu = 51.395$

f. To avoid fines – very expensive, dollars and reputation

g. To avoid putting extra M&M's in bag – If mean and standard deviation are too large, Mars will need lots of extra M&M's to fill the same number of bags or they will be able to fill fewer bags with a given amount of M&M's. Either way, Mars will be cutting into their profit margin.

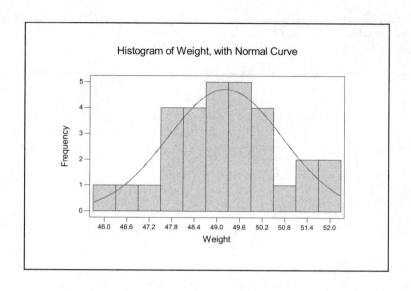

Histogram of Weight, with Normal Curve

6.65 a. $P(x < 525) = \underline{0.056241}$

b. $P(525 < x < 590) = 0.561785 - 0.056241 = \underline{0.505544}$

c. $P(x \geq 590) = 1.000000 - 0.561785 = \underline{0.438215}$

d. $P(x < 525) = P[z < (525 - 584.2)/37.3]$
$$= P[z < -1.59] = 0.5000 - 0.4441 = \underline{0.0559}$$

$P(525 < x < 590) = P[-1.59 < z < (590 - 584.2)/37.3]$
$$= P[-1.59 < z < 0.16]$$
$$= 0.4441 + 0.0636 = \underline{0.5077}$$

$P(x > 590) = P(z > 0.16) = 0.5000 - 0.0636 = \underline{0.4364}$

e. Round-off errors; specifically in (d) when z is calculated to the nearest hundredth and the rounded z score is used with Table 3.

Additional information on computer and calculator commands regarding the normal distribution, can be found in ES10-pp336, 327-330.

6.67 Everyone's generated values will be different, but should have a mean and standard deviation close to 100 and 16, respectively, and be approximately normally distributed.

6.69 Everyone's generated values will be different.

SECTION 6.5 EXERCISES

Z - NOTATION

$z(\alpha) = z_\alpha$ = the z value that has α area to the right of it

1. Draw a picture.
2. Shade in desired α area, starting from the far right tail.
3. Based on the diagram and location, determine the z value using Table 3 (Appendix B, ES10-p810).

6.71 a. z(0.03)   b. z(0.14)   c. z(0.75)

d. z(0.22)   e. z(0.87)   f. z(0.98)

6.73 a. z(0.01)   b. z(0.13)   c. z(0.975)

d. z(0.90)

6.75 a.

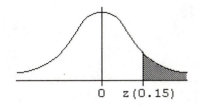

    b.

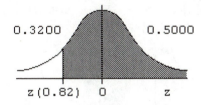

6.77 a. 1.96   b. 1.65   c. 2.33

6.79 a. z(0.08) = <u>1.41</u>

    b. z(0.92) = -<u>1.41</u>

6.81 a. 1.28,   1.65,   1.96,   2.05,   2.33,   2.58

Note the z values. They are the most common occurring z values.
For that reason, Table 4, Part A, has been included in Appendix B
(ES10-p811).   Note Table 4, Part B, for later use.

    b. -2.58,   -2.33,   -2.05,   -1.96,   -1.65,   -1.28

6.83 a. A is an area. z is 0.10 and the area to the right of
z = 0.10; A = 0.5000 - 0.0398 = <u>0.4602.</u>

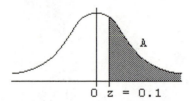

b. B is a z-score. 0.10 is the area to the right of z = B.
Use 0.4000 [0.5000 - 0.1000] to look up the z-score on
Table 3 (Appendix B, ES10-p810). z = B = <u>1.28.</u>

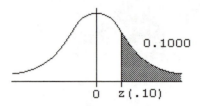

c. C is an area. z is -0.05 and the area to the right of
z = -0.05; C = 0.5000 + 0.0199 = <u>0.5199.</u>

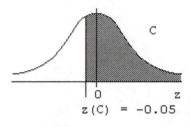

d. D is a z-score. D is to the left of zero [negative], use
   0.4500 [0.5000 - 0.0500] to look it up; z = D = _-1.65._

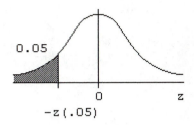

0.05

0

-z(.05)

z

SECTION 6.6 EXERCISES

```
┌─────────────────────────────────────────────────────────────┐
│        Criteria for the Normal Approximation of the Binomial │
│ 1. np and n(1-p) must both be ≥ 5.                           │
│ 2. 0.5 must be added to and/or subtracted from the x values to│
│    allow for an interval.  ex.:  x = 5      ⇒    4.5 < x < 5.5│
│                                  discrete           continuous│
│ 3. μ = np  and  σ = √npq                                     │
│ 4. z = (x - μ)/σ  and Table 3 (Appendix B, ES10-p810) are used to│
│    find probabilities.                                        │
└─────────────────────────────────────────────────────────────┘
```

6.85 np = (100)(0.02) = 2   nq = n(1-p) = (100)(0.98) = 98
No, both np and nq must be greater than or equal to 5.

6.87 Using binomial table (Table 2, Appendix B, ES10-p807;
P(x = 0, 1) = 0.463 + 0.366 = _0.829_
Using normal approximation:
μ = np = (15)(0.05) = 0.75   and
σ = √npq = √(15)(0.05)(0.95) = √0.7125 = 0.844

P(x < 1.5) = P[z < (1.5 - 0.75)/0.844]
           = P(z < 0.89) = 0.5000 + 0.3133 = _0.8133_

6.89 P(x = 4,5) = P(3.5 < x < 5.5)
$\qquad\qquad$ = P[(3.5 − 7.0)/$\sqrt{3.5}$ < z < (5.5 − 7.0)/$\sqrt{3.5}$]
$\qquad\qquad$ = P[−1.87 < z < −0.80]
$\qquad\qquad$ = 0.4693 − 0.2881 = <u>0.1812</u>

$\quad$ P[x = 4,5|B(n = 14,p = 0.5)] = 0.061 + 0.122 = <u>0.183</u>(Table 2,
$\qquad\qquad\qquad\qquad\qquad\qquad\qquad\qquad\qquad$ Appendix B, ES10-p807)

6.91 P(x ≥ 9) = P(x > 8.5)
$\qquad\qquad$ = P[z > (8.5 − 9.1)/1.65]
$\qquad\qquad$ = P[z > −0.36]
$\qquad\qquad$ = 0.5000 + 0.1406 = <u>0.6406</u>

$\quad$ P[x ≥ 9|B(n = 13, p = 0.7)]
$\qquad$ = P(9) + P(10) + P(11) + P(12) + P(13)
$\qquad$ = 0.234 + 0.218 + 0.139 + 0.054 + 0.010
$\qquad$ = <u>0.655</u> from Table 2 (Appendix B, ES10-p807)

6.93 Let x represent the number of patients in the 250 who will
survive melanoma.
$\quad$ x = n(survive)

$\quad$ μ = np = (250)(0.90) = <u>225</u>
$\quad$ σ = $\sqrt{npq}$ = $\sqrt{(250)(0.90)(0.10)}$ = <u>4.74</u>
$\quad$ P(x > 199.5) = P[z > (199.5 − 225.0)/4.74]
$\qquad\qquad\qquad$ = P[z > −5.38]
$\qquad\qquad\qquad$ = 0.5000 + 0.4999997 = <u>0.9999997</u>

6.95 Let x represent the number of female drivers.

$\quad$ μ = np = (50)(0.50) = <u>25</u>
$\quad$ σ = $\sqrt{npq}$ = $\sqrt{(50)(0.50)(0.50)}$ = <u>3.54</u>

$\quad$ a. P(x < 25.5) = P[z < (25.5 − 25)/3.54]
$\qquad\qquad\qquad$ = P[z < 0.14]
$\qquad\qquad\qquad$ = 0.5000 + 0.0557 = <u>0.5557</u>

$\quad$ b. P(x > 37.5) = P[z > (37.5 − 25)/3.54]
$\qquad\qquad\qquad$ = P[z > 3.53]
$\qquad\qquad\qquad$ = 0.5000 − 0.4998 = <u>0.0002</u>

6.97 $\mu = np = (60)[944/(944 + 1,106)] = (60)(0.4605) = 27.63$

$\sigma = \sqrt{npq} = \sqrt{(60)(0.4605)(0.5395)} = 3.86$

At $x = 29.5$: $z = (29.5 - 27.63)/3.86 = 0.48$
Therefore,
$P(x < 29.5) = P(z < 0.48) = 0.5000 + 0.1844 = \underline{0.6844}$

6.99 $\mu = np = (1200)(0.35) = \underline{420}$

$\sigma = \sqrt{npq} = \sqrt{(1200)(0.35)(0.65)} = \underline{16.52}$

a. $P(450 \bullet x \bullet 500) = P(449.5 < x < 500.5)$
$= P[(449.5 - 420)/16.52 < z < (500.5 - 420)/16.52]$
$= P[1.79 < z < 4.87]$ look at these z's
$= 0.5000 - 0.4633 = \underline{0.0367}$

b. Norm. approx. using computer is $1 - 0.965313 = 0.034687$
vs. 0.0367; different by 0.002013

c. Binomial approx. using computer is $1 - 0.967048 = 0.032952$ vs. 0.0367;   different by 0.003748

CHAPTER EXERCISES

6.101 The range from $z = -2$ to $z = +2$ represents two standard deviations on either side of the mean. According to Chebyshev's theorem, there should be *at least 3/4* or *at least 0.75* of a distribution in this interval.

The area under a normal curve is 2(0.4772) or 0.9544.

6.103 a. <u>1.26</u>   b. <u>2.16</u>   c. <u>1.13</u>

6.105 a. $P(|z| > 1.68) = P(z < -1.68) + P(z > +1.68)$
$= 2(0.5000 - 0.4535) = \underline{0.0930}$

b. $P(|z| < 2.15) = P(-2.15 < z < +2.15)$
$= 2(0.4842) = \underline{0.9684}$

6.107 a. <u>1.175</u> or <u>1.18</u>   b. <u>0.58</u>

c. <u>-1.04</u>   d. <u>-2.33</u>

6.109 a. P[x > 25]= P[z > (25-20.9)/4.6]
$\qquad\qquad$ = P[z > 0.89]
$\qquad\qquad$ = 0.5000 - 0.3133 = <u>0.1867</u>

b. P[20 < x < 26]= P[(20-20.9)/4.6 < z < (26-20.9)/4.6]
$\qquad\qquad$ = P[-0.20 < z < 1.11]
$\qquad\qquad$ = 0.0793 + 0.3665 = <u>0.4458</u>

c. P[x < 16]= P[z < (16-20.9)/4.6]
$\qquad\qquad$ = P[z < -1.07]
$\qquad\qquad$ = 0.5000 - 0.3577 = <u>0.1423</u>

6.111 a. P[x < 4000]= P[z < (4000-9600)/2100]
$\qquad\qquad$ = P[z < -2.67]
$\qquad\qquad$ = 0.5000 - 0.4962 = <u>0.0038</u>

b. P[5000 < x < 10000]
$\qquad$ = P[(5000 - 9600)/2100 < z < (10000-9600)/2100]
$\qquad$ = P[-2.19 < z < 0.19]
$\qquad$ = 0.4857 + 0.0753 = <u>0.5610</u>

c. P[x > 16000]= P[z > (16000-9600)/2100]
$\qquad\qquad$ = P[z > 3.05]
$\qquad\qquad$ = 0.5000 - 0.4989 = <u>0.0011</u>

6.113 μ = 2 hours, 50.1 minutes = 170.1 minutes

a. 3 hours = 180 minutes

P[x > 180] = P[z > (180 - 170.1)/20.99]
$\qquad\qquad$ = P[z > 0.47]
$\qquad\qquad$ = 0.5000 - 0.1808 = <u>0.3192</u>

b. P[x < 150] = P[z < (150 - 170.1)/20.99]
$\qquad\qquad$ = P[z < -0.96]
$\qquad\qquad$ = 0.5000 - 0.3315 = <u>0.1685</u>

c. Q_1: $z(0.75) = -0.67$ Q_3: $z(0.25) = +0.67$
   $z = (x - \mu)/\sigma$

   $\pm0.67 = (x - 170.1)/20.99$

   $x = 170.1 + (\pm0.67)(20.99)$

   Bounds for Interquartile range: from 156.0 to 184.2
   minutes

d. Middle 90%: $z(0.95) = -1.65$ $z(0.05) = +1.65$

   $z = (x - \mu)/\sigma$

   $\pm1.65 = (x - 170.1)/20.99$

   $x = 170.1 + (\pm1.65)(20.99)$

   Bounds for Interquartile range: from 135.5 to 204.7
   minutes

6.115

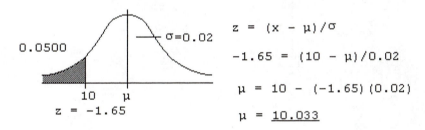

$z = (x - \mu)/\sigma$

$-1.65 = (10 - \mu)/0.02$

$\mu = 10 - (-1.65)(0.02)$

$\mu = \underline{10.033}$

6.117 a. $P[x < 350] = P[z < (350-525)/80] = P[z < -2.19]$
$= 0.5000 - 0.4857 = \underline{0.0143}$

b.

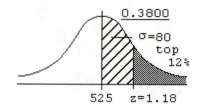

$$z = (x - \mu)/\sigma$$

$$1.18 = (x - 525)/80$$

$$x = 525 + (1.18)(80)$$

$$x = \underline{619.4}$$

c. Q_1 has a z-score of -0.67 and Q_3 has a z-score of +0.67

$$-0.67 = (Q_1 - 525)/80 \qquad\qquad +0.67 = (Q_3 - 525)/80$$

$$Q_1 = 525 + (-0.67)(80) \qquad\qquad Q_3 = 525 + (+0.67)(80)$$

$$Q_1 = \underline{471.4} \qquad\qquad Q_3 = \underline{578.6}$$

Interquartile range = $Q_3 - Q_1$ = 578.6 - 471.4 = $\underline{107.2}$

d.

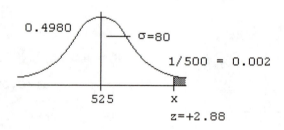

$$z = (x - \mu)/\sigma$$

$$2.88 = (x - 525)/80$$

$$x = 525 + (2.88)(80) = \underline{755.4}$$

6.119 a. The normal approximation is reasonable since both np = 7.5 and nq = 17.5 are greater than 5.

b. μ = np = (25)(0.3) = $\underline{7.5}$

$$\sigma = \sqrt{npq} = \sqrt{(25)(0.3)(0.7)} = \underline{2.29}$$

6.121 a. Use the binomial probablity computer or calculator commands found in ES10-pp292-293.

    b. $P(x \leq 6)$ = 0.005154 + 0.028632 + 0.077943 + 0.138565
                + 0.180904 + 0.184925 + 0.154104 = <u>0.77023</u>

    c. $\mu = np$ = (50)(0.1) = <u>5</u>
       $\sigma = \sqrt{npq} = \sqrt{(50)(0.1)(0.9)}$ = <u>2.12</u>

       Use -0.05 and 6.5 as the data values in the cumulative normal computer or calculator commands found in ES10 pp329-330.

       $P(x \leq 6)$ = 0.760387 - 0.008608 = <u>0.751779</u>

6.123 a. $P[x \leq 75 | B(n = 300, p = 0.2]$ = $P(0)$ + $P(1)$ + $P(2)$ + $P(3)$ +
                                       $P(4)$ + ... + $P(75)$

    b. Use the cumulative binomial probablity computer or calculator commands found in ES10-pp292-293.
       Result: <u>0.9856</u>

    c. $\mu = np$ = (300)(0.2) = <u>60</u>
       $\sigma = \sqrt{npq} = \sqrt{(300)(0.2)(0.8)}$ = <u>6.93</u>

       Use -0.05 and 75.5 as the data values in the cumulative normal computer or calculator commands found in ES10pp329-330

       $P(x \leq 75)$ = 0.987346 - 0.0000 = <u>0.9873</u>

    d. (b) and (c) result in answers that are very close in value.

6.125 $\mu = np$ = (100)(0.80) = <u>80.0</u>
      $\sigma = \sqrt{npq} = \sqrt{(100)(0.80)(0.20)}$ = <u>4.0</u>

     $P(x \leq 70)$ = $P(x < 70.5)$
               = $P[z < (70.5 - 80.0)/4.0]$
               = $P[z < -2.38]$
               = 0.5000 - 0.4913 = <u>0.0087</u>

6.127 $\mu = np = (50)(0.16) = \underline{8}$

$\sigma = \sqrt{npq} = \sqrt{(50)(0.16)(0.84)} = \underline{2.6}$

a. $P(x > 12) = P(x > 12.5)$
$= P[z > (12.5 - 8)/2.6]$
$= P[z > 1.73]$
$= 0.5000 - 0.4582 = \underline{0.0418}$

b. $P(x < 8) = P(x < 7.5)$
$= P[z < (7.5 - 8)/2.6]$
$= P[z < -0.19]$
$= 0.5000 - 0.0753 = \underline{0.4247}$

c. $P(7 \leq x \leq 14) = P(6.5 < x < 14.5)$
$= P[(6.5 - 8)/2.6 < z < (14.5 - 8)/2.6]$
$= P[-0.58 < z < 2.50]$
$= 0.2190 + 0.4938 = \underline{0.7128}$

6.129 $\mu = np = (50)(0.39) = \underline{19.5}$

$\sigma = \sqrt{npq} = \sqrt{(50)(0.39)(0.61)} = \underline{3.45}$

a. $P(x > 20) = P(x > 20.5)$
$= P[z > (20.5 - 19.5)/3.45]$
$= P[z > 0.29]$
$= 0.5000 - 0.1141 = \underline{0.3859}$

b. $P(x < 15) = P(x < 14.5)$
$= P[z < (14.5 - 19.5)/3.45]$
$= P[z < -1.45]$
$= 0.5000 - 0.4265 = \underline{0.0735}$

6.131 $\mu = np = (2500)(0.052) = \underline{130}$

$\sigma = \sqrt{npq} = \sqrt{(2500)(0.052)(0.948)} = \underline{11.1}$

a. $P(x > 150) = P(x > 150.5)$
$= P[z > (150.5 - 130)/11.1]$
$= P[z > 1.85]$
$= 0.5000 - 0.4678 = \underline{0.0322}$

b. $P(x < 125) = P(x < 124.5)$
$$= P[z < (124.5 - 130)/11.1]$$
$$= P[z < -0.50]$$
$$= 0.5000 - 0.1915 = \underline{0.3085}$$

6.133

Nation	Infant Mortality (per 1000)	$\mu = np$	$\sigma = \sqrt{npq}$	$P(x \geq 70)$*
China	25	50	6.98	0.0026
Germany	4	8	2.82	0.0000+
India	58	116	10.45	0.9999+
Japan	3	6	2.45	0.0000+
Mexico	22	44	6.56	0.0001
Russia	17	34	5.78	0.0000+
S. Africa	62	124	10.78	0.9999+
United States	7	14	3.73	0.0000+

*China: $z = (69.5 - 50)/6.98 = 2.79$;
$P(z > 2.79) = 0.5000 - 0.4974 = \underline{0.0026}$
Germany: $z = (69.5 - 8)/2.82 = 21.81$;
$P(z > 21.81) = 0.5000 - 0.4999^{+} = \underline{0.0000+}$
India: $z = (69.5 - 116)/10.45 = -4.45$;
$P(z > -4.45) = 0.5000 + 0.4999^{+} = \underline{0.9999+}$
Japan: $z = (69.5 - 6)/2.45 = 25.92$;
$P(z > 25.92) = 0.5000 - 0.4999^{+} = \underline{0.0000+}$
Mexico: $z = (69.5 - 44)/6.56 = 3.89$;
$P(z > 3.89) = 0.5000 - 0.4999 = \underline{0.0001}$
Russia: $z = (69.5 - 34)/5.78 = 6.14$;
$P(z > 6.14) = 0.5000 - 0.4999^{+} = \underline{0.0000+}$
S. Africa: $z = (69.5 - 124)/10.78 = -5.06$;
$P(z > -5.06) = 0.5000 + 0.4999^{+} = \underline{0.9999+}$
United States: $z = (69.5 - 14)/3.73 = 14.88$;
$P(z > 14.88) = 0.5000 - 0.4999^{+} = \underline{0.0000+}$

c. Only two of the countries had means greater than the 70,
otherwise   means were significantly less than 70.

6.135   a. Middle 95%: z(0.975) = -1.96   z(0.025) = +1.96

$z = (x - \mu)/\sigma$

$\pm 1.96 = (x - 0.00)/0.020$

$x = 0.00 + (\pm 1.96)(0.020)$

Bounds for Middle 95%: from -0.0392 to +0.0392 unit

b. 96/110 = 0.8727 = 87.3%

c. P[-0.030 < x < +0.030]
   = P[(-0.030 - 0.00)/0.020 < z < (0.030 - 0.00)/0.020]
   = P[-1.50 < z < 1.50]
       = 0.4332 + 0.4332 = 0.8664 = 86.6%

CHAPTER 7 ▽ SAMPLE VARIABILITY

Chapter Preview

Chapters 1 and 2 introduced the concept of a sample and its various measures. Measures of central tendency, measures of dispersion, and the shape of the distribution of the data give a single "snapshot" of the population from which the sample was taken. If repeated samples are taken and statistics noted, a clearer picture of the population from which the sample came will develop. These combined statistics will enable us to better predict the population's parameters. Chapter 7 works with this illustration of repeated sampling in the form of a sampling distribution. A sampling distribution is basically a probability distribution for a sample statistic. Therefore, there can be sampling distributions for the sample mean, for the sample standard deviation or for the sample range, to name a few. The significant results that will surface for the probability distribution of the mean specifically, will be contained in the Central Limit Theorem. This theorem justifies the use of the normal distribution in solving a wide range of problems.

Data on the age of the population from the 2000 census taken in the United States is presented in this chapter's opening section, 275 Million Americans.

7.1   a. Histogram

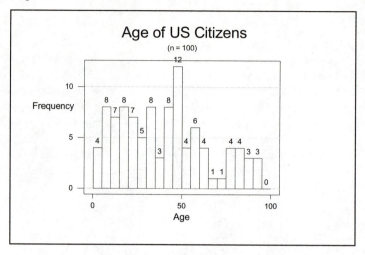

     b. Mounded from 0 to 60 with a tail extending from 60 to
        100 making the distribution skewed to the right.
     c. The sample looks very much like the population.
     d. Not exactly but fairly close.

SECTION 7.2 EXERCISES

7.3   a. No, but hope they are fairly close.
      b. Variability

7.5   a. A sampling distribution of sample means is the distribution
       formed by the means from all possible samples of a fixed
       size that can be taken from a population.

      b. It is one element in the sampling distribution of means for
       samples of size 3.

Use a tree diagram to find all possible samples, for exercise 7.7a. Each sample will have a probability of 1/n, where n is the number of samples. Remember $\sum$P(statistic) = 1.

7.7 a.

000	020	040	060	080	600	620	640	660	680
002	022	042	062	082	602	622	642	662	682
004	024	044	064	084	604	624	644	664	684
006	026	046	066	086	606	626	646	666	686
008	028	048	068	088	608	628	648	668	688
200	220	240	260	280	800	820	840	860	880
202	222	242	262	282	802	822	842	862	882
204	224	244	264	284	804	824	844	864	884
206	226	246	266	286	806	826	846	866	886
208	228	248	268	288	808	828	848	868	888
400	420	440	460	480					
402	422	442	462	482					
404	424	444	464	484					
406	426	446	466	486					
408	428	448	468	488					

b.

$\tilde{x}$	$P(\tilde{x})$
0	13/125
2	31/125
4	37/125
6	31/125
8	13/125
Σ	125/125 ck

c.

$\bar{x}$	$P(\bar{x})$
0/3	1/125
2/3	3/125
4/3	6/125
6/3	10/125
8/3	15/125
10/3	18/125
12/3	19/125
14/3	18/125
16/3	15/125
18/3	10/125
20/3	6/125
22/3	3/125
24/3	1/125
Σ	125/125 ck

7.9   Every student will have different results, however their graphs should be at least similar to these.

a.

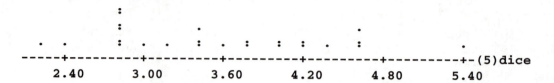

b. The distribution of $\overline{x}$'s is less variable (mostly between 2 and 5) than the distribution of x's (1 to 6) and denser near the middle.

c.

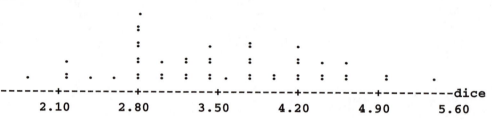

The distribution became more dense and mounded near the middle.

7.11 a. The samples are not all drawn from the same population and they're not the same size - each different type of transit vehicle has a different sample size.

b. The purpose of these repeated samples is for monitoring transit vehicle populations which are continually changing.

Computer commands for repeated sampling can be found in ES10 p368.

7.13 a. Every student will have different results, however the means of the 100 samples, each of size 5, should resemble those listed in (b).

b. sample means

```
6.0    5.8    3.2    6.2    4.8    3.2    5.0    6.2    6.2    6.4
5.0    2.4    5.8    3.4    2.6    3.6    4.6    4.0    5.0    3.8
5.4    5.2    3.6    5.4    4.4    4.2    4.0    6.0    5.8    3.8
4.8    2.8    4.2    5.6    6.4    5.6    2.8    4.8    3.8    3.0
5.2    4.2    5.0    1.6    4.8    3.2    4.2    5.0    4.8    5.0
4.8    4.6    3.4    4.2    3.8    4.2    4.4    5.6    5.2    5.2
6.0    4.0    4.6    3.4    5.2    6.8    4.4    3.6    5.6    4.4
5.0    6.4    5.2    4.8    5.0    6.0    4.0    5.2    5.6    5.8
6.4    7.0    3.2    2.2    4.6    3.2    7.6    3.0    4.6    4.6
5.4    7.2    5.8    3.6    4.4    2.8    4.8    5.4    4.8    4.0
```

c.

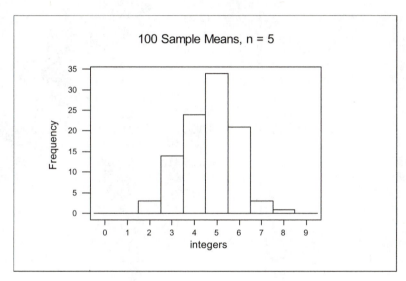

d. The shape is approximately normal, being mounded, approximately symmetrical, and centered near 4.5.

<hr>

Computer commands for repeated sampling can be found in ES10 p.369.

7.15   a. Every student will have different results; however, the means of the 200 samples, each of size 10, should resemble those listed in (b).

b. sample means

91.860	109.866	98.004	86.827	102.329	100.573	98.011
114.985	107.411	96.460	86.894	105.660	96.999	93.765
106.034	94.516	105.026	102.436	103.894	90.158	86.329
102.932	98.434	97.764	95.966	104.754	97.277	96.445
86.349	101.668	101.512	93.906	94.665	95.768	108.146
89.229	109.310	100.526	98.969	101.932	108.459	100.079
104.962	101.606	103.315	91.811	93.948	94.520	106.383
107.004	97.663	101.009	99.208	95.563	92.886	107.509
107.101	105.658	103.223	96.658	95.939	108.203	108.573
90.982	101.484	101.294	103.938	101.708	100.146	97.664
96.985	98.901	97.347	95.219	102.195	95.477	92.300
100.841	112.838	102.455	99.470	96.042	107.587	95.336
95.144	92.109	109.939	96.739	91.548	102.796	94.154
109.485	102.767	101.439	88.470	101.822	104.096	87.914
91.202	95.397	94.413	99.975	94.377	89.091	94.101
104.353	90.017	96.030	104.042	94.446	99.678	89.653
96.662	92.773	98.930	102.418	107.959	98.822	101.498
105.041	96.652	105.297	102.878	96.347	104.832	94.467
109.017	105.082	89.613	110.447	115.052	102.291	90.511
92.783	93.481	102.061	100.769	102.865	104.078	87.550
97.086	100.175	89.797	122.981	99.870	104.534	99.702
99.968	98.243	98.681	105.884	96.934	98.235	97.535
103.662	107.472	82.100	97.276	94.818	101.765	99.148
85.795	107.241	104.025	88.025	104.061	97.676	88.778
102.363	110.035	103.318	110.855	97.457	89.955	110.174
95.746	94.286	106.459	101.952	99.420	94.427	103.308
96.455	102.120	95.651	103.804	111.829	102.809	105.131
101.082	89.315	101.396	101.146	101.080	96.246	94.004
103.730	100.669	101.901	97.239			

c.

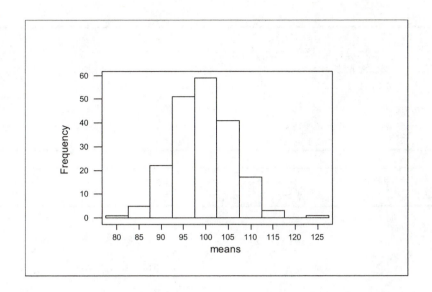

d. The shape is approximately normal, being mounded, approximately symmetrical, and centered near 100.

SECTION 7.3 EXERCISES

The most important sampling distribution is the sampling distribution of the sample means. It provides the information that makes up the sampling distribution of sample means and the central limit theorem.

SAMPLING DISTRIBUTION OF SAMPLE MEANS & CENTRAL LIMIT THEOREM

If all possible random samples, each of size n, are taken from any population with a mean μ and standard deviation σ, the sampling distribution of sample means ($\overline{x}$'s) will result in the following:

1. The mean of the sample means (x bars) will equal the mean of the population, $\mu_{\overline{x}} = \mu$.

2. The standard deviation of the sample means (x bars) will be equal to the population standard deviation divided by the square root of the sample size, $\sigma_{\overline{x}} = \sigma / \sqrt{n}$.   ...

> 3. A normal distribution when the parent population is normally distributed or becomes approximately normal distributed as the sample size increases when the parent population is not normally distributed.

In essence, $\overline{x}$ is normally distributed when n is large enough, no matter what shape the population is. The further the population shape is from normal, the larger the sample size needs to be.

$\sigma_{\overline{x}}$ = the standard deviation of the $\overline{x}$'s is now referred to as the standard error of the mean.

7.17   b. Answers will vary, but should resemble the following results from one simulation:
Mean of xbars = $\overline{\overline{x}}$ = 65.27; very close to μ = 65.15

c. Standard deviation of xbars = $s_{\overline{x}}$ = 1.383; less than σ
$s_{\overline{x}}$ is approximately $2.754/\sqrt{4}$ = 1.377.

d. The shape of the histogram is approximately normal.

e. Took many (1001) samples of size 4 from an approximately normal population and
1. mean of the xbars ≈ μ
2. $s_{\overline{x}}$ ≈ $\sigma / \sqrt{n}$
3. approximately normal distribution

7.19   a. 1.0 or one

b. $\sigma_{\overline{x}} = \sigma / \sqrt{n}$; as n increases the value of this fraction, the standard deviation of sample mean, gets smaller.

7.21 a. <u>500</u>   b. $30/\sqrt{36}$ = <u>5</u>   c. approximately normal

7.23 a.   Approximately normal
b.   4.0 hours
c. $2.1/\sqrt{250}$ = 0.133

-- 240 --

7.25 a. The mean value for the sampling distribution is the same as the population mean, 17.71 pounds/person.
   b. The standard error of the mean is $\sigma / \sqrt{n} = 6.3 / \sqrt{150} = 0.514$ pounds per person.
   c. The sampling distribution will be bell-shaped, approximately normal.

7.27 a. Every student will have different results; however, the means of the 100 samples, each of size 6, should resemble those listed in (b).

   b. sample means

17.0148	21.0960	18.8767	18.4458	23.7957	17.4572	19.9137
22.9338	21.0164	18.9116	15.8072	23.1245	20.1439	20.8047
18.7836	19.5104	16.7224	18.1819	19.7173	19.4121	19.7335

   (your 100 sample means should resemble the above)

   c.

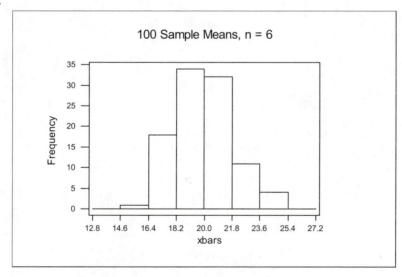

   Mean of xbars $= \overline{\overline{x}} = 19.924$
   Standard deviation of xbars $= s_{\overline{x}} = 1.8023$

   d. $\overline{\overline{x}}$ is approximately 20.
   $s_{\overline{x}}$ is approximately $4.5 / \sqrt{6} = 1.84$.
   The shape of the histogram is approximately normal.

It is helpful to draw a normal curve, locating μ and shading in the desired portion for each problem. A new z formula must now be used to determine probabilities about $\overline{x}$.

$$z = \frac{\overline{x} - \mu}{\sigma / \sqrt{n}}$$

7.29 $z = (46.5 - 43)/(5.2/\sqrt{16}) = \underline{2.69}$

7.31   Both 90 and 110 are two standard errors away from the mean and the $P(0 < z < 2) = 0.4772$ or $P(-2 < z < 0) = 0.4772$.

7.33   a. <u>approximately</u> <u>normal</u>

   b. <u>50</u>

   c. $10/\sqrt{36} = \underline{1.667}$

   d. $P(45 < \overline{x} < 55) = P[(45 - 50)/1.667 < z < (55 - 50)/1.667]$
   $= P[-3.00 < z < +3.00]$
   $= 0.4987 + 0.4987 = \underline{0.9974}$

   e. $P(\overline{x} > 48) = P[z > (48 - 50)/1.667]$
   $= P[z > -1.20]$
   $= 0.5000 + 0.3849 = \underline{0.8849}$

   f. $P(47 < \overline{x} < 53) = P[(47 - 50)/1.667 < z < (53 - 50)/1.667]$
   $= P[-1.80 < z < +1.80]$
   $= 0.4641 + 0.4641 = \underline{0.9282}$

7.35   a. Heights are approximately normally distributed with a $\mu = 69$ and $\sigma = 4$.

   b. $P(x > 70) = P[z > (70 - 69)/4]$
   $= P[z > 0.25]$
   $= 0.5000 - 0.0987 = \underline{0.4013}$

c. The distribution of $\bar{x}$'s will be approximately normally distributed.

d. $\mu_{\bar{x}} = \underline{69}$; $\sigma_{\bar{x}} = 4/\sqrt{16} = \underline{1.0}$

e. $P(\bar{x} > 70) = P[z > (70 - 69)/1.0]$
                  $= P[z > +1.00]$
                  $= 0.5000 - 0.3413 = \underline{0.1587}$

f. $P(\bar{x} < 67) = P[z < (67 - 69)/1.0]$
                  $= P[z < -2.00]$
                  $= 0.5000 - 0.4772 = \underline{0.0228}$

Watch the wording of the various probability problems.

If the probability for an individual item or person (x) is desired, use $z = (x - \mu)/\sigma$.

If the probability for a sample mean ($\bar{x}$) is desired, use $z = (\bar{x} - \mu)/(\sigma/\sqrt{n})$.

7.37   a. $P(38 < x < 40) = P[(38 - 39)/2 < z < (40 - 39)/2]$
                               $= P[-0.50 < z < +0.50]$
                               $= 0.1915 + 0.1915 = \underline{0.3830}$

   b. $P(38 < \bar{x} < 40) =$
                    $= P[(38 - 39)/(2/\sqrt{30}) < z < (40 - 39)/(2/\sqrt{30})]$
                    $= P[-2.74 < z < +2.74]$
                    $= 0.4969 + 0.4969 = \underline{0.9938}$

   c. $P(x > 40) = P[z > (40 - 39)/2]$
                   $= P[z > 0.50]$
                   $= 0.5000 - 0.1915 = \underline{0.3085}$

   d. $P(\bar{x} > 40) = P[z > (40 - 39)/(2/\sqrt{30})]$
                     $= P[z > 2.74]$
                     $= 0.5000 - 0.4969 = \underline{0.0031}$

7.39 $\mu = 11.3$ mph, $\sigma = 3.5$ mph

   a. $P(x > 13.5) = P[z > (13.5 - 11.3)/3.5] = P(z > 0.63)$
                      $= 0.5000 - 0.2357 = \underline{0.2643}$

b. $P(\overline{x} > 13.5) = P[z > (13.5 - 11.3)/(3.5/\sqrt{9}]$
$= P[z > 1.89]$
$= 0.5000 - 0.4706 = \underline{0.0294}$

c. No, especially for (a). The wind speed distribution will be skewed to the right, not normal. If they were normally distributed, most of the wind speeds would be between 0 and 22 mph. That does not allow for the high wind speeds that are known to occur. Samples of size 9 are not large enough for the sampling distribution to approach being approximately normal since the population is strongly skewed.

d. The actual probabilities are most likely not nearly as high as those found in (a) and (b).

7.41 a. $P(550 < \overline{x} < 700)$
$= P[(550-638)/(175/\sqrt{36}) < z < (700-638)/(175/\sqrt{36})]$
$= P[-3.02 < z < 2.13]$
$= 0.4987 + 0.4834 = \underline{0.9821}$

b. $P(\overline{x} > 750)$
$= P[z > (750-638)/(175/\sqrt{36})]$
$= P[z > 3.84]$
$= 0.5000 - 0.4999 = \underline{0.0001}$

c. Answers may vary, but if the normality assumption does not hold, the normal distribution still should allow for reasonable estimates for the probabilities since n > 30.

7.43 If z = -0.67, then
$-0.67 = (\overline{x} - 39.0)/(2/\sqrt{25})$
$-0.268 = \overline{x} - 39.0$
$\overline{x} = 38.732 = \underline{38.73 \text{ inches}}$

7.45 a. Using the computer commands in ES10 p.383:
$$P(4 < \bar{x} < 6) = 0.841345 - 0.158655 = \underline{0.68269}$$

Using Table 3: $P(4 < \bar{x} < 6) =$
$$= P[(4 - 5)/(2/\sqrt{4}) < z < (6 - 5)/(2/\sqrt{4})]$$
$$= P[-1.00 < z < +1.00]$$
$$= 0.3413 + 0.3413 = \underline{0.6826}$$

Everybody will get different answers, however, the results should be similar to the results below.

b. sample means

4.45967	4.04628	4.56959	4.74174	5.00541	4.99137	5.27322
4.18889	6.16806	5.10562	4.21561	5.61553	4.73023	3.23896
5.14986	5.83925	3.87701	4.15342	6.95931	6.69219	4.92261
4.00410	5.33392	5.69277	4.04213	6.00010	5.13663	7.04077

(your 100 sample means should resemble the above)

c.

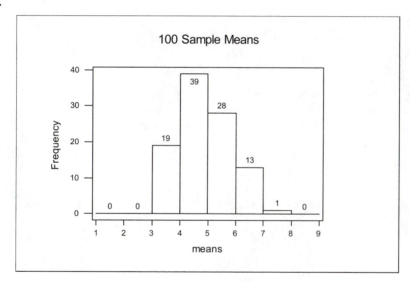

Inspecting the histogram, we find <u>67</u> of the sample means are between 4 and 6; 67/100 = 67%

d. 67% is very close to the expected 0.6826.

7.47 $\sigma_{\overline{x}} = \sigma / \sqrt{n}$: $18.2 / \sqrt{9} = 6.067$ $18.2 / \sqrt{25} = 3.64$

$18.2 / \sqrt{49} = 2.6$ $18.2 / \sqrt{100} = 1.82$

7.49   a Normally distributed with a mean = \$675 and a standard deviation = \$85.

b. P(550 < x < 825) =

$$= P[(550-675)/85 < z < (825-675)/85]$$
$$= P[-1.47 < z < +1.76]$$
$$= 0.4292 + 0.4608 = \underline{0.8900}$$

c. Approximately normally distributed with a mean = \$675 and a standard error = $85/\sqrt{25}$ = \$17

d. P(650 < $\overline{x}$ < 705) =

$= P[(650-675)/(85/\sqrt{25}) < z < (705-675)/(85/\sqrt{25})]$
$= P[-1.47 < z < +1.76]$
$= 0.4292 + 0.4608 = \underline{0.8900}$

e. z is used in (b) and (d) since the distribution of x is given to be normal, which would also cause the sampling distribution of x-bar to be normal.

7.51   a. P(2.63 − e < x < 2.63 + e) = 0.95

P(−1.96 < z < +1.96) = 0.95, using Table 3 (Appendix B, ES10-p810)

z = (x − μ)/σ

+1.96 = [(2.63 + e) − 2.63]/0.25   therefore e = $\underline{0.49}$

b. P(2.63 − E < $\overline{x}$ < 2.63 + E) = 0.95

P(−1.96 < z < +1.96) = 0.95, using Table 3 (Appendix B, ES10-p810)

z = ($\overline{x}$ − $\mu_{\overline{x}}$)/$\sigma_{\overline{x}}$

+1.96 = [(2.63 + E) − 2.63]/(0.25/$\sqrt{100}$)
therefore E = $\underline{0.049}$

7.53   a. $P(x > 1000) = P(z > (1000 - 586)/165) = P(z > 2.51)$
$$= 0.5000 - 0.4940 = \underline{0.0060}$$

   b. $P(\overline{x} < 550)$
$$= P[z < (550 - 586)/(165/\sqrt{20}\,)]$$
$$= P[z < -0.98]$$
$$= 0.5000 - 0.3365 = \underline{0.1635}$$

   c. Daily number of customers can be a very skewed to the right side of the distribution taking into effect holidays, promotions, etc.

7.55   a. $P(245 < x < 255)$
$$= P[(245 - 235)/\sqrt{400} < z < (255 - 235)/\sqrt{400}]$$
$$= P[+0.50 < z < +1.00]$$
$$= 0.3413 - 0.1915 = \underline{0.1498}$$

   b. $P(\overline{x} > 250) = P[z > (250 - 235)/(20/\sqrt{10}\,)]$
$$= P[z > +2.37]$$
$$= 0.5000 - 0.4911 = \underline{0.0089}$$

7.57 $P(\overline{x} < 680) = P[z < (680 - 700)/(120/\sqrt{144}\,)]$
$$= P[z < -2.00]$$
$$= 0.5000 - 0.4772 = \underline{0.0228}$$

7.59 $P(\Sigma x > 38,000) = P[z > (\Sigma x - n\mu)/(\sigma\sqrt{n}\,)]$
$$= P[z > (38,000 - (50)(750))/(25\sqrt{50}\,)]$$
$$= P[z > 500/176.777]$$
$$= P[z > +2.83]$$
$$= 0.5000 - 0.4977 = \underline{0.0023}$$

7.61   a. Let Σx represent the total weight:
$$P(\Sigma x > 4000) = P(\Sigma x/n > 4000/25)$$
$$= P(\overline{x} > 160)$$
$$= P[z > (160 - 300)/(50/\sqrt{25}\,)]$$
$$= P[z > -14.0]$$
$$= \text{approximately } \underline{1.000}$$

   b. $P(\Sigma x < 8000) = P(\overline{x} < 320)$
$$= P[z < (320 - 300)/(50/\sqrt{25}\,)]$$
$$= P[z < 2.00]$$
$$= 0.5000 + 0.4772 = \underline{0.9772}$$

7.63 a. Every student will have different results, however the totals and means of the 50 samples, each of size 10, should resemble those listed below.

sums
1357.86	1344.27	1357.23	1323.87	1335.53	1357.25	1339.91
1365.17	1308.99	1357.99	1346.36	1367.63	1315.55	1354.79
1378.82	1392.20	1320.65	1333.62	1405.80	1373.16	1400.35
1399.88	1414.90	1316.60	1337.11	1309.65	1339.62	1295.95
1350.44	1333.20	1338.42	1389.96	1304.70	1312.90	1312.14
1343.50	1315.46	1390.69	1387.76	1403.76	1339.83	1314.40
1416.16	1355.12	1299.10	1352.39	1367.93	1354.54	1305.35
1325.34						

xbars
135.786	134.427	135.723	132.387	133.553	135.725	133.991
136.517	130.899	135.799	134.637	136.763	131.555	135.479
137.882	139.220	132.065	133.362	140.580	137.316	140.035
139.988	141.490	131.660	133.711	130.965	133.962	129.595
135.044	133.320	133.842	138.996	130.470	131.290	131.214
134.350	131.546	139.069	138.776	140.376	133.983	131.440
141.616	135.512	129.910	135.239	136.793	135.454	130.535
132.534						

b.

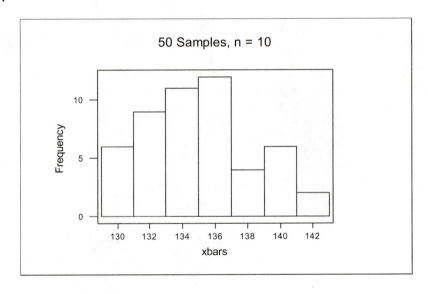

Mean of xbars   =   134.93
Standard deviation of xbars   =   3.2776

c.

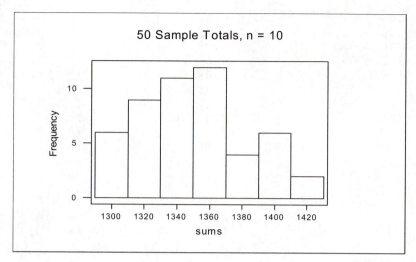

50 Sample Totals, n = 10

Mean of sums   =   1349.3
Standard deviation of sums   =   32.776

d. The histograms in (b) and (c) are identical in shape.
   Using the DESCribe command, it is evident that the xbar
   results are just the sum results divided by 10, the sample
   size.

Variable	N	Mean	Median	TrMean	StDev	SEMean
sums	50	1349.3	1345.3	1348.3	32.8	4.6
xbars	50	134.93	134.53	134.83	3.28	0.46

Variable	Min	Max	Q1	Q3
sums	1296.0	1416.2	1319.6	1369.2
xbars	129.60	141.62	131.96	136.92

7.65 a. $\mu = np = 200(0.3) = \underline{60}$

$\sigma = \sqrt{npq} = \sqrt{200(0.3)(0.7)} = \sqrt{42} = \underline{6.48}$

b. (all other x values have a probability equal to zero)

x	probability	x	probability	x	probability
28	0.0000001	51	0.0238909	74	0.0061875
29	0.0000002	52	0.0293386	75	0.0044550
30	0.0000004	53	0.0351114	76	0.0031403
31	0.0000009	54	0.0409633	77	0.0021673
32	0.0000021	55	0.0466024	78	0.0014647
33	0.0000045	56	0.0517144	79	0.0009694
34	0.0000096	57	0.0559916	80	0.0006284
35	0.0000194	58	0.0591635	81	0.0003990
36	0.0000382	59	0.0610258	82	0.0002481
37	0.0000725	60	0.0614617	83	0.0001512
38	0.0001334	61	0.0604542	84	0.0000903
39	0.0002374	62	0.0580861	85	0.0000528
40	0.0004096	63	0.0545298	86	0.0000303
41	0.0006850	64	0.0500263	87	0.0000170
42	0.0011114	65	0.0448587	88	0.0000093
43	0.0017501	66	0.0393242	89	0.0000050
44	0.0026763	67	0.0337065	90	0.0000027
45	0.0039762	68	0.0282539	91	0.0000014
46	0.0057421	69	0.0231647	92	0.0000007
47	0.0080633	70	0.0185791	93	0.0000003
48	0.0110151	71	0.0145791	94	0.0000002
49	0.0146440	72	0.0111947	95	0.0000001
50	0.0189535	73	0.0084125		

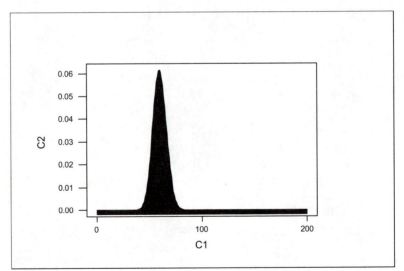

c. Every student will have different results, however the histograms should resemble that in (d)

d.

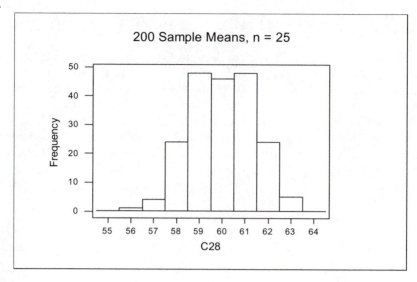

Mean of C28 = $\overline{\overline{x}}$ = 59.979

Standard deviation of C28 = $s_{\overline{x}}$ = 1.368

e. The mean of the sample means, $\overline{\overline{x}}$ = 59.979, is
   approximately equal to μ = 60. The standard deviation of
   the sample means, $s_{\overline{x}}$ = 1.368, is approximately equal to
   $6.48/\sqrt{25}$ = 1.296. The distribution of the sample means is
   approximately normally distributed.

CHAPTER 8 ∇ INTRODUCTION TO STATISTICAL

Chapter Preview

Chapter 8 introduces inferential statistics. Generalizations about population parameters are made based on sample data in inferential statistics. These generalizations can be made in the form of hypothesis tests or confidence interval estimations. Each is calculated with a degree of uncertainty. The integral elements and procedure for obtaining a confidence interval and for completing a hypothesis test will be presented in this chapter. They will be performed with respect to the population mean, μ. The population standard deviation, σ, will be considered as a known quantity.

Statistical information on the average height of women reported by The National Center for Health Statistics is presented in the opening section, 'Were They Shorter Back Then?'.

SECTION 8.1 EXERCISES

8.1   a. female health professionals

b. $\overline{x}$ = 64.7, s = 3.5
Distribution is mounded about center, approximately symmetrical

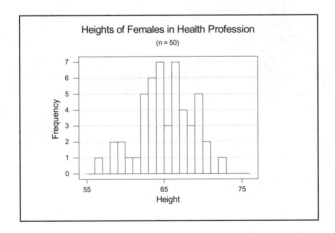

ESTIMATION

Point Estimate for a Population Parameter - the value of the
                                        corresponding sample
                                        statistic

    ex.) $\overline{x}$ is the point estimate of μ
         s is the point estimate of σ
         s^2 is the point estimate of σ^2

Confidence intervals are used to estimate a population parameter on
an interval with a degree of certainty. We could begin by taking a
sample and finding $\overline{x}$ to just estimate μ. The sample statistic, $\overline{x}$,
is a <u>point estimate</u> of the population parameter μ. How good an
estimate it is depends not only on the sample size and variability
of the data, but also whether or not the sample statistic is
unbiased.

Unbiased Statistic - a sample statistic whose sampling distribution
                     has a mean value equal to the corresponding
                     population parameter

One would assume that $\overline{x}$ is not exactly equal to μ, but hopefully
relatively close. It is by this reasoning that we work with
interval estimates of population parameters.

Level of Confidence = $1 - \alpha$ = the probability that the interval
constructed, based on the sample, contains the true population
parameter.

8.3   A point estimate is a single number, usually the sample
       statistic. An interval estimate is an interval of some
       width centered at the point estimate

8.5   n = 15, Σx = 271, Σx^2 = 5015
       a. $\overline{x} = \Sigma x/n = 271/15 = 18.0667 = $ <u>18.1 dollars</u>
       b. $s^2 = (\Sigma x^2 - (\Sigma x)^2/n)/(n - 1)$
            $= (5015 - (271)^2/15)/14 = 8.4952 = $ <u>8.5</u>
       c. $s = \sqrt{s^2} = \sqrt{8.4952} = 2.9146 = $ <u>2.9 dollars</u>

8.7   a. II has the lower variability; both have a mean value equal to the parameter. II would be the better estimator.

b. II has a mean value equal to the parameter, I does not. II would be the better estimator.

c. Neither is a good choice; I is negatively biased with less variability, while II is not biased with a larger variability. II would be the better estimator.

8.9   Difficulty and collector fatigue in obtaining and also in evaluating a very large sample; cost of sampling; destruction of product in cases like the rivets illustration.

8.11 $\sigma_{\bar{x}} = \sigma / \sqrt{n} = 18 / \sqrt{36} = 18 / 6 = 3$

The level of confidence depends on the number of standard errors from the sample mean.

For $\bar{x} \pm z\sigma_{\bar{x}}$: $\bar{x}$ = mean (given)

$\sigma_{\bar{x}}$ = standard error of the mean (given)

z = number of standard errors

Find the corresponding probability for the z-value using Table 3, Appendix B. Multiply the probability by 2 to cover both parts of the interval. This total probability is equal to the level of confidence.

8.13   a. 0.3997 + 0.3997 = 0.7994 = 80.0%
b. 0.4251 + 0.4251 = 0.8502 = 85.0%
c. 0.4750 + 0.4750 = 0.9500 = 95.0%
d. 0.4901 + 0.4901 = 0.9802 = 98.0%

8.15   19($17,320)   = $ 329080
    6($20,200)   =   121200
                    $ 450280 for 25 projects;
therefore 174($450280/25) = $3,133,948.80

8.17   a. Between 11:17 AM and 11:37 AM, the next eruption should occur.

b. Yes; the snapshot was recorded on 8/20/05 at 11:20:48 AM which is within the interval from 11:17 to 11:37 AM.

c. 90% of the eruptions occur within predicted interval

ESTIMATION OF THE POPULATION MEAN - μ

Point estimate of μ: $\overline{x}$

Interval estimate of μ = confidence interval

1 - α = level of confidence, the probability or degree of
        certainty desired (ex.: 95%, 99% ...)

A (1-α) confidence interval estimate for μ is:

$$\overline{x} - z(\alpha/2)\cdot\sigma/\sqrt{n} \ \text{ to } \ \overline{x} + z(\alpha/2)\cdot\sigma/\sqrt{n} \quad **$$

$\overline{x} - z(\alpha/2)\cdot\sigma/\sqrt{n}$ = lower confidence limit

$\overline{x} + z(\alpha/2)\cdot\sigma/\sqrt{n}$ = upper confidence limit

$E = z(\alpha/2)\cdot\sigma/\sqrt{n}$ = maximum error of the estimate

**To find z(α/2):
    (suppose for example that a 95% confidence interval is desired)
                    95% = 1 - α , that is,
                    .95 = 1 - α
                solving for alpha, α, gives
                    α = .05
                dividing both sides by 2 gives
                    α/2 = .025
Now determine the probability associated with z(.025) using Table
4B (Appendix B, ES10-p811), the Critical Values of Standard Normal
Distribution for Two-Tailed Situations. (This table conveniently
gives the most popular critical values for z.)

8.19 Either the sampled population is normally distributed or the
        random sample is sufficiently large for the Central Limit
        Theorem to hold.

8.21   a. $\alpha = 0.02$; $z(\alpha/2) = z(0.01) = \underline{2.33}$

   b. $\alpha = 0.01$; $z(\alpha/2) = z(0.005) = \underline{2.58}$

THE CONFIDENCE INTERVAL: A FIVE-STEP PROCEDURE

Step 1: The Set-Up:
   Describe the population parameter of concern.
Step 2: The Confidence Interval Criteria:
   a. Check the assumptions.
   b. Identify the probability distribution and the formula
      to be used.
   c. Determine the level of confidence, $1 - \alpha$.
Step 3: The sample evidence:
   Collect the sample information.
Step 4: The Confidence Interval:
   a. Determine the confidence coefficient.
   b. Find the maximum error of estimate.
   c. Find the lower and upper confidence limits.
Step 5:   The Results:
   State the confidence interval.

8.23   a. Step 1: The mean, μ
         Step 2: a. normality indicated
                 b. z, $\sigma = 6$ c. $1 - \alpha = 0.95$
         Step 3: $n = 16$, $\overline{x} = 28.7$
         Step 4: a. $\alpha/2 = 0.05/2 = 0.025$; $z(0.025) = 1.96$
                 b. $E = z(\alpha/2) \cdot \sigma/\sqrt{n} = (1.96)(6/\sqrt{16})$
                    $= (1.96)(1.5) = 2.94$
                 c. $\overline{x} \pm E = 28.7 \pm 2.94$
         Step 5: $\underline{25.76 \text{ to } 31.64}$, the 0.95 confidence interval for μ

   b. <u>Yes</u>; the sampled population is normally distributed.

8.25   a. Step 1: The mean, μ
         Step 2: a. normality assumed because of CLT with $n = 86$.
                 b. z, $\sigma = 16.4$ c. $1 - \alpha = 0.90$
         Step 3: $n = 86$, $\overline{x} = 128.5$
         Step 4: a. $\alpha/2 = 0.10/2 = 0.05$; $z(0.05) = \underline{1.65}$
                 b. $E = z(\alpha/2) \cdot \sigma/\sqrt{n} = (1.65)(16.4/\sqrt{86})$
                    $= (1.65)(1.76845) = 2.92$
                 c. $\overline{x} \pm E = 128.5 \pm 2.92$
         Step 5: $\underline{125.58 \text{ to } 131.42}$, the 0.90 confidence interval
                 for μ

   b. <u>Yes</u>; the sample size is sufficiently large to satisfy the
      CLT.

8.27 a. point estimate = $\overline{x}$ = 128.5
b. confidence coefficient = $z(\alpha/2)$ = $z(0.05)$ = 1.65
c. Standard error of the mean = $\sigma/\sqrt{n}$ = $(16.4/\sqrt{86})$ = 1.76845
d. max. error of estimate, $E = z(\alpha/2)\cdot\sigma/\sqrt{n}$
$$= (1.65)(1.76845) = 2.92$$
e. lower confidence limit = $\overline{x}$ - E = 128.5 - 2.92 = 125.58
f. upper confidence limit = $\overline{x}$ + E = 128.5 + 2.92 = 131.42

8.29 Answers will vary.

8.31 a. 15.9; ≈ 68%
b. 31.4; ≈ 95%
c. 41.2; ≈ 99%
d. higher level makes for a wider width; to be more
certain the parameter is contained in the interval

8.33 a. 75.92

b. $E = z(\alpha/2)\cdot\sigma/\sqrt{n}$ = $(2.33)(0.5/\sqrt{10})$ = 0.3684 = 0.368

c. $\overline{x}$ ± E = 75.92 ± 0.368
75.552 to 76.288, the 0.98 confidence interval for μ

8.35 a. Step 1: The mean length of fish caught in Cayuga Lake
Step 2: a. normality assumed, CLT with n = 200.
b. z, σ = 2.5   c. 1-α = 0.90
Step 3: n = 200, $\overline{x}$ = 14.3
Step 4: a. $\alpha/2$ = 0.10/2 = 0.05; $z(0.05)$ = 1.65
b. $E = z(\alpha/2)\cdot\sigma/\sqrt{n}$ = $(1.65)(2.5/\sqrt{200})$
$$= (1.65)(0.17678) = 0.29$$
c. $\overline{x}$ ± E = 14.3 ± 0.29
Step 5: 14.01 to 14.59, the 0.90 confidence interval for μ

b. Step 1-3: as shown in (a), except 2c. 1-α = 0.98
Step 4:   a. $\alpha/2$ = 0.02/2 = 0.01; $z(0.01)$ = 2.33
b. $E = z(\alpha/2)\cdot\sigma/\sqrt{n}$ = $(2.33)(2.5/\sqrt{200})$
$$= (2.33)(0.17678) = 0.41$$
c. $\overline{x}$ ± E = 14.3 ± 0.41
Step 5: 13.89 to 14.71, the 0.98 confidence interval for μ

8.37   Step 1: The mean mathematics score for all eighth-grade
                    students in the US
          Step 2: a. normality assumed, standard error given
                    b. z, $\sigma/\sqrt{n}$ = 8.4   c. 1-α = 0.95
          Step 3: $\overline{x}$ = 504
          Step 4: a. $\alpha/2$ = 0.05/2 = 0.025;   z(0.025) = 1.96
                    b. E = z($\alpha/2$)$\cdot\sigma/\sqrt{n}$ = (1.96)(8.4) = 16.464
                    c. $\overline{x}$ ± E = 504 ± 16.5
          Step 5: <u>487.5 to 520.5</u>, the 0.95 confidence interval for μ

Computer and/or calculator commands to calculate a confidence
interval for μ, provided σ is known, can be found in ES10-pp409-
410. The output will also contain the sample mean and standard
deviation.

8.39   a. The mean length of parts being produced after adjustment.

           b. $\overline{x}$ = $\Sigma x/n$ = 759.2/10 = <u>75.92</u>

           c. Step 1: See part 'a'
              Step 2: a. normality indicated
                        b. z, σ = 0.5   c. 1-α = 0.99
              Step 3: n = 10, $\overline{x}$ = 75.92
              Step 4: a. $\alpha/2$ = 0.01/2 = 0.005;   z(0.005) = 2.58
                        b. E = z($\alpha/2$)$\cdot\sigma/\sqrt{n}$ = (2.58)(0.5/$\sqrt{10}$)
                             = (2.58)(0.158) = 0.408
                        c. $\overline{x}$ ± E = 75.92 ± 0.408
              Step 5: <u>75.512 to 76.328</u>, the 0.99 confidence interval
                        for μ

8.41   a. Step 1: The mean extraction force required for a No.
                        9 cork.
              Step 2: a. normality indicated.
                        b. z, σ = 36   c. 1-α = 0.98
              Step 3: n = 20, $\overline{x}$ = $\Sigma x / n$ = 5913/20 = 295.65
              Step 4: a. $\alpha/2$ = 0.02/2 = 0.01;   z(0.01) = 2.33
                        b. E = z($\alpha/2$)$\cdot\sigma/\sqrt{n}$ = (2.33)(36/$\sqrt{20}$)
                             = (2.33)(8.050) = 18.7565
                        c. $\overline{x}$ ± E = 295.65 ± 18.7565
              Step 5: <u>276.9 to 314.4</u>, the 0.98 confidence interval for μ

b. Step 1: The mean extraction force required for a No. 9 cork.
     Step 2: a. normality indicated.
          b. z, $\sigma = 36$
          c. $1-\alpha = 0.98$
     Step 3 $n = 8$, $\bar{x} = \sum x / n = 2514.8/8 = 314.35$
     Step 4: a. $\alpha/2 = 0.02/2 = 0.01$; $z(0.01) = 2.33$
          b. $E = z(\alpha/2)\cdot\sigma/\sqrt{n} = (2.33)(36/\sqrt{8})$
              $= (2.33)(12.728) = 29.65624$
          c. $\bar{x} \pm E = 314.35 \pm 29.65624$
    Step 5: <u>284.69 to 344.01</u>, the 0.98 confidence interval for μ

c. The two different sample means gave different center points for the confidence intervals.

d. The decrease in sample size, increased the width of the second interval.

e. The 310 Newtons is contained in both intervals. Neither interval gives reason to doubt the truthfulness of the claim.

8.43 The numbers are calculated for samples of snow for some of the months; it would be impossible to gather this information from every inch of snow.

To find the <u>sample size n</u> required for a $1-\alpha$ confidence interval, use the formula: $n = [z(\alpha/2)\cdot\sigma/E]^2$, where

    z = standard normal distribution
    α = calculated from the $1-\alpha$ confidence interval desired
    σ = population standard deviation
    E = maximum error of the estimate

The maximum error of the estimate, E, is the amount of error that is tolerable or allowed. Quite often, finding the word "within" in an exercise will locate the acceptable value for E.

8.45 $n = [z(\alpha/2)\cdot\sigma/E]^2 = [(2.33)(3)/1]^2 = 48.8601 = \underline{49}$

8.47 $n = [z(\alpha/2)\cdot\sigma/E]^2 = [(2.58)(1.0)/0.5]^2 = 26.6 = \underline{27}$

8.49 $n = [z(\alpha/2) \cdot \sigma/E]^2 = [(1.96)(\sigma)/(0.4\sigma)]^2 = 24.01 = \underline{25}$

SECTION 8.4 EXERCISES

DEFINITIONS FOR HYPOTHESIS TESTS

<u>Hypothesis</u> - a statement that something is true.

<u>Null Hypothesis,</u> H_O - a statement that specifies a value for a
                          population parameter
                             ex.: H_O: The mean weight is 40 pounds

<u>Alternative Hypothesis,</u> H_a - opposite of H_O, a statement that
                               specifies an "opposite" value for a
                               population parameter
                               ex.: H_a: The mean weight is not 40
                               pounds

<u>Type I Error</u> - the error resulting from rejecting a true null
               hypothesis

<u>α (alpha)</u> - the probability of a type I error, that is the
                   probability of rejecting H_O when it is true.

<u>Type II Error</u> - the error resulting from not rejecting a false
               null hypothesis

<u>β (beta)</u> - the probability of a type II error, that is the
                 probability of not rejecting H_O when it is false.

Keep α and β as small as possible, depending on the severity of the
respective error.

8.51 H_O: The system is reliable
      H_a: The system is not reliable

8.53   a. H_O: Special delivery mail does not take too much time
          H_a: Special delivery mail takes too much time

b. H_o: The new design is not more comfortable
   H_a: The new design is more comfortable

c. H_o: Cigarette smoke has no effect on the quality of a
       person's life
   H_a: Cigarette smoke has an effect on the quality of a
       person's life

d. H_o: The hair conditioner is not effective on "split ends"
   H_a: The hair conditioner is effective on "split ends"

8.55 Type A correct decision:
Truth of situation: the party will be a dud.
Conclusion: the party will be a dud.
Action: did not go [avoided dud party]

Type B correct decision:
Truth of situation: the party will be a great time.
Conclusion: the party will be a great time.
Action: did go [party was great time]

Type I error:
Truth of situation: the party will be a dud.
Conclusion: the party will be a great time.
Action: did go [party was a dud]

Type II error:
Truth of situation: the party will be a great time.
Conclusion: the party will be a dud.
Action: did not go [missed great party]

Remember; the truth of the situation is not known before the
decision is made, the conclusion reached and the resulting
actions take place. Only after the party is over can the
evaluation be made.

8.57 a. H_o: The victim is alive
        H_a: The victim is not alive

b. Type A correct decision: The victim is alive and is
   treated as though alive.

   Type I error: The victim is alive, but is treated
   as though dead.

   Type II error: The victim is dead, but treated as
   if alive.

Type B correct decision: The victim <u>is</u> <u>dead</u> and treated <u>as</u> <u>dead</u>.

   c. The type I error is very serious. The victim may very
      well be dead shortly without the attention that is not
      being received.

      The type II error is not as serious. The victim is
      receiving attention that is of no value. This would be
      serious only if there were other victims that needed this
      attention.

8.59 You missed a great time.

8.61 a. Type A correct decision: The majority of Americans do
           favor laws against assault weapons and it is decided that
           they do favor the laws.
           Type B correct decision: The majority of Americans do not
           favor laws against assault weapons and it is decided that
           they do not favor the laws.

        b. Type A correct decision: The fast food menu is not low
           salt and it is decided that it is not low salt.
           Type B correct decision: The fast food menu is low salt
           and it is decided that it is low salt.

        c. Type A correct decision: The building must not be
           demolished and it is decided that it should not be
           demolished.
           Type B correct decision: The building must be demolished
           and it is decided that it should be demolished.

        d. Type A correct decision: There is no waste in government
           spending and it is decided that there is no waste.
           Type B correct decision: There is waste in government
           spending and it is decided that there is waste.

8.63 Type I error: Teaching techniques have no significant effect
        on students' exam scores and it is decided that teaching
        techniques do have a significant effect on students' exam
        scores.

        Action: Teaching techniques which have no significant effect
        will be believed to be effective and used accordingly.

Type II error: Teaching techniques have a significant effect on students' exam scores and it is decided that teaching techniques have no significant effect on students' exam scores.

Action: Teaching techniques which have an significant effect will be believed to be ineffective and not used.

```
┌─────────────────────────────────────────────────────────────────┐
│          TERMINOLOGY FOR DECISIONS IN HYPOTHESIS TESTS            │
│                                                                   │
│ Reject H_O: use when the evidence disagrees with the null         │
│         hypothesis.                                                │
│                                                                   │
│ Fail to reject H_O: use when the evidence does not disagree with  │
│               the null hypothesis.                                │
│                                                                   │
│ Note: The purpose of the hypothesis test is to allow the evidence │
│ a chance to discredit the null hypothesis.                        │
│ Remember: If one believes the null hypothesis to be true,         │
│ generally there is no test.                                       │
└─────────────────────────────────────────────────────────────────┘
```

8.65 a. Type I   b. Type II   c. Type I   d. Type II

8.67 a. Commercial is not effective.

b. Commercial is effective.

8.69 a. The type I error is very serious and, therefore, we are willing to allow it to occur with a probability of 0.001; that is, only 1 chance in 1000.

b. The type I error is somewhat serious and, therefore, we are willing to allow it to occur with a probability of 0.05; that is, 1 chance in 20.

c. The type I error is not at all serious and, therefore, we are willing to allow it to occur with a probability of 0.10; that is, 1 chance in 10.

8.71 a. α b. β

8.73 α is the probability of rejecting a TRUE null hypothesis; $1-\beta$ is the probability of rejecting a FALSE null hypothesis; they are two distinctly different acts that both result in rejecting the null hypothesis.

8.75 a. When the test procedure begins, the experimenter is thoroughly convinced the alternative hypothesis can be shown to be true; thus when the decision *reject* H_O is attained, the experimenter will want to say something like "see I told you so." Thus the statement of the conclusion is a fairly strong statement like; "the evidence shows beyond a shadow of a doubt (is significant) that the alternative hypothesis is correct."

b. When the test procedure begins, the experimenter is thoroughly convinced the alternative hypothesis can be shown to be true; thus when the decision *fail to reject* H_O is attained, the experimenter is disappointed and will want to say something like "okay so this evidence was not significant, I'll try again tomorrow." Thus the statement of the conclusion is a fairly mild statement like, "the evidence was not sufficient to show the alternative hypothesis to be correct."

8.77 a. $\alpha = P(\text{rejecting } H_O \text{ when the } H_O \text{ is true})$
$$= P(x \geq 86 | \mu=80) = P(z > (86 - 80)/5) = P(z > 1.20)$$
$$= 0.5000 - 0.3849 = \underline{0.1151}$$

b. $\beta = P(\text{accepting } H_O \text{ when the } H_O \text{ is false})$
$$= P(x < 86 | \mu=90) = P(z < (86 - 90)/5) = P(z < -0.80)$$
$$= 0.5000 - 0.2881 = \underline{0.2119}$$

8.79 a. Acceptable diameters: from 23.5 mm to 24.5 mm
(1 unacceptable: Cork 1, 24.51 mm)
Acceptable ovalization: • 0.7 mm
(2 unacceptable: Cork 2, 0.88 mm and cork 21, 0.76 mm)
Acceptable lengths: from 44.3 to 45.7
(1 unacceptable: Cork 21, 44.27 mm)
Therefore 29 corks pass Part 1 inspection.

b. The batch will be refused, there are 3 corks that do not meet the limits of specification.

c. Answers will vary.

SECTION 8.5 EXERCISES

THE PROBABILITY-VALUE HYPOTHESIS TEST: A FIVE-STEP PROCEDURE

Step 1: The Set-Up:
    a. Describe the population parameter of concern.
    b. State the null hypothesis (H_O) and the alternative hypothesis (H_a).
Step 2: The Hypothesis Test Criteria:
    a. Check the assumptions.
    b. Identify the probability distribution and the test statistic formula to be used.
    c. Determine the level of significance, α.
Step 3: The Sample Evidence:
    a. Collect the sample information.
    b. Calculate the value of the test statistic.
Step 4: The Probability Distribution:
    a. Calculate the p-value for the test statistic.
    b. Determine whether or not the p-value is smaller than α.
Step 5: The Results:
    a. State the decision about H_O.
    b. State the conclusion about H_a.

The Null and Alternative Hypotheses, H_O and H_a

H_O: $\mu = 100$ versus H_a: $\mu \neq 100$ (= and $\neq$ form the opposite ofeach other)

    H_a is a two-sided alternative

    possible wording for this combination:
        a) mean is different from 100 ($\neq$)
        b) mean is not 100 ($\neq$)
        c) mean is 100 (=)   . . .

OR ——

H_O: μ = 100 ($\leq$) versus H_a: μ > 100   ($\leq$ and > form the opposite of
each other)

H_a is a one-sided alternative

possible wording for this combination:
   a) mean is greater than 100 (>)
   b) mean is at most 100 ($\leq$)
   c) mean is no more than 100 ($\leq$)

OR ——

H_O: μ = 100 ($\geq$) versus H_a: μ < 100   ($\geq$ and < form the opposite of
each other)

H_a is a one-sided alternative
possible wording for this combination:
   a) mean is less than 100 (<)
   b) mean is at least 100 ($\geq$)
   c) mean is no less than 100 ($\geq$)

——

Always show equality (=) in the null hypothesis, since the null
hypothesis must specify a single specific value for μ
(like μ = 100).

The null hypothesis could be rejected in favor of the alternative
hypothesis for three different reasons; 1) $\mu \neq$ 100 or 2) μ > 100 or
3) μ < 100. Together, the two opposing statements, H_O and H_a, must
contain or account for all numerical values around and including μ.
This allows for the addition of $\geq$ or $\leq$ to the null hypothesis.
Therefore, if the alternative hypothesis is < or >, $\geq$ or $\leq$,
respectively, may be added to the null hypothesis. If $\geq$ or $\leq$ is
being tested, the appropriate symbol should be written in
parentheses after the amount stated for μ.
Sometimes, depending on the wording, it is easier to write the
alternative hypothesis first. The alternative hypothesis can only
contain >, < or $\neq$.

                                                        . . .

8.81 H_O: The mean shearing strength is at least 925 lbs.
H_a: The mean shearing strength is less than 925 lbs.

8.83 a. H_O: $\mu = 1.25$ ($\leq$)
H_a: $\mu > 1.25$

b. H_O: $\mu = 335$ ($\geq$)
H_a: $\mu < 335$

c. H_O: $\mu = 230,000$
H_a: $\mu \neq 230,000$

d. H_O: $\mu = 210$ ($\leq$)
H_a: $\mu > 210$

e. H_O: $\mu = 9.00$ ($\leq$)
H_a: $\mu > 9.00$

ERRORS

Type I error - occurs when H_O is rejected and it is a true statement.

Type II error - occurs when H_O is accepted and it is a false statement.

8.85 Type A correct decision: The mean shearing strength is at least 925 lbs and it is decided that it is.
Type I error: The mean shearing strength is at least 925 lbs and it is decided that it is less than 925 lbs.
Type II error: The mean shearing strength is less than 925 lbs and it is decided that it is greater than or equal to 925 lbs.
Type B correct decision: The mean shearing strength is less than 925 lbs and it is decided that it is less than 925 lbs.

Type II error; you buy and use weak rivets.

Since we work under the assumption that H_O is a true statement, all decisions are made <u>based on</u> or <u>pertaining to</u> H_O.

If we are unable to reject H_O, the terminology "fail to reject H_O" is used; whereas if we are able to reject H_O, "reject H_O" is used. After this decision statement, include an additional statement, explaining how the test results support or did not support the experimenter's convictions, to form the conclusion.

8.87 A type I error would be committed if a decision of reject H_O was reached and interpreted as 'mean hourly charge is less than \$60 per hour' when in fact the mean hourly charge is at least \$60 per hour.

A type II error would be committed if a decision of fail to reject H_O was reached and interpreted as 'mean hourly charge is at least \$60 per hour' when in fact the mean hourly charge is less than \$60 per hour.

Use the sample information (sample mean and size) and the population parameter in the null hypothesis (μ) to calculate the test statistic z^*.

$$z^* = (\overline{x} - \mu)/(\sigma/\sqrt{n})$$

8.89 a. $z = (\overline{x} - \mu)/(\sigma/\sqrt{n})$

$z^* = (10.6 - 10)/(3/\sqrt{40}) = \underline{1.26}$

b. $z = (\overline{x} - \mu)/(\sigma/\sqrt{n})$

$z^* = (126.2 - 120)/(23/\sqrt{25}) = \underline{1.35}$

c. $z = (\overline{x} - \mu)/(\sigma/\sqrt{n})$

$z^* = (18.93 - 18.2)/(3.7/\sqrt{140}) = \underline{2.33}$

d. $z = (\overline{x} - \mu)/(\sigma/\sqrt{n})$

$z^* = (79.6 - 81)/(13.3/\sqrt{50}) = \underline{-0.74}$

DECISIONS AND CONCLUSIONS

Since the null hypothesis, H_O, is usually thought to be the statement whose truth is being challenged by the experimenter, all decisions are about the null hypothesis. The alternative hypothesis, H_a, however is usually thought to express the experimenter's viewpoint. Thus, the interpretation of the decision or conclusion is expressed from the experimenter and alternative hypothesis point of view.

Decision:
  1) If the p-value is less than or equal to the specified level of significance (α), the null hypothesis will be rejected
  (if **P** $\leq \alpha$, reject H_O).
  2) If the p-value is greater than the specified level of significance (α), fail to reject the null hypothesis
  (if **P** $> \alpha$, fail to reject H_O).

Conclusion:
  1) If the decision is "reject H_O," the conclusion should read "There is sufficient evidence at the α level of significance to show that ... (the meaning of the alternative hypothesis)."

  2) If the decision is "fail to reject H_O," the conclusion should read "There is not sufficient evidence at the α level of significance to show that ... (the meaning of the alternative hypothesis)."

8.91 a. *Reject H_O* or *Fail to reject H_O*
     b. When the calculated p-value is smaller than or equal to α, the decision will be *reject H_O*.

     When the calculated p-value is larger than α, the decision will be *fail to reject H_O*.

8.93 a. Reject H_O, **P** $< \alpha$
     b. Fail to reject H_O, **P** $> \alpha$
     c. Reject H_O, **P** $< \alpha$
     b. Reject H_O, **P** $< \alpha$

8.95 a. Fail to reject H_O b. Reject H_O

8.97 a. Directions - no answer.
   b. ≈ 0.0000
   c. no means down by 1451 or lower; The probability of taking a sample of 24 and having a mean less than 1451 when the true mean is equal to 1500 is 0.0000.
   d. Reject Ho.

The p-value approach uses the calculated test statistic to find the area under the curve that contains the calculated test statistic and any values "beyond" it, in the direction of the alternative hypothesis. This probability (area under the curve), based on the position of the calculated test statistic, is compared to the level of significance (α) for the test and a decision is made.

8.99 The p-value measures the likeliness of the sample results based on a true null hypothesis.

Rules for calculating the p-value

1) If H_a contains <, then the p-value = $P(z < z^*)$.
2) If H_a contains >, then the p-value = $P(z > z^*)$.
3) If H_a contains $\neq$, then the p-value = $2P(z > |z^*|)$.

The p-value can then be calculated by using the z* value with Table 3, or it can be found directly using Table 5 (Appendix B, in ES10-p812), or it can be found using a computer and/or calculator.

8.101 p-value = $2 \cdot P(z > 1.1)$ = 2(0.5000 - 0.3643) = 2(0.1357) = $\underline{0.2714}$

8.103 a. p-value = $P(z > 1.48)$ = 0.5000 - 0.4306 = $\underline{0.0694}$

   b. p-value = $P(z < -0.85)$ = 0.5000 - 0.3023 = $\underline{0.1977}$

   c. p-value = $2 \cdot P(z > 1.17)$ = 2(0.5000 - 0.3790) = $\underline{0.2420}$

   d. p-value = $P(z < -2.11)$ = 0.5000 - 0.4826 = $\underline{0.0174}$

   e. p-value = $2 \cdot P(z > 0.93)$ = 2(0.5000 - 0.3238) = $\underline{0.3524}$

8.105 a. **P** = P(z > z*) = 0.0582

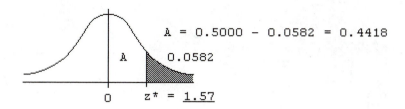

A = 0.5000 − 0.0582 = 0.4418

0.0582

z* = <u>1.57</u>

b. **P** = P(z < z*) = 0.0166

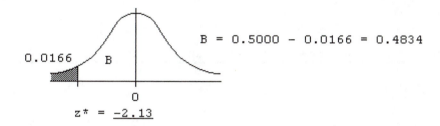

B = 0.5000 − 0.0166 = 0.4834

0.0166

z* = <u>−2.13</u>

c. **P** = P(z < −z*) + P(z > +z*) = 2·P(z > +z*) = 0.0042

P(z > +z*) = 0.0021

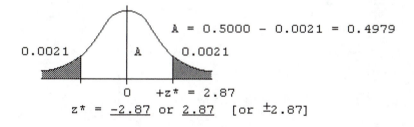

A = 0.5000 − 0.0021 = 0.4979

0.0021   0.0021

+z* = 2.87

z* = <u>−2.87</u> or <u>2.87</u>   [or <u>±2.87</u>]

8.107 $\mathbf{P}$ = P(z > z*) = 0.0170   0.5000 - 0.0170 = 0.4830

$\quad$ z* = -2.12

$\quad$ z = ($\overline{x}$ - μ)/(σ/$\sqrt{n}$)

$\qquad$ -2.12 = (14 - 16)/(σ/$\sqrt{50}$)

$\qquad$ (-2.12)(σ/$\sqrt{50}$) = -2

$\qquad$ σ = (-2)($\sqrt{50}$/(-2.12) = <u>6.67</u>

Computer and calculator commands to complete a hypothesis test for a mean μ with a standard deviation σ can be found in ES10-p438-439.

Compare the calculated p-value to the given level of significance, α. Using the rules for comparison as stated in SSM*-p269, a decision can be made about the null hypothesis.

(*SSM denotes this manual, the Student Solutions Manual)

8.109 a. H$_O$: μ = 525   vs.   H$_a$: μ < 525

$\quad$ b. Fail to reject H$_O$; the population mean is not significantly less than 525.

$\quad$ c. $\sigma_{\overline{x}}$ = σ/$\sqrt{n}$ = 60.0/$\sqrt{38}$ = 9.7333 = <u>9.733</u>

8.111 a. H$_O$: μ = 6.25   vs.   H$_a$: μ ≠ 6.25

$\quad$ b. Reject H$_O$; the population mean is significantly different than 6.25.

$\quad$ c. $\sigma_{\overline{x}}$ = σ/$\sqrt{n}$ = 1.4/$\sqrt{78}$ = <u>0.1585</u>

$\quad$ d. $\overline{x}$ = Σx/n;   therefore Σx = n·$\overline{x}$ = (78)(6.596) = <u>514.488</u>

$\qquad$ s² = [Σx² - ((Σx)²/n)]/(n - 1)

$\qquad$ (1.273)² = [Σx² - ((514.488)²/78]/(78 - 1)

$\qquad$ 1.62053 = [Σx² - 3393.5628]/77

$\qquad$ Σx² = 3393.5628 + (1.62053)(77) = <u>3518.3437</u>

```
┌─────────────────────────────────────────────────────────────────┐
│                    WORD   PROBLEMS                                │
│ 1. Look for the key words that indicate the need for a hypothesis │
│      test for μ using a z-test.  Statements that mention; "testing│
│      a mean amount" or "make a decision about the mean value" and │
│      the fact that the population standard deviation (σ) or       │
│      variance (σ²) are given are examples of these key words. ... │
│                                                                   │
│ 2. Write down the values needed: $\mu, \sigma, \bar{x}, n, \alpha$.│
│      If the mean is mentioned in a sentence with no reference to a│
│      sample, then it is most likely μ (population mean).  If the  │
│      mean is mentioned in a sentence involving the sample or      │
│      thereafter, then it is usually $\bar{x}$ (sample mean).  Often the│
│      sample size (n) is mentioned in the same sentence.  If the   │
│      standard deviation or variance is given in a sentence without│
│      any reference to a sample, then it also is usually the       │
│      population σ or σ² respectively.                             │
│                                                                   │
│ 3. Proceed with the hypothesis steps as outlined in:             │
│      ES10-pp427&444-445   or   SSM-p265.                          │
└─────────────────────────────────────────────────────────────────┘
```

8.113 a. The mean price for all laptops.

    b. H_o: μ = \$1240 (≥)
       H_a: μ < \$1240

    c. z* = $(1211 - 1240)/(66.75/\sqrt{35})$ = -2.57
       p-value = P(z < -2.57) = P(z > 2.57) = 0.5000 - 0.4949
            = 0.0051

    d. Fail to reject H_o, there is not sufficient evidence to
       conclude the mean price for laptops is less than \$1240, at
       the 0.001 level of significance.

8.115 Step 1: a. The mean checkout time at a local grocery store
            b. H_o: μ = 12 (≥)
               H_a: μ < 12
      Step 2: a. normality is indicated
            b. z, σ = 2.3;   c. α = 0.02
      Step 3: a. n = 28, $\bar{x}$ = 10.9
            b. z = $(\bar{x} - \mu)/(\sigma/\sqrt{n})$
               z* = $(10.9 - 12)/(2.3/\sqrt{28})$ = -2.53
      Step 4: a. **P** = P(z < -2.53) = P(z > 2.53);
             Using Table 3, Appendix B, ES10-p810:
             P = 0.5000 - 0.4943 = 0.0057
             Using Table 5, Appendix B, ES10-p812:
             0.0054 < **P** < 0.0062
            b. p-value is smaller than α

Step 5. a. Reject Ho
         b. There is sufficient evidence to conclude that the
            mean checkout time this week was less than 12
            minutes, at the 0.02 level of significance.

8.117 Step 1: a. The average amount spent on Mother's Day.
                b. $H_{o:}$ $\mu = \$104.63$ ($\geq$)
                   $H_{a:}$ $\mu < \$104.63$
         Step 2: a. normality is assumed, CLT with n = 60
                 b. z, $\sigma = 29.50$;   c. $\alpha = 0.05$
         Step 3: a. n = 60, $\overline{x} = 94.27$
                 b. $z = (\overline{x} - \mu)/(\sigma/\sqrt{n})$
                    $z^* = (94.27 - 104.63)/(29.50/\sqrt{60}) = -2.72$
         Step 4: a. **P** = P(z < -2.72) = P(z > 2.72);
                    Using Table 3, Appendix B, ES10-p810:
                    P = 0.5000 - 0.4967 = 0.0033
                    Using Table 5, Appendix B, ES10-p812:
                    0.0030 < **P** < 0.0035
                 b. p-value is smaller than α
         Step 5. a. Reject Ho
                  b. There is sufficient evidence to conclude that the
                     average amount spent on Mother's Day is less than
                     $104.63, at the 0.05 level of significance.

8.119 a. The mean accuracy of quartz watches measured in seconds in
             error per month.
          b. $H_{o:}$ $\mu = 20$ ($\leq$)
             $H_{a:}$ $\mu > 20$
          c. normality is assumed, CLT with n = 36
             use z with $\sigma = 9.1$; an $\alpha = 0.05$ is given
          d. n = 36, $\overline{x} = 22.7$
          e. $z = (\overline{x} - \mu)/(\sigma/\sqrt{n})$
             $z^* = (22.7 - 20)/(9.1/\sqrt{36}) = 1.78$

             P = P(z > 1.78);
                  Using Table 3, Appendix B, ES10-p810:
                  P = 0.5000 - 0.4625 = 0.0375
                  Using Table 5, Appendix B, ES10-p812:
                  0.0359 < **P** < 0.0401
          f. **P** < α; Reject H_o
             At the 0.05 level of significance, there is sufficient
             evidence to support the contention that the wrist watches
             priced under $25 exhibit greater error (less accuracy)
             than watches in general.

8.121 Results will vary; however, expect your results to be similar to those shown in Table 8.8.

SECTION 8.6 EXERCISES

THE CLASSICAL HYPOTHESIS TEST: A FIVE-STEP PROCEDURE

Step 1: The Set-Up:
    a. Describe the population parameter of concern.
    b. State the null hypothesis (H_O) and the alternative hypothesis (H_a).

Step 2: The Hypothesis Test Criteria:
    a. Check the assumptions.
    b. Identify the probability distribution and the test statistic formula to be used.
    c. Determine the level of significance, α.

Step 3: The Sample Evidence:
    a. Collect the sample information.
    b. Calculate the value of the test statistic.

Step 4: The Probability Distribution:
    a. Determine the critical region and critical value(s).
    b. Determine whether or not the calculated test statistic is in the critical region.
Step 5: The Results:
    a. State the decision about H_O.
    b. State the conclusion about H_a.

Review "The Null and Alternative Hypotheses, H_O and H_a" in SSM-pp265-266, if necessary.

8.123 H_O: The mean shearing strength is at least 925 lbs.
    H_a: The mean shearing strength is less than 925 lbs.

8.125 a. H_O: $\mu = 1.25$ ($\geq$)
        H_a: $\mu < 1.25$

                                               b. H_O: $\mu = 335$
                                                 H_a: $\mu \neq 335$

    c. H_O: $\mu = 230,000$ ($\leq$)
        H_a: $\mu > 230,000$

Review "Errors" and "H$_O$ Decisions", if necessary, in:
ES10-pp421&422, or SSM-pp267&268.

8.127 H$_O$: $\mu = 350$ ($\leq$) vs. H$_a$: $\mu > 350$

    a. It is decided that the average salt content is more than 350 mg when in fact, it is not more than 350 mg.

    b. It is decided that the average salt content is less than or equal to 350 mg when in fact it is greater than 350 mg.

8.129 H$_O$: $\mu = 85$ ($\leq$) vs. H$_a$: $\mu > 85$

Type I error: It is decided that the mean minimum plumber's call is greater than \$85 when in fact it is not more than \$85.

Type II error: It is decided that the mean minimum plumber's call is at most \$85 when in fact it is greater than \$85.

Critical region - that part under the curve where H$_O$ will be rejected (size based on α)
Noncritical region - the remaining part under the curve where H$_O$ will not be rejected
Critical value(s) - the z(α) or boundary point values of z, separating the critical and noncritical regions

See ES10-pp450-451 for a visual display of these regions and value(s).

8.131 a. The critical region is the set of all values of the test statistic that will cause us to reject H$_O$.

    b. The critical value(s) is the value(s) of the test statistic that forms the boundary between the critical region and the non-critical region. The critical value is in the critical region.

8.133 Because alpha and beta are interrelated; if one is reduced, the other one becomes larger. Alpha is area of critical region when H_o is true, beta is related to the noncritical region when H_o is false.

Determining the test criteria

1. Draw a picture of the standard normal (z) curve.
     (0 is at the center)

2. Locate the critical region (based on α and H_a)
     a) if H_a contains <, all of the α is placed in the left tail
     b) if H_a contains >, all of the α is placed in the right tail
     c) if H_a contains ≠, place $\alpha/2$ in each tail.

3. Shade in the critical region (the area where you will reject H_o).

4. Find the appropriate critical value(s) using the $z(\alpha)$ concept and the Standard Normal Distribution (Table 3, Appendix B, ES10-p810, or Table 4(a) for one-tail and Table 4(b) for two-tails, Appendix B, ES10-p811).

     If H_a contains >, the critical value is $z(\alpha)$.
     If H_a contains <, the critical value is $z(1-\alpha)$ or $-z(\alpha)$.
     If H_a contains ≠, the critical values are $\pm z(\alpha/2)$ or $z(\alpha/2)$ with $z(1-\alpha/2)$.

     Remember this boundary value divides the area under the curve into critical and noncritical regions and is part of the critical region.

8.135 $z \le -2.33$

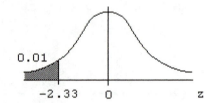

8.137 a.

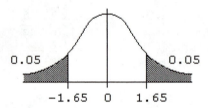

b.

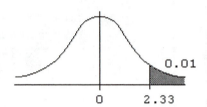

c.

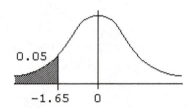

d.

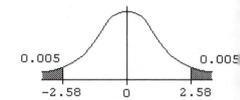

For exercises 8.139 and 8.140, use $z* = (\overline{x} - \mu)/(\sigma/\sqrt{n})$, substituting the given values, then solving for the required unknown.

8.139 $z* = (\overline{x} - \mu)/(\sigma/\sqrt{n})$

$-1.18 = (\overline{x} - 250)/(22.6/\sqrt{85})$

$-1.18 = (\overline{x} - 250)/2.451314$

$-2.89255 = \overline{x} - 250$

$\overline{x} = 247.107449 = \underline{247.1}$

$\overline{x} = \Sigma x/n$

$247.107449 = \Sigma x/85$

$\Sigma x = \underline{21,004.133}$

NOTE: Standard Error $= \sigma_{\overline{x}} = \sigma / \sqrt{n}$ and $z = (\overline{x} - \mu)/(\sigma/\sqrt{n})$

8.141 a. $z = n$(standard errors from mean):

$z = (\overline{x} - \mu)/(\sigma/\sqrt{n})$

$z = (4.8 - 4.5)/(1.0/\sqrt{100}) = \underline{3.0}$

$\underline{\overline{x} = 4.8}$ is 3.0 standard errors **above** the mean $\mu = 4.5$

b. If $\alpha = 0.01$, the critical region is $z \geq 2.33$. Since z* is equal to 3.00, it is in the critical region. Therefore, yes, $\underline{\text{reject } H_O}$.

RESULTS, DECISIONS AND CONCLUSIONS

Since the null hypothesis, H_O, is usually thought to be the statement whose truth is being challenged by the experimenter, all decisions are about the null hypothesis. The alternative hypothesis, H_a, however is usually thought to express the experimenter's viewpoint. Thus, the conclusion (interpretation of the decision) is expressed from the experimenter and alternative hypothesis point of view.

The two possible outcomes are:
1. z^* falls in the critical region or
2. z^* falls in the noncritical region.   . . .

Decision and Conclusion:

If z^* falls in the critical region, we **reject H_0**. The conclusion is very strong and proclaims the alternative to be the case, that is, there is sufficient evidence to overturn H_0 in favor of H_a. It should read something like "There is sufficient evidence at the α level of significance to show that ...(the meaning of the H_a)."

If z^* falls in the noncritical (acceptance) region, we **fail to reject H_0**. The conclusion is much weaker, that is, it suggests that the data does not provide sufficient evidence to overturn H_0. This does not necessarily mean that we have to accept H_0 at this point, but only that this sample did not provide sufficient evidence to reject H_0. It should read something like "There is not sufficient evidence at the α level of significance to show that ...(the meaning of the H_a)."

8.143 a. *Reject H_0* or *Fail to reject H_0*

      b. When the calculated test statistic falls in the critical region, the decision will be *reject H_0*.

        When the calculated test statistic falls in the noncritical region, the decision will be *fail to reject H_0*.

Computer and calculator commands to complete a hypothesis test using the classical approach can be found in ES10-pp438-439. It is the same command used for the probability approach. Compare the calculated z value (test statistic) with the corresponding critical value(s). The locations of z*, relative to the critical value of z, will determine the decision you must make about the null hypothesis.

8.145 a. H_0: $\mu = 15.0$ vs. H_a: $\mu \neq 15.0$

      b. Critical values: $\pm z(0.005) = \pm 2.58$
        Decision: reject H_0
        Conclusion: There is sufficient evidence to conclude that the mean is different than 15.0.

      c. $\sigma_{\bar{x}} = \sigma/\sqrt{n} = 0.5/\sqrt{30} = 0.091287 = \underline{0.0913}$

8.147 a. H_O: μ = 72 ($\leq$)   vs. H_a: μ > 72

  b. Critical values: z(0.05) = 1.65
      Decision: Fail to reject H_O
      Conclusion: There is not sufficient evidence to conclude
      the mean is greater than 72.

  c. $\sigma_{\overline{x}}$ = σ / $\sqrt{n}$ = 12.0 / $\sqrt{36}$ = <u>2.00</u>

See SSM-p273 for information on "Word Problems", if necessary.

<u>Hint for writing the hypotheses for exercise 8.149</u>
Look at the fourth sentence of the exercise, "Is there sufficient
evidence to conclude ... scored <u>higher than</u> the state average?''.
"higher than" indicates <u>greater than</u> (>). Therefore, the
alternative hypothesis is <u>greater than</u> (>). Equality (=) is used
in the null hypothesis (as usual), but it stands for less than or
equal to ($\leq$).

8.149 Step 1: a. The mean score for the Emergency Medical Services
                     Certification Examiniation
               b. H_O: μ = 79.68 ($\leq$)
                   H_a: μ > 79.68
        Step 2: a. normality is assumed, CLT with n = 50
               b. z, σ = 9.06;   c. α = 0.05
        Step 3: a. n = 50, $\overline{x}$ = 81.05
               b. z = ($\overline{x}$ - μ)/($\sigma/\sqrt{n}$)
                   z* = (81.05 - 79.68)/(9.06/$\sqrt{50}$) = 1.07
        Step 4: a. z(0.05) = 1.65

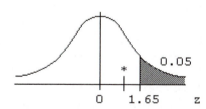

               b. z* falls in the noncritical region, see Step 4a
        Step 5. a. Fail to reject Ho
               b. There is not sufficient evidence conclude that
                   the sample average is higher than the state
                   average, at the 0.05 level of significance.

8.151 Step 1: a. The current mean age of race track fans

      b. $H_o:$ $\mu = 55$

        $H_a:$ $\mu \neq 55$

Step 2: a. normality assumed, CLT with n = 35

      b. z,   σ = 8   c. α = 0.10

Step 3: a. n = 35, $\overline{x}$ = 52.7

      b. $z = (\overline{x} - \mu)/(\sigma/\sqrt{n})$

        $z* = (52.7 - 55)/(8/\sqrt{35}) = -1.70$

Step 4: a. $\pm z(0.05) = \pm 1.65$

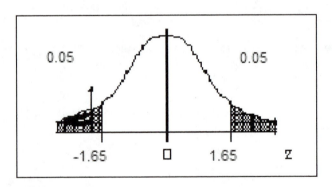

      b. z* falls in the critical region, see Step 4a

Step 5: a. Reject H_o

      b. At the 0.10 level of significance, the sample does provide sufficient evidence to conclude the mean age of race track fans is different than 55 years old.

8.153   Step 1: a. The mean time of Yankee baseball games
            b. H_0: $\mu = 170.1$ (≤)
                H_a: $\mu > 170.1$
        Step 2: a. normality indicated
            b. z, $\sigma = 21.0$ c. $\alpha = 0.05$
        Step 3: a. n = 8, $\overline{x} = 190.375$
            b. $z = (\overline{x} - \mu)/(\sigma/\sqrt{n})$
                $z* = (190.375 - 170.1)/(21.0/\sqrt{8}) = 2.73$
        Step 4: a. $z(0.05)' = 1.65$

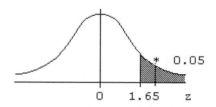

            b. z* falls in the critical region, see Step 4a
        Step 5: a. Reject H_0
            b. At the 0.05 level of significance, there is
                sufficient evidence to support the contention
                that Yankee games, on the average, last longer
                than games of Major League Baseball.

8.155   Step 1: a. The average amount of water Americans drink per
                day
            b. H_0: $\mu = 36.8$ (≤)
                H_a: $\mu > 36.8$
        Step 2: a. normality assumed, CLT with n = 42
            b. z, $\sigma = 11.2$ c. $\alpha = 0.05$

Step 3: a. n = 42, $\overline{x}$ = 39.3

b. z = $(\overline{x} - \mu)/(\sigma/\sqrt{n})$

z* = $(39.3 - 36.8)/(11.2/\sqrt{42})$ = 1.45

Step 4: a. z(0.05) = 1.65

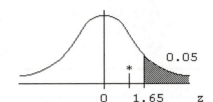

b. z* falls in the noncritical region, see Step 4a

Step 5: a. Fail to reject H_o

b. At the 0.05 level of significance, the sample does not provide sufficient evidence to show that education professionals consume, on the average, more water daily than the national average.

8.157 Results will vary, however expect your results to be similar to those shown in Table 8.12.

CHAPTER EXERCISES

8.159 a. $\overline{x}$ = <u>32.0</u>

b. σ = <u>2.4</u>

c. n = <u>64</u>

d. 1 - α = <u>0.90</u>

e. z(α/2) = z(0.05) = <u>1.65</u>

f. $\sigma_{\overline{x}}$ = $2.4/\sqrt{64}$ = <u>0.3</u>

g. E = z(α/2)·$\sigma/\sqrt{n}$ = (1.65)(0.3) = <u>0.495</u>

h. UCL = $\overline{x}$ + E = 32.0 + 0.495 = <u>32.495</u>

i. LCL = $\overline{x}$ - E = 32.0 - 0.495 = <u>31.505</u>

8.161   Step 1: The mean age of volunteer ambulance members in upstate New York

Step 2: a. normality assumed, CLT with n = 80

        b. z, σ = 7.8   c. 1-α = 0.95

Step 3: n = 80, $\overline{x}$ = 45

Step 4: a. $\alpha/2$ = 0.05/2 = 0.025; z(0.025) = 1.96

        b. E = z($\alpha/2$)·$\sigma/\sqrt{n}$ = (1.96)(7.8/$\sqrt{80}$)

          = (1.96)(0.872) = 1.709

        c. $\overline{x}$ ± E = 45 ± 1.7

Step 5: <u>43.3 to 46.7</u>, the 0.95 confidence interval for μ

Exercise 8.163 shows the effect of the level of confidence (1 - α) on the width of a confidence interval.

8.163 a. Step 1: The mean weights of full boxes of a certain kind of cereal

Step 2: a. normality indicated

        b. z, σ = 0.27   c. 1-α = 0.95

Step 3: n = 18, $\overline{x}$ = 9.87

Step 4: a. $\alpha/2$ = 0.05/2 = 0.025; z(0.025) = 1.96

        b. E = z($\alpha/2$)·$\sigma/\sqrt{n}$ = (1.96)(0.27/$\sqrt{18}$)

          = (1.96)(0.0636) = 0.12

        c. $\overline{x}$ ± E = 9.87 ± 0.12

Step 5: <u>9.75 to 9.99</u>, the 0.95 confidence interval for μ

b. Step 1: The mean weights of full boxes of a certain kind Of cereal

Step 2: a. normality indicated

        b. z, σ = 0.27   c. 1-α = 0.99

Step 3: n = 18, $\overline{x}$ = 9.87

Step 4: a. $\alpha/2$ = 0.01/2 = 0.005; z(0.005) = 2.58

        b. E = z($\alpha/2$)·$\sigma/\sqrt{n}$ = (2.58)(0.27/$\sqrt{18}$)

          = (2.58)(0.0636) = 0.16

        c. $\overline{x}$ ± E = 9.87 ± 0.16

Step 5: <u>9.71 to 10.03</u>, the 0.99 confidence interval for μ

c. The increased confidence level widened the interval.

8.165 a. Step 1: The mean score for a clerk-typist position
Step 2: a. normality assumed, CLT with n = 100
b. z, σ = 10.5   c. 1-α = 0.99
Step 3: n = 100, $\overline{x}$ = 72.6
Step 4: a. $\alpha/2$ = 0.01/2 = 0.005; z(0.005) = 2.58
b. E = z($\alpha/2$)·$\sigma/\sqrt{n}$ = (2.58)(10.5/$\sqrt{100}$)
= (2.58)(1.05) = 2.71
c. $\overline{x}$ ± E = 72.6 ± 2.71
Step 5: <u>69.89 to 75.31</u>, the 0.99 confidence interval for μ

b. <u>Yes.</u>   75.0 falls within the interval.

8.167   Step 1: The mean time for National League baseball games
Step 2: a. normality assumed, CLT with n = 48
b. z, σ = 21   c. 1-α = 0.98
Step 3: n = 48, $\overline{x}$ = 169.1
Step 4: a. $\alpha/2$ = 0.02/2 = 0.01; z(0.01) = 2.33
b. E = z($\alpha/2$)·$\sigma/\sqrt{n}$ = (2.33)(21/$\sqrt{48}$)
= (2.33)(3.031) = 7.06
c. $\overline{x}$ ± E = 169.1 ± 7.06
Step 5: <u>162.04 to 176.16</u>, the 0.98 confidence interval for

8.169 a. 125.10 to 132.95, the 0.95 confidence interval for μ

b. $\sigma_{\overline{x}}$ = $\sigma / \sqrt{n}$ = 10.0 / $\sqrt{25}$ = 2.00

$\overline{x}$ ± z($\alpha/2$)·$\sigma/\sqrt{n}$ = 129.02 ± (1.96)(2.00)
129.02 ± 3.92

<u>125.10 to 132.94</u>, the 0.95 confidence interval for μ

8.171 n = [z($\alpha/2$)·σ/E]^2 = [(2.58)(3.7)/1]^2 = 91.126 = <u>92</u>

8.173 n = [z($\alpha/2$)·σ/E]^2 = [(2.58)(σ)/($\sigma/3$)]^2 = 59.9 = <u>60</u>

8.175 When using probability-value approach:

a. The 0.01 is used as a 'boundary' separating the values of P that will lead to a rejection of the null hypothesis from the values of P that do not lead to a rejection of the null hypothesis.

b. The value of the test statistic would have to be less extreme to lead to a rejection of the null hypothesis.

8.177 a. H_O: $\mu = 100$ b. H_a: $\mu \neq 100$

c. $\alpha = \underline{0.01}$ d. $\mu = \underline{100}$

e. $\bar{x} = \underline{96}$ f. $\sigma = \underline{12}$

g. $\sigma_{\bar{x}} = 12/\sqrt{50} = 1.697 = \underline{1.70}$

h. $z* = (\bar{x} - \mu)/(\sigma/\sqrt{n}) = (96-100)/1.7 = \underline{-2.35}$

i. p-value $= 2 \cdot P(z < -2.35) = 2 \cdot P(z > 2.35)$
$$= 2(0.5000 - 0.4906)$$
$$= 2(0.0094) = \underline{0.0188}$$

j. Fail to reject H_O

k. p-value $= 0.0188$

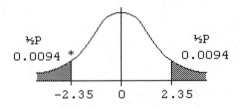

8.179 Step 1: a. The mean delay time in a garden sprinkler system
     b. H_o: $\mu = 45$ ($\leq$)
      H_a: $\mu > 45$
  Step 2: a. normality indicated
     b. z, $\sigma = 8$ c. $\alpha = 0.02$
  Step 3: a. $n = 15$, $\overline{x} = 50.1$
     b. $z = (\overline{x} - \mu)/(\sigma/\sqrt{n})$
      $z* = (50.1 - 45)/(8/\sqrt{15}) = 2.47$
  Step 4: -- using p-value approach ----------------------
     a. **P** $= P(z > 2.47)$;
        Using Table 3, Appendix B, ES10-p810:
        P $= 0.5000 - 0.4932 = 0.0068$
        Using Table 5, Appendix B, ES10-p812:
        $0.0062 < P < 0.0071$
    b. **P** $< \alpha$
   --- using classical approach ----------------------
     a.

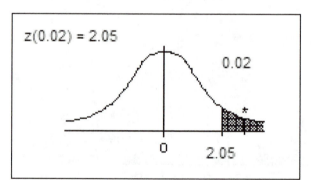

    b. z* falls in the critical region, see Step 4a

  Step 5: a. Reject H_o
     b. At the 0.02 level of significance, the sample
      does provide sufficient evidence to conclude that
      the mean delay time is more than 45 seconds.

8.181 a. H_O: $\mu = 0.50$
H_a: $\mu \neq 0.50$

b. $z* = (0.51 - 0.50)/(0.04/\sqrt{25}) = 1.25$

$P = 2 \cdot P(z > 1.25) = 2(0.5000 - 0.3944) = \underline{0.2112}$

c. $z = \pm 2.33$

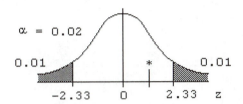

$\alpha = 0.02$

0.01   0.01

-2.33 0   2.33   z

NOTE: When both methods of hypothesis testing are asked for; steps 1, 2, 3 and 5 are the same, and step 4 is shown twice, p-value approach followed by the classical. Dashed lines are used to separate the answers.

8.183 Step 1: a. The mean weight of one load of pollen and nectar being carried by a worker bee to the hive after collecting it

b. H_O: $\mu = 0.0113$ ($\leq$)
H_a: $\mu > 0.0113$

Step 2: a. normality indicated, CLT with n = 200
b. z, $\sigma = 0.0063$ c. $\alpha = 0.01$

Step 3: a. n = 200, $\overline{x} = 0.0124$
b. $z = (\overline{x} - \mu)/(\sigma/\sqrt{n})$

$z* = (0.0124 - 0.0113)/(0.0063/\sqrt{200}) = 2.47$

Step 4: -- using p-value approach -----------------------
a. **P** = P(z > 2.47);
Using Table 3, Appendix B, ES10-p810:
P = 0.5000 - 0.4932 = 0.0068
Using Table 5, Appendix B, ES10-p812:
0.0062 < P < 0.0071

b. **P** < α

--- using classical approach ----------------------
a.

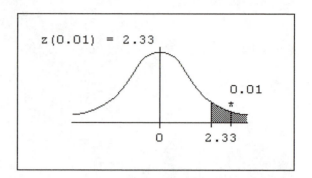

b. z* falls in the critical region, see Step 4a

Step 5:   a. Reject H_O
          b. At the 0.01 level of significance, the sample
             does provide sufficient evidence to conclude that
             the mean load of pollen and nectar carried by
             Fuzzy's strain of Italian worker bees is greater
             than the rest of the honey bee population.

8.185 Step 1: a. The mean customer checkout time at a large
                  supermarket
          b. $H_{O:}$ $\mu = 9$ $(\leq)$
             $H_{a:}$ $\mu > 9$
Step 2: a. normality indicated
        b. z, $\sigma = 2.5$ c. $\alpha = 0.02$
Step 3: a. n = 24, $\bar{x}$ = 10.6
        b. $z = (\bar{x} - \mu)/(\sigma/\sqrt{n})$
           $z* = (10.6 - 9.0)/(2.5/\sqrt{24}) = 3.14$
Step 4: -- using p-value approach ----------------------
         a. **P** = P(z > 3.14);
                 Using Table 3, Appendix B, ES10-p810:
                 P = 0.5000 - 0.4992 = 0.0008
                 Using Table 5, Appendix B, ES10-p812:
                 0.0008 < P < 0.0010
         b. **P** < α

a.

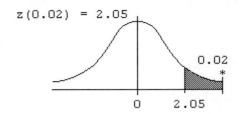

z(0.02) = 2.05

0.02
*

0   2.05

   b. z* falls in the critical region, see Step 4a

Step 5: a. Reject H_o

   b. At the 0.02 level of significance, the sample does provide sufficient evidence to conclude the mean waiting time is more than the claimed 9 minutes.

8.187 a. Step 1: a. The mean annual consumption of natural gas by residential customers

   b. $H_{o:}$ $\mu = 129.2$ ($\geq$)
      $H_{a:}$ $\mu < 129.2$

Step 2: a. normality assumed, CLT with n = 300

   b. z, $\sigma = 18$ c. $\alpha = 0.01$

Step 3: a. n = 300, $\bar{x} = 127.1$

   b. $z = (\bar{x} - \mu)/(\sigma/\sqrt{n})$

      $z* = (127.1 - 129.2)/(18/\sqrt{300}) = -2.02$

Step 4. a. **P** = P(z < -2.02) = P(z > 2.02);
            Using Table 3, Appendix B, ES10-p810:
            P = 0.5000 - 0.4783 = 0.0217
            Using Table 5, Appendix B, ES10-p812:
            0.0202 < P < 0.0228

   b. **P** > α

Step 5: a. Fail to reject H_o

   b. At the 0.01 level of significance, the sample does not provide sufficient evidence to conclude that the mean annual consumption has declined.

b. The p-value indicates the likelihood (approximately 0.02) of being wrong when you state that a significant reduction has not occurred.

8.189 a. H_a: $\mu \neq 18$;   Fail to reject H_o;   The population mean is not significantly different from 18.

b. $\sigma_{\overline{x}} = \sigma / \sqrt{n} = 4.00 / \sqrt{28} = 0.756$

$z = (\overline{x} - \mu)/(\sigma/\sqrt{n})$

$z* = (17.217 - 18)/(0.756) = -1.04$

p-value $= 2 \cdot P(z < -1.04) = 2 \cdot P(z > 1.04)$
$= 2(0.5000 - 0.3508) = 2(0.1492) = 0.2984 = \underline{0.30}$

8.191 a. Step 1: The mean, μ
Step 2: a. normality assumed, CLT with n = 100
   b. z, $\sigma = 5.0$ c. $1-\alpha = 0.95$
Step 3: n = 100, $\overline{x} = 40.6$
Step 4: a. $\alpha/2 = 0.05/2 = 0.025$;   z(0.025) = 1.96
   b. E = $z(\alpha/2) \cdot \sigma/\sqrt{n} = (1.96)(5/\sqrt{100})$
      $= (1.96)(0.50) = 0.98$
   c. $\overline{x} \pm E = 40.6 \pm 0.98$
Step 5: $\underline{39.6}$ to $\underline{41.6}$, the 0.95 confidence interval for μ

b. Step 1: a. The mean, μ
   b. H_o: $\mu = 40$
      H_a: $\mu \neq 40$
Step 2: a. normality assumed, CLT with n = 100
   b. z, $\sigma = 5.0$ c. $\alpha = 0.05$
Step 3: a. n = 100, $\overline{x} = 40.6$
   b. z = $(\overline{x} - \mu)/(\sigma/\sqrt{n})$
      $z* = (40.6 - 40)/(5/\sqrt{100}) = 1.20$
Step 4: a. **P** = 2P(z > 1.20);
      Using Table 3, Appendix B, ES10-p810:
      P = 2(0.5000 - 0.3849) = 2(0.1151) = 0.2302
      Using Table 5, Appendix B, ES10-p812:
      P = 2(0.1151) = 0.2302
   b. **P** > α

Step 5: a. Fail to reject H_o
b. At the 0.05 level of significance, there is not sufficient evidence to support the contention that the mean is not equal to 40.

c. Step 1: a. The mean, μ
b. H_o: $\mu = 40$
   H_a: $\mu \neq 40$
Step 2: a. normality assumed, CLT with n = 100
b. z, $\sigma = 5.0$ c. $\alpha = 0.05$
Step 3: a. n = 100, $\overline{x} = 40.6$
b. $z = (\overline{x} - \mu)/(\sigma/\sqrt{n})$
   $z* = (40.6 - 40)/(5/\sqrt{100}) = 1.20$
Step 4: a. $\pm z(0.025) = \pm 1.96$

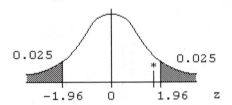

b. z* falls in the noncritical region, see * in Step 4a.
Step 5: a. Fail to reject H_o
b. At the 0.05 level of significance, there is not sufficient evidence to support the contention that the mean is not equal to 40.

d.

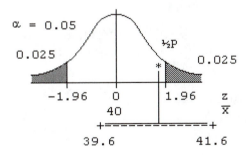

$z* = 1.20$ is in the noncritical region or P = 0.2302 is greater than α, and $\mu = 40$ is within the interval estimate of 39.6 to 41.6.

8.193 a. Step 1: The mean, μ
Step 2: a. normality assumed, CLT with n = 100
b. z, σ = 5.0   c. 1-α = 0.95
Step 3: n = 100, $\overline{x}$ = 40.9
Step 4: a. $\alpha/2$ = 0.05/2 = 0.025; z(0.025) = 1.96
b. E = z($\alpha/2$)·$\sigma/\sqrt{n}$ = (1.96)(5/$\sqrt{100}$)
    = (1.96)(0.50) = 0.98
c. $\overline{x}$ ± E = 40.9 ± 0.98
Step 5: <u>39.9 to 41.9</u>, the 0.95 confidence interval for μ

b. Step 1: a. The mean, μ
b. H_0: μ = 40 ($\leq$)
   H_a: μ > 40
Step 2: a. normality assumed, CLT with n = 100
b. z, σ = 5.0   c. α = 0.05
Step 3: a. n = 100, $\overline{x}$ = 40.9
b. z = ($\overline{x}$ - μ)/($\sigma/\sqrt{n}$)
   z* = (40.9 - 40)/(5/$\sqrt{100}$) = 1.80
Step 4: a. **P** = P(z > 1.80);
       Using Table 3, Appendix B, ES10-p810:
       P = (0.5000 - 0.4641) = 0.0359
       Using Table 5, Appendix B, ES10-p812:
       P = 0.0359
b. **P** < α
Step 5: a. Reject H_0
b. At the 0.05 level of significance, there is
sufficient evidence to support the contention
that the mean is greater than 40.

c. Step 1: a. The mean, μ
b. H_0: μ = 40 ($\leq$)
   H_a: μ > 40
Step 2: a. normality assumed, CLT with n = 100
b. z, σ = 5.0   c. α = 0.05
Step 3: a. n = 100, $\overline{x}$ = 40.9
b. z = ($\overline{x}$ - μ)/($\sigma/\sqrt{n}$)
   z* = (40.9 - 40)/(5/$\sqrt{100}$) = 1.80

Step 4: a. z(0.05) = 1.65

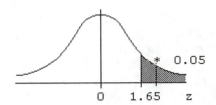

b. z* falls in the critical region, see Step 4a
Step 5: a. Reject H_o
b. At the 0.05 level of significance, there is sufficient evidence to support the contention that the mean is greater than 40.

d.

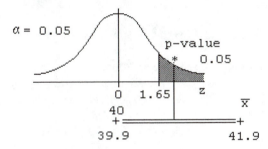

z* = 1.80 is in the critical region and P = 0.0359 is less than α, and μ = 40 is within the interval estimate of 39.9 to 41.9. Since the interval is two-sided and the hypothesis tests are one-sided, it is hard to compare the estimate and the hypothesis tests.

8.195 a. (2)$H_{a:}$ r > A   Failure to reject H_o will result in the drug being marketed. Because of the high current mortality rate, burden of proof is on the old ineffective drug.

b. (1)$H_{a:}$ r < A   Failure to reject H_o will result in the new drug not being marketed. Because of the low current mortality rate, burden of proof is on the new drug.

8.197 Every student will have different results, but they should be similar to the following.

   a. Using Minitab;

     The commands needed to obtain 50 rows/samples of 28 data per row/sample:

     Calc > Random Data > Normal >

       50 rows in columns **C1-C28**; Mean **18**; Standard deviation **4**

     The commands needed to obtain the 50 sample means:

     Calc > Row Statistic … >

       Mean of **C1-C28**; Store results in **C29**

     The commands to calculate z*:

     Calc > Calculator … >

       Store result (z*) in **C30**;

       Using expression **(C29-18)/(4/SQRT(28))**

     To sort the z* values into ranked order:

     Data > Sort … >

       Sort column **C30**; Store in **C31**; Sort by **C30**

     Using Excel:

     The commands needed to obtain 50 rows/samples of 28 data per row/sample:

     Tools > Data Analysis > Random Number Generation > OK

       Number of Variables **28**; Number of Random Numbers **50**;

       Distribution Normal; Mean **18**; Standard deviation **4**;

       Output Range **A1**

     The commands needed to obtain the 50 sample means:

     Activate cell **AC1**

     Insert function f$_x$ > Statistical > AVERAGE > OK

       Number1 **A1:AB1** > OK

       Drag right corner of average value box down to give other averages

     The commands to calculate z*:

     Activate cell **AD1**

     Edit Formula (=) **> (AC1-18)/(4/SQRT(28))**

   b. Examine column C31 or AD1, count the z*'s that are less than -1.04 and greater than 1.04. In one run of the commands 17/50 or 34% of the values were more extreme than the given z-values; On average, 30% should be more extreme (see 8.165b); an empirical p-value.

c. Critical values = ±1.65; Examine column C31 or AD1, count the z*'s that are less than -1.65 and greater than 1.65. 6/50 = 12%; On an average 10% (α=0.10) should fall in the critical region; an empirical level of significance.

CHAPTER 9 ▽ INFERENCES INVOLVING ONE POPULATION

Chapter Preview

Chapter 9 continues the work of inferential statistics started in Chapter 8. The concepts of hypothesis tests and confidence intervals will still be presented but on samples where the population standard deviation (σ) is unknown. Also inferences regarding the population binomial probability (p) and the population variance/standard deviation ($σ^2$/σ) will be introduced.

An article published by the U.S. Department of Health and Human Services about exercise is the topic of the chapter's opening section 'Get Enough Daily Exercise?'.

SECTION 9.1 EXERCISES

9.1   a. Mean amount of physical exercise time per week for women.
          b.

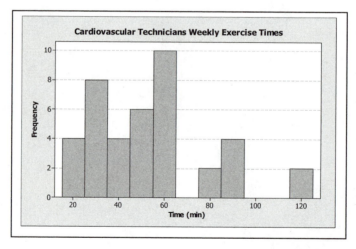

          c. Skewed to the right or maybe, bi-modal.
          d. Yes, amount of time does not appear to be normally distributed.

9.3 Pick any 3 numbers; the fourth must be the negative of the sum of the first three. For example, 4, 3, 1, whose sum is 8; the fourth is required to be -8 for the sum to be zero. Three numbers were chosen freely, however there was no choice for the last number.

---

t-Distribution
(used when σ is unknown)

Key facts about the t-distribution:

1. The total area under the t-distribution is 1.

2. It is symmetric about 0.

3. Its shape is a more "spread out" version of the normal shape.

4. A different curve exists for each sample size.

5. The shape of the distribution approaches the normal distribution shape as n increases [For df > 100, t is approximately normal.].

6. Critical values are determined based on α and degrees of freedom(df) - Table 6 (Appendix B, ES10-p813).

7. Degrees of freedom is abbreviated as *df*, where df = n - 1 for this application.

Explore the t-distribution for different degrees of freedom using the Chapter 9 Skillbuilder Applet 'Properties of t-distribution on your CD.

Notation: t(df,α) = t(degrees of freedom, area to the right)
                    ↑   ↑   ↑
              Table 6   row id #   column id #

 ex.: t(13,.025) means df = 13 (row) and α = .025 (column), using

       Table 6, t(13,.025) = 2.16   (df = n-1)

For t(df,α), consider the α given as the amount in one tail and use
the top row label - "Amount of α in One-Tail". For two-tailed
tests, an additional row label is given - "Amount of α in Two-
Tails". Note that it is twice the amounts in the one-tail row,
therefore α does not have to be divided by two.
For α > 0.5000, use the 1-α amount and negate the t-value.

ex.: t(14,0.90); α = 0.90, 1-α = 0.10,
                    t(14,0.90) = -t(14,0.10) = -1.35
(Table 6 is in Appendix B, ES10-p813.)

9.5   a. t(12, 0.01) = <u>2.68</u>   b. t(22, 0.025) = <u>2.07</u>

          b. t(50, 0.10) = <u>1.30</u>   d. t(8, 0.005) = <u>3.36</u>

9.7   a. t(18, 0.90) = -t(18, 0.10) = <u>-1.33</u>
          b. t(9, 0.99) = -t(9, 0.01) = <u>-2.82</u>
          c. t(35, 0.975) = -t(35, 0.025) = <u>-2.03</u>
          d. t(14, 0.98) = -t(14, 0.02) = -2.26 (computer calculation)
          Using Table 6: 2.14 < t < 2.62; Interpolation: 2.30

For a two-sided test:
      1. divide α by 2 and use the top row of column labels of
         Table 6 (Appendix B, ES10-p813) identified as "Amount of α
         (α/2, in this case) in One-Tail"
      or
      2. use the second row of column labels of Table 6 identified
         as "Amount of α in Two-Tails."

9.9   a. t(19,0.05) = 1.73   b. ±t(3,0.025) = ±3.18

          c. -t(18,0.01) = -2.55   d. t(17,0.10) = 1.33

9.11 $\alpha/2 = 0.05/2 = 0.025$; $\pm t(12, 0.025) = \pm 2.18$

9.13   a. -2.49   b. 1.71   c. -0.685

9.15   df = 7

9.17   Use the cumulative probability distribution function, Student's t distribution with 18 DF

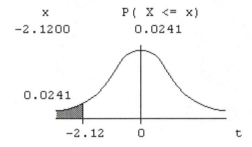

```
        x              P( X <= x)
    -2.1200             0.0241
```

9.19   a.   Symmetric about mean: mean is 0

b.   Standard deviation of t-distribution is greater than 1; t-distribution is different for each different sample size while there is only one z-distribution, t has df.

```
┌─────────────────────────────────────────────────────────────────────────┐
│                Estimating μ - the population mean                         │
│                          (σ unknown)                                      │
│                                                                           │
│   1. point estimate: x̄                                                    │
│                                                                           │
│   2. confidence interval:  x̄ ± t(df,α/2)·(s/√n), where df = n-1           │
│                                                                           │
│   Review steps for constructing a confidence interval for μ: ES10-        │
│   p404, SSM-p256.  The t-distribution is used when σ is unknown, and      │
│   sampling is from an approximately normal distribution or the sample     │
│   size is large.                                                          │
└─────────────────────────────────────────────────────────────────────────┘
```

9.21   Step 1: The mean, μ
           Step 2: a. normality assumed
                   b. t   c. $1-\alpha = 0.95$
           Step 3: n = 24, $\bar{x}$ = 16.7,   s = 2.6
           Step 4: a. $\alpha/2 = 0.05/2 = 0.025$; df = 23; t(23, 0.025) = 2.07
                   b. E = $t(df,\alpha/2)\cdot(s/\sqrt{n})$ = (2.07)(2.6/$\sqrt{24}$)
                         = 1.0986 = 1.10
                   c. $\bar{x} \pm E$ = 16.7 ± 1.10
           Step 5: <u>15.60 to 17.8</u>, the 0.95 confidence interval for μ

9.23   Step 1: The mean amount spent on prescription drugs by older
                   Minnesotans
           Step 2: a. normality assumed, CLT with n = 3000
                   b. t   c. $1-\alpha = 0.99$
           Step 3: n = 3000, $\bar{x}$ = $85,   s = $50.35
           Step 4: a. $\alpha/2 = 0.01/2 = 0.005$;   df = 2999;
                         t(2999, 0.005) = 2.58
                   b. E = $t(df,\alpha/2)\cdot(s/\sqrt{n})$ = (2.58)(50.35 / $\sqrt{3000}$)
                         = 2.37169 = 2.37
                   c. $\bar{x} \pm E$ = 85 ± 2.37
           Step 5: <u>$82.63 to $87.37</u>, the 0.99 confidence interval for μ

9.25   a. $\bar{x} = \Sigma x/n$ = 3582.17/41 = <u>$87.37</u>
           b. s = $\sqrt{[\Sigma(x - \bar{x})^2 / (n - 1)]}$ = $\sqrt{9960.336 / 40}$ = <u>$15.78</u>

c. Step 1: The mean textbook cost per semester
   Step 2: a. normality assumed, CLT with n = 41
           b. t   c. 1-α = 0.90
   Step 3: n = 41, $\overline{x}$ = 87.37, s = 15.78
   Step 4: a. α/2 = 0.10/2 = 0.05; df = 40; t(40, 0.05) = 1.68
           b. E = t(df,α/2)·(s/√n) = (1.68)(15.78/√41) = 4.14
           c. $\overline{x}$ ± E = 87.37 ± 4.14
   Step 5: $83.23 to $91.51, the 0.90 confidence interval
           for μ

9.27 a. 16(5.5) = 88 ounces; 74.9/16 = 4.68 lbs

  b. The data appears to follow a straight line,
     indicating an approximately normal distribution.

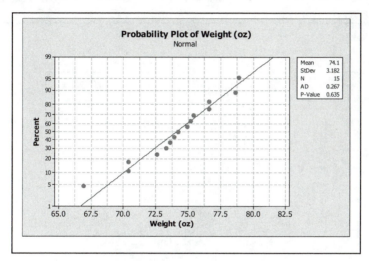

  c. $\overline{x}$ = Σx/n = 3582.17/41 = 74.10; s = 3.182

  d. Step 1: The mean weight of chicken broilers
     Step 2: a. normality indicated in part b
             b. t   c. 1-α = 0.98
     Step 3: n = 15, $\overline{x}$ = 74.10, s = 3.182
     Step 4: a. α/2 = 0.02/2 = 0.01; df = 14;
                 t(14, 0.01) = 2.62
             b. E = t(df,α/2)·(s/√n) = (2.62)(3.182 / √15)
                 = 2.1525 = 2.15
             c. $\overline{x}$ ± E = 74.10 ± 2.15
     Step 5: 71.95 to 76.25, the 0.98 confidence interval
             for μ

e. 1000(71.95) and 1000(76.25) = from 71950 to 76250
   ounces total.

9.29 Verify - answers given in exercise.

9.31

Variable	N	Mean	StDev	SE Mean	98.0 % CI
C1	12	7.750	2.137	0.617	(6.073, 9.427)

Computer and/or calculator commands to calculate a confidence
interval for the population mean, μ, when the population standard
deviation, σ, is unknown can be found in ES10-p481.

9.33 a.

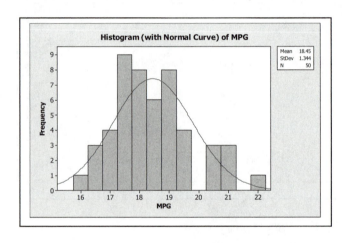

Distribution is mounded in center, approximately
symmetrical. With n = 50, the assumptions are
satisfied.

b. Step 1: The mean mileage per gallon for SUV's.
    Step 2: a. The sampled population appears to be normally
               distributed.
            b. t   c. $1-\alpha = 0.95$
    Step 3: n = 50, $\overline{x}$ = 18.45,   s = 1.3440

Step 4: a. $\alpha/2 = 0.05/2 = 0.025$;   df = 49;
          $t(49, 0.025) = 2.02$
      b. $E = t(df,\alpha/2)\cdot(s/\sqrt{n}) = (2.02)(1.3440/\sqrt{50})$
          $= 0.3839$
      c. $\overline{x} \pm E = 18.45 \pm 0.38$
Step 5: <u>18.07 to 18.83</u>, the 0.95 confidence interval
          for μ

c. One can expect at least 18 miles per gallon.

Hypotheses are written the same way as before. Sample size and standard deviation have no effect on the stating of hypotheses.

9.35   a. H_O: $\mu = 11$ ($\geq$)   vs. H_a: $\mu < 11$

b. H_O: $\mu = 54$ ($\leq$)   vs. H_a: $\mu > 54$

c. H_O: $\mu = 75$ vs. H_a: $\mu \neq 75$

9.37 $\mu = 32$, n = 16, $\overline{x} = 32.93$,   s = 3.1
$t = (\overline{x} - \mu)/(s/\sqrt{n})$
$t* = (32.93 - 32)/(3.1/\sqrt{16}) = \underline{1.20}$

Calculating the **P**-value using the t-distribution

Table 6 or Table 7 (Appendix B, ES10-pp813&814) or a computer and/or calculator, can be used to <u>estimate</u> the p-value

1. Using Table 6 to place bounds on the value of **P**
      a) locate df row
      b) locate the absolute value of the calculated t-value between two critical values in the df row
      c) the p-value is in the interval between the two corresponding probabilities at the top of the columns; read the bounds from either the one-tail or two-tailed column headings as per Ha.

                                        . . .

---

2. Using Table 7 to estimate or place bounds on the value of **P**
   a) locate the absolute value of the calculated t-value and the df directly for the corresponding probability value

   OR

   b) locate the absolute value of the calculated t-value and its df between appropriate bounds. From the box formed, use the upper left and lower right values for the interval. (see ES10-p484)

3. Using a computer and/or calculator
   a) the p-value is calculated directly and given in the output when completing the hypothesis test (see ES10-p487)

   OR

   b) the p-value is calculated using the cumulative probability commands:
   Subtract the probability value from 1 or multiply it by 2, depending on the exercise. The cumulative probability given is $P(t \leq t$-value). (see ES10-pp478&479)

9.39 a. **P** = P(t < -2.01|df=10) = P(t > 2.01|df=10)
using Table 6: 0.025 < **P** < 0.05
using Table 7: 0.031 < **P** < 0.037
using computer: **P** = 0.036

b. **P** = P(t > +2.01|df=10);
using Table 6: 0.025 < **P** < 0.05
using Table 7: 0.031 < **P** < 0.037
using computer: **P** = 0.036

c. **P** = P(t < -2.01|df=10) + P(t > +2.01|df=10)
= 2P(t > 2.01|df=10)
using Table 6:   0.05 < **P** < 0.10
using Table 7:   0.062 < **P** < 0.074
using computer: **P** = 0.072

d. **P** = P(t < -2.01|df=10) + P(t > +2.01|df=10)
= 2P(t > 2.01|df=10)
using Table 6:   0.05 < **P** < 0.10
using Table 7:   0.062 < **P** < 0.074
using computer: **P** = 0.072

9.41 a. $P = P(t > 1.20 | df = 15)$;
                Using Table 6, Appendix B, ES10-p813:
                $0.10 < P < 0.25$
                Using Table 7, Appendix B, ES10-p814:
                $P = 0.124$
            Using computer: $P = 0.124$
        $\alpha = 0.05$; $P > \alpha$; <u>Fail to reject H_o</u>

Draw a picture as before, of an "approximately" normal distribution
curve.   Shade in the critical regions based on the alternative
hypothesis (H_a).   Using α and df, find the critical value(s) using
Table 6 (Appendix B, ES10-p813).

b. $t(15, 0.05) = 1.75$; $t^* = 1.20$

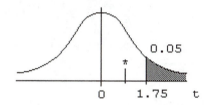

t^* falls in the noncritical region; <u>Fail to reject H_o</u>

9.43   Test of mu = 32 vs mu > 32

N	Mean	StDev	T	P
16	32.9300	3.1000	1.20	0.124

9.45   a. $P = 2P(t > 1.60 | df = 14)$
            using Table 6: $0.10 < P < 0.20$
            using Table 7: $0.065 < \tfrac{1}{2}P < 0.068$; $0.130 < P < 0.136$
            using computer: $P = 0.132$
            $\alpha = 0.05$; $P > \alpha$; Fail to reject H_o

b. $P = P(t > 2.16 | df = 24)$
   using Table 6: $0.01 < P < 0.025$
   using Table 7: $0.019 < P < 0.024$
   using computer: $P = 0.0205$
   $\alpha = 0.05$; $P < \alpha$; Reject H_0

$\alpha = 0.05$
$t(24, 0.05) = 1.71$

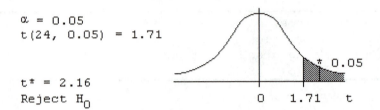

$t* = 2.16$
Reject H_0

c. $P = P(t < -1.73 | df = 44) = P(t > 1.73 | df = 44)$
   using Table 6: $0.025 < P < 0.05$
   using Table 7: $0.039 < P < 0.049$
   using computer: $P = 0.045$
   $\alpha = 0.05$; $P < \alpha$; Reject H_0

$\alpha = 0.05$
$-t(44, 0.05) = -1.68$

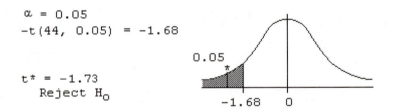

$t* = -1.73$
   Reject H_0

d. Results for each of the decisions are identical.

Hint for writing the hypotheses for exercise 9.47
Look at the first sentence of the exercise, "A student group
maintains that the average student must travel for at least 25
minutes ...". "At least" indicates greater than or equal to ($\geq$).
Since the greater than or equal to includes the equality, it
belongs in the null hypothesis. Continue to write the null
hypothesis with the equal sign (=), but include the greater-than or
equal-to sign ($\geq$) in parentheses after it. The negation is "..less
than", hence the alternative hypothesis must have a less-than sign
($<$).

Hypothesis tests, (p-value approach), will be completed in the same
format as before. You may want to review: ES10-p427, SSM-p265.
The only differences are in:

    1. the calculated test statistic, which is t, where
        $t = (\overline{x} - \mu)/(s/\sqrt{n})$

    2. using Table 6 or Table 7 (Appendix B, ES10-pp813&814)
       to <u>estimate</u> the p-value

Remember: σ (population standard deviation) is unknown, therefore
        s (sample standard deviation) is used.

9.47   a. P-value approach:

Step 1: a. The mean travel time to college
       b. H_o: $\mu = 25$ (at least) ($\geq$)
          H_a: $\mu < 25$ (less than)
Step 2: a. Travel times are mounded; CLT is satisfied
          with n = 31
       b. t   c. $\alpha = 0.01$
Step 3: a. n = 31, $\overline{x} = 19.4$, s = 9.6
       b. $t = (\overline{x} - \mu)/(s/\sqrt{n})$
          $t^* = (19.4 - 25.0)/(9.6/\sqrt{31}) = -3.25$
Step 4: a. **P** = $P(t < -3.25 | df = 30) = P(t > 3.25 | df = 30)$;
          Using Table 6, Appendix B, ES10-p813:
          P < 0.005
          Using Table 7, Appendix B, ES10-p814:
          0.001 < **P** < 0.002
          Using computer: **P** = 0.0014
       b. **P** < α
Step 5: a. Reject H_o
       b. At the 0.01 level of significance, the sample
          does provide sufficient evidence to justify the
          contention that mean travel time is less than 25
          minutes.

Hypothesis tests (classical approach) will be completed using the same format as before. You may want to review: ES10-pp444-445, SSM-p275. The only differences are in:
1. finding the critical value(s) of t, remember you need α (column) and degrees of freedom (df = n - 1)(row) for the t-distribution

2. the calculated test statistic, which is t, where
$$t = (\overline{x} - \mu)/(s/\sqrt{n}).$$

b. Classical approach:

Step 1: a. The mean travel time to college
        b. H_o: $\mu = 25$ (at least) ($\geq$)
           H_a: $\mu < 25$ (less than
Step 2: a. Travel times are mounded; assume normality,
           CLT with n = 31
        b. t   c. $\alpha = 0.01$
Step 3: a. n = 31, $\overline{x} = 19.4$,   s = 9.6
        b. $t = (\overline{x} - \mu)/(s/\sqrt{n})$
           $t* = (19.4 - 25.0)/(9.6/\sqrt{31}) = -3.25$
Step 4: a. $-t(30, 0.01) = -2.46$

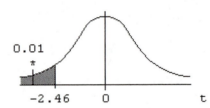

0.01

-2.46   0   t

        b. t* falls in the critical region, see Step 4a
Step 5: a. Reject H_o
        b. At the 0.01 level of significance, the sample
           does provide sufficient evidence to justify the
           contention that mean travel time is less than 25
           minutes.

9.49 Step 1: a. The mean number of pairs of jeans for any given
                      size offered at Levi stores
                   b. H_o: $\mu = 130$ (≤)
                      H_a: $\mu > 130$
         Step 2: a. normality indicated
                 b. t   c. $\alpha = 0.01$
         Step 3: a. n = 24, $\bar{x}$ = 141.3,   s = 36.2
                 b. t = $(\bar{x} - \mu)/(s/\sqrt{n})$
                    t* = $(141.3 - 130)/(36.2/\sqrt{24}) = 1.53$
         Step 4: -- using p-value approach --------------------
                 a. P = P(t > 1.53|df = 23);
                      Using Table 6, Appendix B, ES10-p813:
                      0.05 < P < 0.10
                      Using Table 7, Appendix B, ES10-p814:
                      0.061 < P < 0.074
                      Using computer: P = 0.0698
                 b. P > α
                 -- using classical approach ------------------
                 a. t(23, 0.01) = 2.50

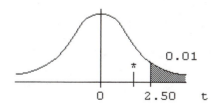

                 b. t* falls in the noncritical region, see Step 4a
         ----

Step 5:   a. Fail to reject Ho
          b. The sample does not provide sufficient evidence
             to justify the contention that this sample of
             stores carries a greater selection of jeans, at
             the 0.01 level of significance.

9.51 Verify - answers given in exercise.

Computer and/or calculator commands to perform a hypothesis test
for a population mean if the population standard deviation (σ) is
unknown can be found in ES10-pp486-487.

The alternative command works the same as in the z-distribution (σ
known) command. The output will also look the same except a t-
value will be calculated in place of the z-value.

9.53 Test of mu = 52.00 vs mu < 52.00

Variable	N	Mean	StDev	SE Mean	T	P
C1	12	49.75	5.48	1.58	-1.42	0.091

$\alpha = 0.01$; $P > \alpha$; Fail to reject H_O
OR
$-t(11, 0.01) = -2.72$;
$t* = -1.42$ falls in the noncritical region; Fail to reject H_O

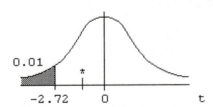

9.55 Sample statistics: n = 6, $\Sigma x = 222$, $\Sigma x^2 = 8330$,
                              $\overline{x} = 37.0$, s = 4.817

Step 1: a. The mean test score at a certain university
        b. H_O: $\mu = 35$ (reasonable)
           H_a: $\mu \neq 35$ (not reasonable)

Step 2: a. normality indicated
         b. t   c. α = 0.05
Step 3: a. n = 6, x̄ = 37.0,   s = 4.817
         b. t = (x̄ - μ)/(s/√n)
            t* = (37.0 - 35.0)/(4.817/√6) = 1.02
Step 4: -- using p-value approach --------------------
         a. P = 2P(t > 1.02|df = 5);
                 Using Table 6, Appendix B, ES10-p813:
                 0.20 < P < 0.50
                 Using Table 7, Appendix B, ES10-p814:
                 0.161 < ½P < 0.182]   0.322 < P < 0.364
             Using computer: P = 0.3545
         b. P > α
         -- using classical approach ------------------
         a. ±t(5, 0.025) = ±2.57

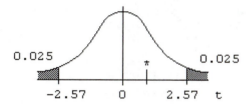

$$0.025 \qquad\qquad 0.025$$
$$-2.57 \qquad 0 \qquad 2.57 \quad t$$

         b. t* falls in the noncritical region, see Step 4a

    Step 5: a. Fail to reject H_o
             b. The sample does not provide sufficient
                evidence to reject the claim that the mean
                score is 35, at the 0.05 level of
                significance.

9.57 a.

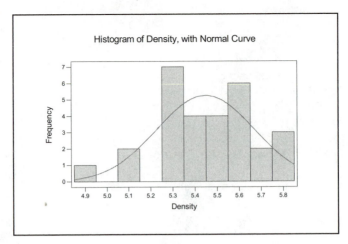

Histogram of Density, with Normal Curve

The distribution is mounded and appears to have an approximately normal distribution.

b. Step 1: a. The mean density of the earth to the density of water
              b. H_{0}: μ = 5.517
                 H_a: μ < 5.517 (less than)
    Step 2: a. normality indicated above
              b. t   c. α = 0.05
    Step 3: a. n = 29, $\overline{x}$ = 5.4479,   s = 0.2209
              b. t = $(\overline{x} - \mu)/(s/\sqrt{n})$

              $t^* = (5.4479 - 5.517)/(0.2209/\sqrt{29}) = -1.68$

    Step 4: -- using p-value approach --------------------
              a. **P** = P(t < -1.68|df = 28);
                 Using Table 6, Appendix B, ES10-p813:
                 0.05 < **P** < 0.10
                 Using Table 7, Appendix B, ES10-p814:
                 0.05 < **P** < 0.061
                 Using computer: **P** = 0.052
              b. **P** > α

-- using classical approach ------------------
  a. $-t(28, 0.05) = -1.70$

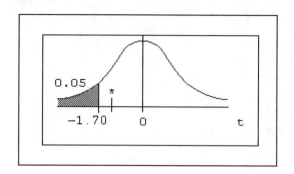

  b. t* falls in the noncritical region, see Step 4a

Step 5: a. Fail to reject H_o
        b. There is not sufficient evidence to show,
           the mean of Cavendish's data is
           significantly less than today's recognized
           standard of 5.517 g/cm^3, at the 0.05 level
           of significance.

9.59 a. Yes, the assumption of normality seems reasonable.
         Manufacturers are required to keep manufacturing
         processes within specifications.

  b. Step 1: The mean amount of acetaminophen per tablet.
     Step 2: a. Normality assumed, CLT with n = 30.
             b. t   c. $1-\alpha = 0.99$
     Step 3: n = 30, $\overline{x}$ = 596.3, s = 4.7
     Step 4: a. $\alpha/2 = 0.01/2 = 0.005$; df = 29;
                t(29, 0.005) = 2.76
             b. $E = t(df, \alpha/2) \cdot (s/\sqrt{n}) = (2.76)(4.7/\sqrt{30}) = 2.368$
             c. $\overline{x} \pm E = 596.3 \pm 2.37$
     Step 5: 593.93 to 598.67, the 0.99 confidence
             intervalfor μ

  c. The confidence interval in part b suggests that the mean
     amount of acetaminophen in one pill is less than 600 mg.
     The interval did not contain the 600 mg.

9.61 a. Step 1: a. The mean length of corks in a shipment
$\qquad$ b. H_O: $\mu = 45.0$
$\qquad$ H_a: $\mu \neq 45.0$
$\quad$ Step 2: a. normality indicated
$\qquad$ b. t $\qquad$ c. $\alpha = 0.02$
$\quad$ Step 3: a. n = 12, $\overline{x}$ = 44.8775, s = 0.2176
$\qquad$ b. $t = (\overline{x} - \mu)/(s/\sqrt{n})$

$\qquad$ $t* = (44.8775 - 45.0)/(0.2176/\sqrt{12}) = -1.95$
$\quad$ Step 4: -- using p-value approach --------------------
$\qquad$ a. **P** = 2P(t < -1.95|df = 11)
$\qquad\qquad$ = 2P(t > 1.95|df = 11);
$\qquad$ Using Table 6, Appendix B, ES10-p813:
$\qquad$ 0.05 < **P** < 0.10
$\qquad$ Using Table 7, Appendix B, ES10-p814:
$\qquad$ 2(0.034 < **P** < 0.043) = (0.068 < P < 0.086)
$\qquad$ Using computer: **P** = 0.0771
$\qquad$ b. **P** > α
$\quad$ -- using classical approach ------------------
$\quad$ a. $\pm$t(11, 0.01) = $\pm$2.72
$\quad$ b. t* falls in the noncritical region, see Step 4a
$\quad$ ---
$\quad$ Step 5: a. Fail to reject H_O
$\qquad$ b. There is not sufficient evidence to show
$\qquad$ that the mean length of the corks is
$\qquad$ different from 45.0mm, at the 0.02 level of
$\qquad$ significance.

$\quad$ b. Step 1: a. The mean length of corks in a shipment
$\qquad$ b. H_O: $\mu = 45.0$
$\qquad$ H_a: $\mu \neq 45.0$
$\quad$ Step 2: a. normality indicated
$\qquad$ b. t $\qquad$ c. $\alpha = 0.02$
$\quad$ Step 3: a. n = 18, $\overline{x}$ = 45.0183, s = 0.3199
$\qquad$ b. $t = (\overline{x} - \mu)/(s/\sqrt{n})$
$\qquad$ $t* = (45.0183 - 45.0)/(0.3199/\sqrt{18}) = 0.24$
$\quad$ Step 4: -- using p-value approach --------------------
$\qquad$ a. **P** = 2P(t > 0.24|df = 17);
$\qquad$ Using Table 6, Appendix B, ES10-p813:
$\qquad$ **P** > 0.500
$\qquad$ Using Table 7, Appendix B, ES10-p814:
$\qquad$ 2(0.384 < **P** < 0.422) = (0.768 < P < 0.844)
$\qquad$ Using computer: **P** = 0.813
$\qquad$ b. **P** > α

a. ±t(17, 0.01) = ±2.57
b. t* falls in the noncritical region, see Step 4a

Step 5: a. Fail to reject H_o
b. There is not sufficient evidence to show that the
second sample mean length of the corks is
different from 45.0mm, at the 0.02 level of
significance.

c. The test statistic in part d was much smaller since the
sample mean was closer to the hypothesized mean.

d. The increase in the sample size made up for the difference
in standard deviations between the two samples.   The
larger standard deviation was with the larger sample size.

e. The standard deviation in part d was smaller that that in
part c.   In combination with a sample mean closer to the
claimed mean produced a very small test statistic,
resulting in a Fail to Reject Ho decision.

SECTION 9.2 EXERCISES

---
p' = sample proportion   p' = x/n
x = number of successes
n = sample size (number of independent trials)

9.63 a. x = number of successes = 45 (with only two outcomes,
"success' and "failure")
n = sample size = number of independent trials = 150
b. p' = 45/150 = # of successes/# of trials = 0.30
p' = sample proportion of success
c. p' = x/n = 24/250 = <u>0.096</u>
d. p' = x/n = 640/2050 = 0.312195 = <u>0.312</u>
e. p' = x/n = 892/1280 = 0.696875 = <u>0.697</u>

9.65 a. Yes, it seems likely that the mean of the observed
proportions would be the true proportion for the
population.

b. Unbiased because the mean of the p' distribution is p,
the parameter being estimated.

9.67 a. $\alpha = 0.10$, $\alpha/2 = 0.05$; $z(\alpha/2) = 1.65$ or $-z(\alpha/2) = -1.65$
b. $\alpha = 0.05$, $\alpha/2 = 0.025$; $z(\alpha/2) = 1.96$ or $-z(\alpha/2) = -1.96$
c. $\alpha = 0.01$, $\alpha/2 = 0.005$; $z(\alpha/2) = 2.58$ or $-z(\alpha/2) = -2.58$

Estimating p - the population proportion

1. point estimate: $p' = x/n$

2. confidence interval: $p' \pm z(\alpha/2) \cdot \sqrt{p'q'/n}$

        ↑

     point   maximum error
   estimate   of estimate

Computer and/or calculator commands to calculate a confidence interval for the population proportion, p, can be found in ES10-pp499-500.

Review: ES10-p404, SSM-p256, "The Confidence Interval: A Five-Step Procedure" if necessary.
Explore the relationship between a z* value and its corresponding confidence interval using the Skillbuilder Applet 'z* & Confidence Level' on your CD.

9.69 a. $\sqrt{p'q'/n} = \sqrt{(0.23)(0.77)/400} = \underline{0.02104}$

b. Step 1: The proportion of convertibles driven by students
Step 2: a. The sample was randomly selected and each subject's response was independent of those of the others surveyed.
b. $n = 400$; $n > 20$, $np = (400)(92/400) = 92$, $nq = (400)(308/400) = 308$, np and nq both > 5
c. $1 - \alpha = 0.95$
Step 3: $n = 400$, $x = 92$, $p' = x/n = 92/400 = 0.23$
Step 4: a. $z(\alpha/2) = z(0.025) = 1.96$
b. $E = z(\alpha/2) \cdot \sqrt{p'q'/n} = 1.96\sqrt{(0.23)(0.77)/400}$
$= (1.96)(0.02104) = 0.041$
c. $p' \pm E = 0.23 \pm 0.041$
Step 5: $\underline{0.189 \text{ to } 0.271}$ is the 0.95 interval for p = P(drives convertible)

9.71 a. n = 3,003 people surveyed, a trial is the surveying
         of each person, success is when they "do not know
         that caffeine dehydrates", p = P(did not know), x =
         number of surveyed 3,003 people who say they did not
         know that caffeine dehydrates
      b. 0.20 (20%). It is a statistic. It is being used to
         estimate the parameter.
      c. $1.96 \cdot \sqrt{(0.20)(0.80)/3003}$ = 1.96(0.0073) = 0.0143
      d. Maximum error of 1.4% is smaller than the quoted
         margin of error = 1.8%.
      e. 0.20 ± 0.014 or 0.186 to 0.214

9.73 Step 1: The proportion of students that support the proposed
               budget amount
       Step 2: a. The sample was randomly selected and each
                  subject's response was independent of those of
                  the others surveyed.
               b. n = 60; n > 20, np = (60)(22/60) = 22,
                  nq = (60)(38/60) = 38, np and nq both > 5
               c. 1 - α = 0.99
       Step 3: n = 60, x = 22, p' = x/n = 22/60 = 0.367
       Step 4: a. z(α/2) = z(0.005) = 2.58
               b. E = z(α/2)· $\sqrt{p'q'/n}$ = $2.58\sqrt{(0.367)(0.633)/60}$
                     = (2.58)(0.0622) = 0.161
               c. p' ± E = 0.367 ± 0.161
       Step 5: 0.206 to 0.528, the 0.99 interval for p = P(favor
                                                       budget)

9.75 Step 1: The proportion of managers and professionals who work
               late five days a week
       Step 2: a. The sample was randomly selected and each
                  subject's response was independent of those of
                  the others surveyed.
               b. n = 1742; n > 20, np = (1742)(0.278) = 484.3,
                  nq = (1742)(0.722) = 1257.7, np and nq both > 5
               c. 1 - α = 0.99
       Step 3: n = 1742, x not given, p' = 0.278 (given)
       Step 4: a. z(α/2) = z(0.005) = 2.58
               b. E = z(α/2)· $\sqrt{p'q'/n}$ = $2.58 \cdot \sqrt{(0.278)(0.722)/1742}$
                     = (2.58)(0.0107) = 0.028
               c. p' ± E = 0.278 ± 0.028
       Step 5: 0.250 to 0.306, the 0.99 interval for
                          p = P(working late five days a week)

9.77 There are many possible reasons why the results could be biased. Here are a few:
1. Many people will not reveal information about credit cards to telephone callers.
2. Many people will not ever admit to being "easy prey" to offers like this.
3. Many people answer questions with answers they think the caller wants to hear hoping to end the call quickly.

9.79 a. $E = 1.96\sqrt{(0.70)(0.30)/1020} = 1.96(0.01435) = 0.028$

$E = 1.96\sqrt{(0.57)(0.43)/1019} = 1.96(0.01551) = 0.030$

$E = 1.96\sqrt{(0.15)(0.85)/1021} = 1.96(0.01117) = 0.022$

b. The variation was caused by the differing values of p. More explicitly, the differing product of pq: 0.2100, 0.2451, and 0.1275.

c. Yes.

d. The "margin of error" (MoE) is typically reported as the maximum error of estimate calculated using p = 0.5 because this yields the maximum value. Rounding up yields a slightly larger error, in turn a wider interval. "Conservative" equates with a less restrictive (narrower) interval.

e. p = 0.5

9.81 Verify - answers given in exercise

9.83 a. $p = P(\text{head}) = 12{,}012/24{,}000 = \underline{0.5005}$

b. $\sqrt{p'q'/n} = \sqrt{(0.5005)(0.4995)/24000} = \underline{0.003227}$

c. $E = z(\alpha/2)\cdot\sqrt{p'q'/n} = (1.96)(0.003227) = 0.006325 = 0.0063$

$p \pm E = 0.5005 \pm 0.0063$

$\underline{0.4942 \text{ to } 0.5068}$, the 0.95 interval for p = P(head)

d. - f. Each student will have different results. Each set of results will yield an empirical probability whose value is very close to 0.50; in fact, you should expect 95% of such results to be within 0.0063 of 0.50.

9.85 a. – e. The distributions do not look normal, they are skewed
right. The gaps are caused by working with a discrete
distribution, the binomial distribution. Both histograms
look the same because they are both showing the same set
of data, one as a proportion and the other as a
standardized proportion.

f. The distribution is not normal and the normal distribution
should not be used to calculate probabilities.

Sample Size Determination Formula for a Population Proportion

$$n = \frac{[z(\alpha / 2)]^2 \cdot p^* \cdot q^*}{E^2}$$

Sample Size - A Four-Step Prodecure

Step 1: Use the level of confidence, 1-α, to find z(α/2)
Step 2: Find the maximum error of estimate
Step 3: Determine p* and q* = 1 - p* (if not given, use p* = 0.5)
Step 4: Use formula to find n

9.87 Step 1: 1 - α = 0.95; z(α/2) = z(0.025) = 1.96
Step 2: E = 0.02
Step 3: no estimate given, p* = 0.5 and q* = 0.5
Step 4: n = {[z(α/2)]²·p*·q*}/E²
            = (1.96²)(0.5)(0.5)/(0.02²) = <u>2401</u>

9.89 a. Step 1: 1 - α = 0.90; z(α/2) = z(0.05) = 1.65
Step 2: E = 0.02
Step 3: p* = 0.81 and q* = 0.19
Step 4: n = {[z(α/2)]²·p*·q*}/E²
            = (1.65²)(0.81)(0.19)/(0.02²) = 1047.48 = 1048

b. Step 1: 1 - α = 0.90; z(α/2) = z(0.05) = 1.65
Step 2: E = 0.04
Step 3: p* = 0.81 and q* = 0.19
Step 4: n = {[z(α/2)]²·p*·q*}/E²
            = (1.65²)(0.81)(0.19)/(0.04²) = 261.87 = 262

c. Step 1: $1 - \alpha = 0.98;\quad z(\alpha/2) = z(0.01) = 2.33$
   Step 2: $E = 0.02$
   Step 3: $p* = 0.81$ and $q* = 0.19$
   Step 4: $n = \{[z(\alpha/2)]^2 \cdot p* \cdot q*\}/E^2$
   $= (2.33^2)(0.81)(0.19)/(0.02^2) = 2088.77 = 2089$

d. Increasing the maximum error decreases the required sample size. The maximum error is located in the denominator of formula 9.8 and therefore an increase will reduce the resulting value for n.

e. Increasing the level of confidence increases the required sample size. The level of confidence determines the value of z used, and it is located in the numerator of formula 9.8 and therefore an increase in 1-α will increase z and will increase the resulting value for n.

Hypotheses are written with the same rules as before. Now replace μ, the population mean, with p, the population proportion.

(ex.: H_O: p = P(driving a convertible) = 0.45 vs.
   H_a: p = P(driving a convertible) $\neq$ 0.45,
if driving a convertible is considered the success)

Review: ES10-pp417-419, SSM-pp265-266, if necessary.

9.91   a. H_O: p = P(work) = 0.60 ($\leq$)   vs. H_a: p > 0.60

b. H_O: p = P(interested in quitting) = 1/3 ($\leq$)
        vs. H_a: p > 1/3

c. H_O: p = P(vote for) = 0.50 ($\leq$)   vs. H_a: p > 0.50

d. H_O: p = P(seriously damaged) = 3/4 ($\geq$)
        vs. H_a: p < 3/4

e. H_O: p = P(H|tossed fairly) = 0.50   vs. H_a: p $\neq$ 0.50

9.93   **a.** p = 0.70, n = 300, x = 224, p' = x/n = 224/300 = 0.747
   $z = (p' - p)/\sqrt{pq/n}$
   $z* = (0.747 - 0.70)/\sqrt{(0.7)(0.3)/300} = 1.78$

b. $p = 0.50$, $n = 450$, $x = 207$, $p' = x/n = 207/450 = 0.46$

$z = (p' - p)/\sqrt{pq/n}$

$z* = (0.46 - 0.50)/\sqrt{(0.50)(0.50)/450} = -1.70$

c. $p = 0.35$, $n = 280$, $x = 94$, $p' = x/n = 94/280 = 0.336$

$z = (p' - p)/\sqrt{pq/n}$

$z* = (0.336 - 0.35)/\sqrt{(0.35)(0.65)/280} = -0.49$

d. $p = 0.90$, $n = 550$, $x = 508$, $p' = x/n = 508/550 = 0.924$

$z = (p' - p)/\sqrt{pq/n}$

$z* = (0.924 - 0.90)/\sqrt{(0.90)(0.10)/550} = 1.88$

Determining the p-value will follow the same procedures as before. We are again working with the normal distribution. Review: ES10-p433, Table 8.7; SSM-p270, if necessary.

9.95 a. $\mathbf{P} = 2P(z > 1.48) = 2(0.5000 - 0.4306) = 2(0.0694) = \underline{0.1388}$

b. $\mathbf{P} = 2P(z < -2.26) = 2P(z > 2.26) = 2(0.5000 - 0.4881)$
$= 2(0.0119) = \underline{0.0238}$

c. $\mathbf{P} = P(z > 0.98) = (0.5000 - 0.3365) = \underline{0.1635}$

e. $\mathbf{P} = P(z < -1.59) = P(z > 1.59) = (0.5000 - 0.4441)$
$= \underline{0.0559}$

Determining the test criteria will also follow the same procedures as before. We are again working with the normal distribution. Review: ES10-pp450-451, SSM-p277, if necessary.

9.97 a.   b.

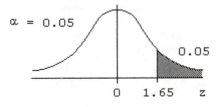

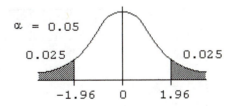

c.

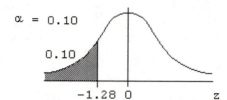

$\alpha = 0.10$

0.10

−1.28 0   z

d.

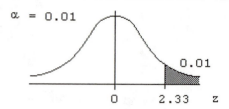

$\alpha = 0.01$

0.01

0   2.33   z

Since n ≤ 15 and x is discrete, Table 2 (Appendix B, ES10-pp807-809) will be used to determine the level of significance, α.
x can be any value, 0 through n, for each experiment.
List all values in numerical sequence.
Draw a vertical line separating the set of values that belong in the critical region and the set of values that belong in the noncritical region.
Add all of the probabilities associated with those numbers in the critical region to find α.

9.99   a. $\alpha = P[x = 12, 13, 14, 15 | B(n=15, p=0.5)]$
$= 0.014 + 0.003 + 2(0+) = \underline{0.017}$

b. $\alpha = P[x = 0, 1 | B(n=12, p=0.3)]$
$= 0.014 + 0.071 = \underline{0.085}$

c. $\alpha = P[x = 0, 1, 2, 3, 9, 10 | B(n=10, p=0.6)]$
$= (0+) + 0.002 + 0.011 + 0.042 + 0.040 + 0.006 = \underline{0.101}$

d. $\alpha = P[x = 4, 5, 6, \ldots, 14 | B(n=14, p=0.05)]$
$= 0.004 + 10(0+) = \underline{0.004}$

List all x values in numerical order, as before. Based on H_a[#], add consecutive probabilities until the sum is as close to the given α as possible, without exceeding it. Draw a vertical line at this point, separating the critical and noncritical regions.

[#] If H_a contains <, begin adding from x = 0 towards x = n.
If H_a contains >, begin adding from x = n towards x = 0.
If H_a contains ≠, begin adding simultaneously from x = 0 and x = n toward the center.

9.101 a. (1) Correctly fail to reject H_O

b. $\alpha = P[x = 14,15|B(n = 15,p = 0.7)]$
$= 0.031 + 0.005 = \underline{0.036}$

c. (4) Commit a type II error

d. $\mathbf{P} = P[x = 13,14,15|B(n = 15,p = 0.7)]$
$= 0.092 + 0.031 + 0.005 = \underline{0.128}$

Review: "The Probability-Value Hypothesis Test: A Five-Step Procedure"; ES10-p427, SSM-p265 and/or "The Classical Hypothesis Test: A Five-Step Procedure"; ES10-pp444-445, SSM-p275, if necessary.

Use $z = \dfrac{p' - p}{\sqrt{\dfrac{pq}{n}}}$, for calculating the test statistic.

$p' = x/n$, if not given directly.

Computer and/or calculator commands to perform a hypothesis test for a population proportion can be found in ES10-p506.

Hint for writing the hypotheses for exercise 9.103

Look at the last sentence in the exercise, "If the consumer group ... that less than 90% ...?" The words "less than", (<.) do not indicate any equality, therefore the alternative is less than (<). The negation is "NOT less than", which is > or =. Equality (=) is used in the null hypothesis, but stands for greater than or equal to (≥). Remember to use *p* as the population parameter in the hypotheses.

9.103 Step 1: a. The proportion of claims settled within 30 days

b. H_O:p = P(claim is settled within 30 days)=0.90 (≥)
H_a: p < 0.90

Step 2: a. independence assumed

b. z; n = 75; n > 20, np = (75)(0.90) = 67.5,
nq = (75)(0.10) = 7.5, both np and nq > 5

c. α = 0.05

Step 3: a. n = 75, x = 55, p' = x/n = 55/75 = 0.733

b. $z = (p' - p)/\sqrt{pq/n}$
$z* = (0.733 - 0.900)/\sqrt{(0.9)(0.1)/75} = -4.82$

Step 4: -- using p-value approach --------------------
        a. $P = P(z < -4.82) = P(z > 4.82)$;
              Using Table 3, Appendix B, ES10-p810:
              $P = 0.5000 - 0.499997 = 0.000003$
              Using Table 5, Appendix B, ES10-p812:
              $P > 0+$
        b. $P < \alpha$
        -- using classical approach ------------------
        a. $-z(0.05) = -1.65$
          b. z* falls in the critical region, see Step 4a

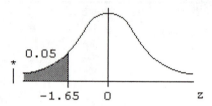

Step 5: a. Reject H_o
         b. The sample provides sufficient evidence that p is
             significantly less than 0.90; it appears that
             less than 90% are settled within 30 days as
             claimed, at the 0.05 level of significance.

Hint for writing the hypothesis for exercise 9.105

Look at the first sentence of the exercise, "A politician claims
she will receive 60% of the vote ..." The words "will receive"
indicate equality (=). Since a politician would be interested in a
majority, anything equal to or greater than the stated percentage
would be acceptable. Greater than or equal to (≥) includes the
equality, therefore it belongs in the null hypothesis. Continue to
write the null hypothesis with the equal sign (=), but include the
greater-than or equal-to sign(≥) in parentheses after it. The
negation is "less than," hence the alternative hypothesis must have
a < sign.

9.105 Step 1: a. The proportion of vote for a politician in an
                       upcoming election
                   b. H_o: p = P(vote for) = 0.60 [will receive 60% of
                                                                    vote]
                       H_a: p < 0.60 [will receive less than 60%]

Step 2: a. independence assumed
          b. z; n = 100; n > 20, np = (100)(0.60) = 60,
             nq = (100)(0.40) = 40, both np and nq > 5
          c. α = 0.05
Step 3: a. n = 100, x = 50, p' = x/n = 50/100 = 0.500
          b. z = (p' - p)/$\sqrt{pq/n}$
             z* = (0.500 - 0.600)/$\sqrt{(0.6)(0.4)/100}$ = -2.04
Step 4: -- using p-value approach -------------------
          a. **P** = P(z < -2.04) = P(z > 2.04);
                Using Table 3, Appendix B, ES10-p810:
                P = 0.5000 - 0.4793 = 0.0207
                Using Table 5, Appendix B, ES10-p812:
                0.0202 < **P** < 0.0228
          b. **P** < α
          -- using classical approach ------------------
          a. -z(0.05) = -1.65

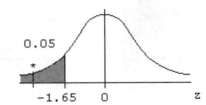

          b. z* falls in the critical region, see Step 4a

Step 5: a. Reject H_O
          b. The sample provides sufficient evidence that the
             proportion is significantly less than 0.60, at
             the 0.05 level; it appears that less than 60%
             support her.

9.107 Step 1: a. The proportion of adults in favor of allowing
                legal use of marijuana for medical purposes
          b. H_O: p = P(in favor) = 0.72 (≥)
             H_a: p < 0.72
Step 2: a. independence assumed
          b. z; n = 200; n > 20, np = (200)(0.72) = 144,
             nq = (200)(0.28) = 56, both np and nq > 5
          c. α = 0.05

Step 3: a. n = 200, x = 134,   p' = x/n = 134/200 = 0.67
        b. $z = (p' - p)/\sqrt{pq/n}$
           $z* = (0.67 - 0.72)/\sqrt{(0.72)(0.28)/200} = -1.57$
Step 4: -- using p-value approach --------------------
        a. **P** = P(z < -1.57);
             Using Table 3, Appendix B, ES10-p810:
             P = 0.5000 - 0.4418 = 0.0582
             Using Table 5, Appendix B, ES10-p812:
             0.0548 < **P** < 0.0606
        b. **P** > α
        -- using classical approach ------------------
        a. -z(0.05) = -1.65

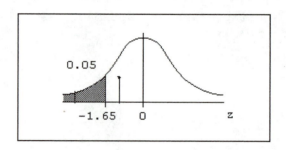

        b. z* falls in the noncritical region, see Step 4a

Step 5: a. Fail to reject H_O
        b. The sample does not provide sufficient evidence
           that the Midwest has a lower proportion in favor
           of legalizing marijuana for medicinal purposes
           than the rest of the country, at the 0.05 level
           of significance.

9.109 a. p' = 324/1000 = 0.324

         $z = (p' - p)/\sqrt{pq/n}$

         $z* = (0.324 - 0.35)/\sqrt{(0.35)(0.65)/1000}$
            = -0.026/0.015083 = -1.72

b & c.

Step 1: a. The proportion of profession women that fear public speaking

        b. H_o: $p = 0.35$ ($\geq$)

           H_a: $p < 0.35$

Step 2: a. independence assumed

        b. z; n = 1000; n > 20, np = (1000)(0.35) = 350, nq = (1000)(0.65) = 650, both np and nq > 5

        c. $\alpha = 0.01$

Step 3: a. n = 1000, x = 324, p' = x/n = 324/1000 = 0.324

        b. z* = -1.72 (from part a above)

Step 4: -- using p-value approach --------------------

        a. **P** = P(z < -1.72) = P(z > 1.72);

           Using Table 3, Appendix B, ES10-p810:

           P = 0.5000 - 0.4573 = 0.0427

           Using Table 5, Appendix B, ES10-p812:

           0.0401 < **P** < 0.0446

        b. **P** > α

        -- using classical approach ------------------

        a. z(0.01) = -2.33

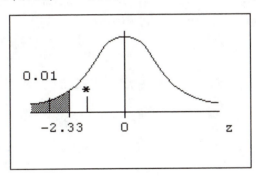

        b. z* falls in the noncritical region, see Step 4a

Step 5: a. Fail to reject H_o

        b. At the 0.01 level of significance there is insufficient evidence to show that less than 35% of the country's professional women fear public speaking.

9.111 a. $P(x \geq 140 | B(250, 0.50)) = 1 - P(x \leq 139)$
$$= 1.0000 - 0.9668 = \underline{0.0332}$$

b. $P(x \leq 110 | B(250, 0.50)) = \underline{0.0332}$

c. $P(\text{more extreme}) = 0.0332 + 0.0332 = \underline{0.0664}$

d. Coenen's claim: "struck evenly or balanced" or
H_O: $P(H) = 0.5$,
Blight's suspicion: "not balanced" or H_a: $P(H) \neq 0.5$;

e. "If the coin were unbiased" is the statement of the null hypothesis. The probability of "getting a result as extreme" is the definition of the p-value.

f. 0.560 ± 0.062 or 0.498 to 0.622

g. the 6% is a maximum error of estimate since the results are from a binomial experiment and not a survey – therefore there is no chance for bias due to an interviewing process

9.113 a. H_O: $p = 0.225$ ($\leq$)
H_a: $p > 0.225$

b. $P < 0.05$
$z(0.05) = 1.65$; $z = 2.71 > 1.65$
Reject Ho   Support Ha at the 0.05 level of significance

c. $p' = x/n = 61/200 = 0.305$

SECTION 9.4 EXERCISES

9.115 a. A: $n = 6$, $\Sigma x = 43$, $\Sigma x^2 = 323$

$$s^2 = (\Sigma x^2 - (\Sigma x)^2/n)/(n - 1) = (323 - 43^2/6)/5$$
$$= 2.96667$$

$$s = \sqrt{s^2} = \sqrt{2.96667} = 1.72$$

B: $n = 6$, $\Sigma x = 48$, $\Sigma x^2 = 448$

$$s^2 = (\Sigma x^2 - (\Sigma x)^2/n)/(n - 1) = (448 - 48^2/6)/5 = 12.8$$

$$s = \sqrt{s^2} = \sqrt{12.8} = 3.58$$

b. Increased standard deviation

c. The 15 is quite different than the rest of the data in this sample. It had a big effect on the standard deviation; it approximately doubled the standard deviation.

χ^2 Distribution
(used for inferences concerning σ and σ^2)

Key facts about the χ^2 curve:
  1) the total area under a χ^2 curve is 1
  2) it is skewed to the right (stretched out to the right side, not symmetrical)
  3) it is nonnegative, starts at zero and continues out towards $+\infty$
  4) a different curve exists for each sample size
  5) uses α and degrees of freedom, df, to determine table values
  6) degrees of freedom is abbreviated as 'df', where, df = n-1
  7) for df > 2, the mean of the distribution is df.

Notation: χ^2(df,α) = χ^2("degrees of freedom","area to the right")
                  ↑   ↑   ↑
            Table 8   row id #   column id #

ex.) Right tail: χ^2(13,.025) = 24.7   (n must have been 14)

   Left tail: χ^2(13,.975) = 5.01

Note: For left tail, "area to right" includes both the area in the "middle" and the area of the "right" tail.

Explore the chi-square distribution for various degrees of freedom using Chapter 9 Skillbuilder Applet "Chi-Square Probabilities".

9.117 a. 23.2   b. 23.3   c. 3.94   d. 8.64

9.119 a. χ^2(19,0.05) = 30.1   b. χ^2(4,0.01) = 13.3
   c. χ^2(17,0.975) = 7.56   d. χ^2(60,0.95) = 43.2

    e. $\chi^2(21, 0.95) = \underline{11.6}$ and $\chi^2(21, 0.05) = \underline{32.7}$

    f. $\chi^2(6, 0.975) = \underline{1.24}$ and $\chi^2(6, 0.025) = \underline{14.5}$

Draw a diagram of a chi square distribution, labeling the given information and shading the specified region(s).
Percentiles: $P_k = k^{th}$ percentile $\Rightarrow$ k% of the data lies <u>below</u> (to left of) this value

---

9.121 a. $\chi^2(5, 0.05) = \underline{11.1}$

    b. $\chi^2(5, 0.05) = \underline{11.1}$

    c. $\chi^2(5, 0.10) = \underline{9.24}$

9.123 $1 - (0.01 + 0.05) = \underline{0.94}$

9.125 Chi-Square with 15 DF

      x   P(X <= x)
  20.2000   0.8356

    a. $P(\chi^2 < 20.2 | df = 15) = 0.8356$
    b. $P(\chi^2 > 20.2 | df = 15) = 1 - 0.8356 = 0.1644$

Hypotheses for variability are written with the same rules as before. Now, in place of μ or p, the population standard deviation, σ, or the population variance, σ^2, will be used.
(ex. H_o: $\sigma = 3.7$ vs. H_a: $\sigma \neq 3.7$) Review: ES10-p417-419, SSM-pp265-266, if necessary.

9.127 a. H_o: $\sigma = 24$ ($\leq$)   vs. H_a: $\sigma > 24$

    b. H_o: $\sigma = 0.5$ ($\leq$)   vs. H_a: $\sigma > 0.5$

    c. H_o: $\sigma = 10$ vs. H_a: $\sigma \neq 10$

    d. H_o: $\sigma^2 = 18$ ($\geq$)   vs. H_a: $\sigma^2 < 18$

    e. H_o: $\sigma^2 = 0.025$ vs. H_a: $\sigma^2 \neq 0.025$

Chi-square test statistic $\quad-\quad \chi^{2}* = (n-1)s^2/\sigma^2$

9.129 a. $\chi^{2}* = (n-1)s^2/\sigma^2 = (17)(785)/(532) = \underline{25.08}$

b. $\chi^{2}* = (n-1)s^2/\sigma^2 = (40)(78.2)/(52) = \underline{60.15}$

Hypothesis tests (p-value approach) will be completed in the same format as before. You may want to review: ES10-p427, SSM-p265. The only differences are in:
1. the calculated test statistic, which is χ^2, where
$$\chi^2 = (n-1)s^2/\sigma^2$$
2. using Table 8 to <u>estimate</u> the p-value
    a) locate df row
    b) locate the calculated χ^2-value between two critical values in the df row, the p-value or ½p-value is in the interval between the two corresponding probabilities of the critical values

Computer and/or calculator probability and cumulative probability commands for values of χ^2 can be found in ES10-p518-519.

9.131 a. **P** $= 2P(\chi^{2}* > 27.8|df = 14)$

    Using Table 8: $0.01 < $ ½**P** $< 0.025;$ $\underline{0.02 < \mathbf{P} < 0.05}$
    Using computer: **P** $= 0.0302$

b. **P** $= P(\chi^{2}* > 33.4|df = 17) = \underline{0.01}$ both are same

c. **P** $= 2P(\chi^{2}* > 37.9|df = 25)$
    Using Table 8: $0.025 < $ ½**P** $< 0.05;$ $\underline{0.05 < \mathbf{P} < 0.10}$
    Using computer: **P** $= 0.0946$

d. **P** $= P(\chi^{2}* < 26.3|df = 40)$
    Using Table 8: $\underline{0.025 < \mathbf{P} < 0.05}$
    Using computer: **P** $= 0.0469$

Hypothesis tests (classical approach) will be completed using the same format as before. You may want to review: ES10-pp444-445, SSM-p275. The only differences are :
1. the χ^2 distribution
 a) draw a skewed right distribution (starting at zero)
 b) locate df value as middle value
 c) shade in the critical region(s) based on the alternative hypothesis (H_a).
2. finding critical value(s) from Table 8
 a) degrees of freedom (n-1) is the row id #
 b) α, area to the right, is the column id #
3. the left tail uses $1-\alpha$ or $1-\alpha/2$ for its probability
4. the calculated test statistic is χ^{2*}, where
$$\chi^{2*} = (n-1)s^2/\sigma^2$$

9.133 a. $P = P(\chi^2 > 25.08 | df = 17)$
 Using Table 8: $0.05 < \mathbf{P} < 0.10$
 Using computer: $\mathbf{P} = 0.0929$
 $\alpha = 0.01$; $\mathbf{P} > \alpha$; Fail to reject H_o

 b. $\chi^2(17, 0.01) = 33.4$ $\chi^2 \geq 33.4$

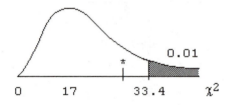

χ^{2*} falls in the noncritical region; Fail to reject $H_\circ$

Hint for writing the hypotheses for Exercise 9.135
Look at the last sentence in the exercise, "Does this sample ... the population standard deviation is <u>not equal to</u> 8 ...?". "Not equal to" ($\neq$) belongs in the alternative hypothesis. The negation becomes "equal to" and the null hypothesis would be written with an equality sign (=). Remember to use the population standard deviation, σ, as the parameter in the hypotheses.

9.135 Step 1: a. The standard deviation, σ
    b. H_{o}: $\sigma = 8$
      H_a: $\sigma \neq 8$

Step 2: a. normality assumed

        b. χ^2 c. $\alpha = 0.05$

Step 3: a. n = 51, $\overline{x}$ = 98.2, s^2 = 37.5

        b. χ^{2*} = (n-1)s^2/σ^2 = (50)(37.5)/8^2 = 29.3

Step 4: -- using p-value approach --------------------

        a. **P** = 2P(χ^2 < 29.3|df = 50);

           Using Table 8: 0.005 < ½**P** < .010; 0.01 < **P** < 0.02

           Using computer: **P** = 0.0172

        b. **P** < α

        -- using classical approach ------------------

        a. χ^2(50, 0.975) = 32.4, χ^2(50, 0.025) = 71.4

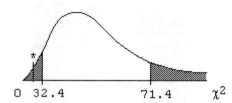

        b. χ^{2*} falls in the critical region, see Step 4a

Step 5: a. Reject H_o

        b. There is sufficient reason to conclude that the
           population standard deviation is not equal to 8,
           at the 0.05 level of significance.

9.137 a. Allows the use of the chi-square distribution to calculate probabilities.

b.

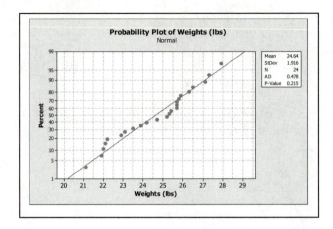

The data values follow very closely to the straight line, indicating an approximately normal distribution.

c & d.

Step 1: a. The standard deviation of plates used in weight lifting

  b. $H_{o}:$ $\sigma = 1.0$ ($\leq$)
     $H_{a}:$ $\sigma > 1.0$

Step 2: a. normality indicated

  b. χ^2 c. $\alpha = 0.01$

Step 3: a. n = 24, s = 1.916

  b. $\chi^{2*} = (n-1)s^2/\sigma^2 = (23)(1.916^2)/(1.0^2) = 84.43$

Step 4: -- using p-value approach --------------------

  a. $\mathbf{P} = P(\chi^2 > 84.43 | df = 23)$;

    Using Table 8: $\mathbf{P} < 0.005$

    Using computer: $\mathbf{P} = 0.0+$

  b. $\mathbf{P} < \alpha$

-- using classical approach ------------------
a. $\chi^2(23, 0.01) = 41.6$

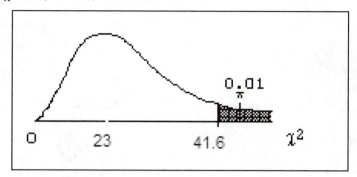

b. $\chi^{2}*$ falls in the critical region, see Step 4a

Step 5: a. Reject H_O
b. There is sufficient reason to conclude that the variability in the weights is greater than the acceptable one-pound standard deviation, at the 0.01 level of significance.

9.139 Step 1: a. The standard deviation of ranchland value in Missouri
b. H_O: $\sigma = 85$ $(\le)$
   H_a: $\sigma > 85$
Step 2: a. normality indicated
b. χ^2 c. $\alpha = 0.05$
Step 3: a. n = 31, s = 125
b. $\chi^{2}* = (n-1)s^2/\sigma^2 = (30)(125^2)/(85^2) = 64.88$
Step 4: -- using p-value approach --------------------
a. **P** = $P(\chi^2 > 64.88 | df = 30)$
   Using Table 8: **P** < 0.005
   Using computer: **P** = 0.0002
b. **P** < α

-- using classical approach ------------------
a. $\chi^2(30, 0.05) = 43.8$

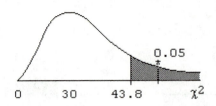

0.05

0   30   43.8 χ^2

b. χ^{2*} falls in the critical region, see Step 4a

Step 5: a. Reject H_o
b. There is sufficient reason to conclude that the
variability in ranchland value in Missouri is
greater than the variablility for the region, at
the 0.05 level of significance.

9.141 a.
Step 1: a. The standard deviation of dry weights of corks
b. H_o: $\sigma = 0.3275$ [not different]
H_a: $\sigma \neq 0.3275$ [differs]
Step 2: a. normality indicated
b. χ^2 c. $\alpha = 0.02$
Step 3: a. n = 10, s = 0.2920, $s^2 = 0.2920^2 = 0.085264$
b. $\chi^{2*} = (n-1)s^2/\sigma^2 = (9)(0.085264)/0.3275^2 = 7.15$
Step 4: -- using p-value approach --------------------
a. $\mathbf{P} = 2 \cdot P(\chi^2 < 7.15 | df = 9)$;
Using Table 8: $2(0.25 < P < 0.50)$; $0.50 < P < 1.00$
Using computer: $\mathbf{P} = 0.7570$
b. $\mathbf{P} > \alpha$

-- using classical approach ------------------
a. $\chi^2(9, 0.99) = 2.09$, $\chi^2(9, 0.01) = 21.7$

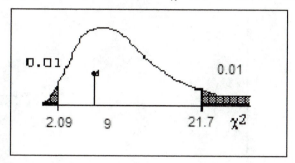

b. χ^{2*} falls in the noncritical region, see Step 4a

Step 5: a. Fail to reject H_o
        b. There is not sufficient reason to show that the
           standard deviation is different from 0.3275
           grams, at the 0.02 level of significance.
b.
Step 1: a. The standard deviation of dry weights of corks
        b. H_o: $\sigma = 0.3275$ [not different]
           H_a: $\sigma \neq 0.3275$ [differs]
Step 2: a. normality indicated
        b. χ^2 c. $\alpha = 0.02$
Step 3: a. $n = 20$, $s = 0.2808$, $s^2 = 0.2808^2 = 0.078849$
        b. $\chi^{2*} = (n-1)s^2/\sigma^2 = (19)(0.078849)/0.3275^2 = 13.97$
Step 4: -- using p-value approach --------------------
        a. $P = 2 \cdot P(\chi^2 < 13.97 | df = 19)$;
           Using Table 8: $2(0.10 < P < 0.25)$; $0.20 < P < 0.50$
           Using computer: $P = 0.4292$
        b. $P > \alpha$

a. $\chi^2(19, 0.99) = 7.63,\quad \chi^2(19, 0.01) = 36.19$

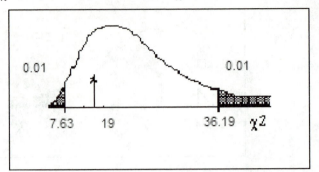

0.01   0.01

7.63   19   36.19 χ^2

b. χ^{2*} falls in the noncritical region, see Step 4a

Step 5: a. Fail to reject H_o
b. There is not sufficient reason to show that the standard deviation is different from 0.3275 gram, at the 0.02 level of significance.

c. Larger sample standard deviations increase the calculated chi-square value.

d. Larger sample sizes increase the number of degree of freedom and in turn increase the calculated chi-square value. With regards to the p-value and critical-value, since each df has a different distribution, it is not possible to determine, in general, how their values are effected.

9.143 Chi-Square with 23 DF

        x   P(X <= x)
    36.59   0.964126

$P = P(\chi 2* > 36.59 | df = 23) = (1 - 0.9641) = 0.0359$

9.145 Results will vary.

Minitab commands:
Choose: Calc > Random Data > Normal
Generate 2000 rows into C1 with mean = 100 and st.dev. = 50.
Choose: Graph > Histogram
Use cutpoints with positions -100:300/25.

A histogram of 2000 random data presents a good picture of the population. Notice, the data ranges from -75 to 275.

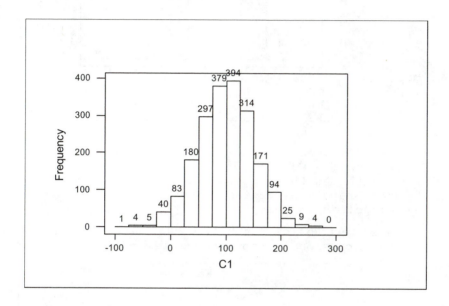

a. Minitab commands:
   Choose: Calc > Random Data > Normal
   Generate 200 rows into C2-C11 with mean = 100 & stdev. = 50.
   Choose: Calc > Row Statistics
   Select Standard deviation for C2-C11 and store in C12.
   Choose: Graph > Histogram
   Use cutpoints with positions 10:100/10.

   The histogram on the next page shows the 200 sample standard deviations from samples of n = 10 from N(100,50). Notice, the s-values range from 20 to 90, and appear to be slightly skewed to the right. Remember, the population standard deviation is 50.

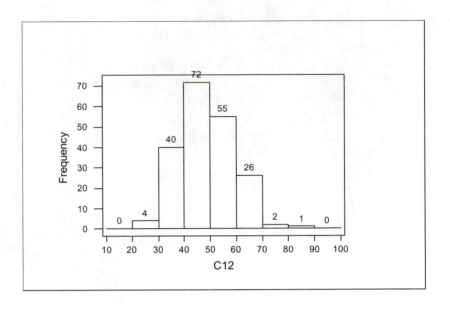

b. Minitab commands:
   Choose: Calc > Calculator
   Store into C13 the expression: (9*C12**2)/(50**2)
   Choose: Graph > Histogram
   Use midpoints (automatic)

   The histogram below shows the 200 calculated chi-square
   values corresponding to the samples of n = 10 above.

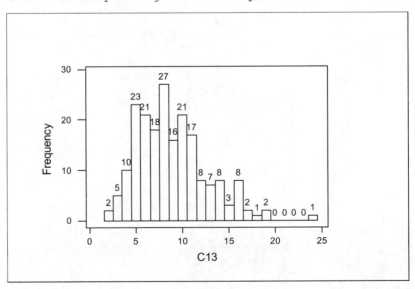

By redrawing the histogram using key critical values from Table 8 (df = 9), we will be able to compare the observed percentage of the chi-square values with the expected percentages.

Minitab commands:
Enter cutpoints of 0, 2.09, 3.33, 4.17, 5.90, 8.34, 11.4, 14.7, 16.9, 21.7, and 25 into C14.
Choose: Graph > Histogram
Use cutpoints with positions in C14.

Table 8 indicates that 1%, 4%, 5%, 15%, 25%, 25%, 15%, 5%, 4%, 1% should occur for the intervals.

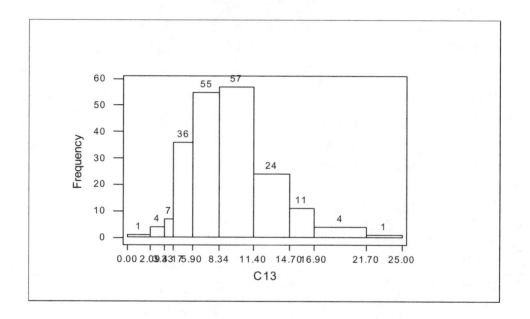

The histogram above shows 1/200 = 0.5%, 4/200 = 2%, 7/200 = 3.5%, 36/200 = 18%,   55/200 = 27.5%, 57/200 = 28.5%, 24/200 = 12%, 11/200 = 5.5%, 4/200 = 2%, and 1/200 = 0.5%. As a set of 10 percentages, they seem very close to what is expected.
c-e.
The results obtained using the exponential distribution are not expected to follow the chi-square distribution as closely. You will need to calculate observed percentages for several intervals in order to detect the true picture.
Minitab commands;
Use the commands above substituting Exponential with a mean of 100 for Normal. Use cutpoints of 0:700/25 for the first histogram. Try cutpoints of 0:40/1 for C13.

CHAPTER EXERCISES

9.147 Step 1: The mean mileage on certain tires for a utility
company
Step 2: a. normality assumed
b. t   c. $1-\alpha = 0.98$
Step 3: $n = 100$, $\overline{x} = 36,000$, $s = 2000$
Step 4: a. $\alpha/2 = 0.02/2 = 0.01$; $df = 99$; $t(99, 0.01) = 2.38$
b. $E = t(df,\alpha/2)\cdot(s/\sqrt{n}) = (2.38)(2000/\sqrt{100}) = 476$
c. $\overline{x} \pm E = 36,000 \pm 476$
Step 5: 35,524 to 36,476, the 0.98 estimate for μ

9.149 a. $\overline{x} = \Sigma x/n = 878.2/100 = \underline{8.782}$

$s^2 = \Sigma(x - \overline{x})^2/(n - 1)$

$= 49.91/99 = 0.5041$

$s = \sqrt{s^2} = \sqrt{0.5041} = \underline{0.710}$

b. Point estimate for μ is $\overline{x} = \underline{8.78}$

c. Step 1: The mean size of oranges
Step 2: a. normality assumed
b. t   c. $1-\alpha = 0.95$
Step 3: $n = 100$, $\overline{x} = 8.78$, $s = 0.710$
Step 4: a. $\alpha/2 = 0.05/2 = 0.025$; $df = 99$; $t(99, 0.025)$
$= 1.99$
b. $E = t(df,\alpha/2)\cdot(s/\sqrt{n}) = (1.99)(0.71/\sqrt{100})$
$= 0.14129 = 0.14$
c. $\overline{x} \pm E = 8.78 \pm 0.14$
Step 5: 8.64 to 8.92, the 0.95 estimate for μ

9.151 Since $df = 71$, therefore $n = df + 1 = 72$

9.153 a. $\overline{x} = \$908.30$, $s = \$118.50$

b. The data values are somewhat mounded towards the center.

c. Step 1: The mean cost of required textbooks at NY private colleges

Step 2: a. normality assumed, see part b above.
b. t   c. $1-\alpha = 0.95$

Step 3: $n = 15$, $\bar{x} = 908.30$, $s = 118.50$

Step 4: a. $\alpha/2 = 0.05/2 = 0.025$; $df = 9$;
$t(9, 0.025) = 2.26$
b. $E = t(df, \alpha/2) \cdot (s/\sqrt{n}) = (2.26)(118.50/\sqrt{10}) = 84.689$
c. $\bar{x} \pm E = 908.30 \pm 84.689$

Step 5: \$823.61 to \$992.99, the 0.95 estimate for μ

d. 95% confident the mean cost of required textbooks at NY private colleges is between \$823.61 and \$992.99.

e. Public colleges have a higher mean cost than private colleges, \$935.86 versus \$908.30.

f. the confidence interval for the public colleges is much wider than that of the private colleges due to the larger standard deviation.

9.155 Step 1: a. The mean yield per acre in a large cherry orchard
b. H_o: $\mu = 4.35$ [no increase] ($\leq$)
H_a: $\mu > 4.35$ [increase]

Step 2: a. normality indicated
b. t   c. $\alpha = 0.05$

Step 3: a. $n = 15$, $\bar{x} = 4.589$, $s = 0.568$
b. $t = (\bar{x} - \mu)/(s/\sqrt{n})$
$t* = (4.589 - 4.35)/(0.568/\sqrt{15}) = 1.63$

Step 4: -- using p-value approach --------------------
a. **P** $= P(t > 1.63 | df = 14)$;
Using Table 6, Appendix B, ES10-p813:
$0.05 <$ **P** < 0.10
Using Table 7, Appendix B, ES10-p814:
$0.055 <$ **P** < 0.068
Using computer: **P** $= 0.0627$
b. **P** $> \alpha$

a. t(14, 0.05) = 1.76

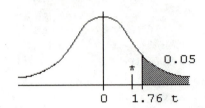

0.05

0   1.76 t

b. t* falls in the noncritical region, see Step 4a

Step 5: a. Fail to reject H_O
        b. The sample does not provide sufficient evidence
           to show an increase, at the 0.05 level.

9.157 a. Summary of data: n = 20, Σx = 629, Σx² = 21,013

$\bar{x}$ = Σx/n = 629/20 = <u>31.45</u>

s² = [Σx² - (Σx)²/n]/(n - 1)

 = [21,013 - (629²/20)]/19 = 64.7868

s = $\sqrt{s^2}$ = $\sqrt{64.7868}$ = <u>8.049</u>

b. Step 1: a. The mean age at which mothers give birth to
               abnormal males
            b. H_O: μ = 28.0 (≤)
               H_a: μ > 28.0   (older)
    Step 2: a. normality indicated
            b. t   c. α = 0.05
    Step 3: a. n = 20, $\bar{x}$ = 31.45,   s = 8.049
            b. t = ($\bar{x}$ - μ)/(s/$\sqrt{n}$)
               t* = (31.45 - 28.0)/(8.049/$\sqrt{20}$) = 1.92
    Step 4:   -- using p-value approach ------------------
            a. **P** = P(t > 1.92|df = 19);
               Using Table 6, Appendix B, ES10-p813:
               0.025 < **P** < 0.05
               Using Table 7, Appendix B, ES10-p814:
               0.029 < **P** < 0.037
               Using computer: **P** = 0.035
            b. **P** < α

        -- using classical approach ------------------
           a. t(19, 0.05) = 1.73

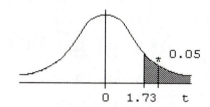

           b. t* falls in the critical region, see * Step 4a

Step 5: a. Reject H_O
        b. The sample does provide sufficient evidence to
           justify the contention that the mean age of
           mothers of abnormal male children is
           significantly greater than the mean age of
           mothers with normal male children, at the 0.05
           level.

9.159

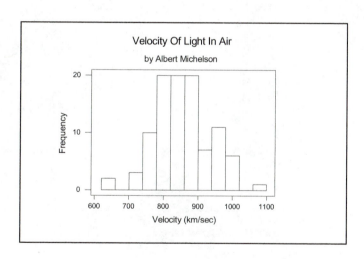

Step 1: a. The mean value for the velocity of light in air
        b. H_O: $\mu = 734.5$
           H_a: $\mu \neq 734.5$
Step 2: a. normality assumed, note graph above; CLT
        b. t   c. $\alpha = 0.01$

                    -- 347 --

Step 3: a. $n = 100$, $\bar{x} = 852.40$, $s = 79.01$
       b. $t = (\bar{x} - \mu)/(s/\sqrt{n})$

         $t\text{*} = (852.40 - 734.5)/(79.01/\sqrt{100}) = 14.92$
Step 4: -- using p-value approach --------------------
       a. $\mathbf{P} = 2P(t > 14.92 | df = 99)$;
             Using Table 6, Appendix B, ES10-p813:
             $\mathbf{P} < 0.005$
             Using Table 7, Appendix B, ES10-p814:
             $\mathbf{P} = 0.00+$
             Using computer: $\mathbf{P} = 0.00+$

       b. $\mathbf{P} < \alpha$
      -- using classical approach ------------------
       a. $\pm t(99, 0.005) = \pm 2.65$;
           critical region: $t \leq -2.65$, $t \geq 2.65$
       b. $t\text{*}$ falls in the critical region

Step 5: a. Reject H_o
       b. At the 0.01 level of significance there is
          sufficient evidence to show that the true
          constant value for the velocity of light in air
          is different from 299,734.5 km/sec.

9.161 Step 1: The proportion of married men who prefer a brand of
          instant coffee
     Step 2: a. The sample was randomly selected and each
          subject's
          response was independent of those of the others
          surveyed.
       b. $n = 100$; $n > 20$, $np = (100)(20/100) = 20$,
          $nq = (100)(80/100) = 80$, np and nq both > 5
       c. $1 - \alpha = 0.95$
     Step 3: $n = 100$, $x = 20$
     Step 4: a. $z(\alpha/2) = z(0.025) = 1.96$
       b. $E = z(\alpha/2) \cdot \sqrt{p'q'/n} = 1.96\sqrt{(0.20)(0.80)/100}$
          $= (1.96)(0.04) = 0.078$
       c. $p' \pm E = 0.200 \pm 0.078$
     Step 5: <u>0.122 to 0.278</u>, the 0.95 interval for p = P(preferred
         this company's brand of instant coffee)

9.163 a.

Step 1: The proportion of dissatisfied customers at a local auto dealership

Step 2: a. The sample was randomly selected and each subject's response was independent of those of the others surveyed.

b. $n = 60$; $n > 20$, $np = (60)(14/60) = 14$, $nq = (60)(46/60) = 46$, np and nq both > 5

c. $1 - \alpha = 0.95$

Step 3: a. $n = 60$, $x = 14$, $p' = x/n = 14/60 = 0.233$

Step 4: a. $z(\alpha/2) = z(0.025) = 1.96$

b. $E = z(\alpha/2) \cdot \sqrt{p'q'/n} = 1.96\sqrt{(0.233)(0.767)/60}$
$= (1.96)(0.054575) = 0.107$

c. $p' \pm E = 0.233 \pm 0.107$

Step 5: <u>0.126 to 0.340</u>, the 0.95 interval for p =
P(dissatisfied)

b. The dealer has overestimated his percent of satisfied customers; it appears to be between 66% and 87%, which is less than 90%.

9.165 a. Step 1: $1 - \alpha = 0.90$; $z(\alpha/2) = z(0.05) = 1.65$

Step 2: $E = 0.03$

Step 3: $p^* = 0.13$ and $q^* = 0.87$

Step 4: $n = \{[z(\alpha/2)]^2 \cdot p^* \cdot q^*\}/E^2$
$= [(1.65^2)(0.13)(0.87)]/(0.03^2) = 342.1275 = \underline{343}$

b. Step 1: $1 - \alpha = 0.95$; $z(\alpha/2) = z(0.025) = 1.96$

Step 2: $E = 0.06$

Step 3: $p^* = 0.13$ and $q^* = 0.87$

Step 4: $n = \{[z(\alpha/2)]^2 \cdot p^* \cdot q^*\}/E^2$
$= [(1.96^2)(0.13)(0.87)]/(0.06^2) = 120.69 = \underline{121}$

c. Step 1: $1 - \alpha = 0.99$; $z(\alpha/2) = z(0.005) = 2.58$

Step 2: $E = 0.09$

Step 3: $p^* = 0.13$ and $q^* = 0.87$

Step 4: $n = \{[z(\alpha/2)]^2 \cdot p^* \cdot q^*\}/E^2$
$= [(2.58^2)(0.13)(0.87)]/(0.09^2) = 92.94 = \underline{93}$

9.167 Step 1: $1 - \alpha = 0.98$; $z(\alpha/2) = z(0.01) = 2.33$

Step 2: $E = 0.02$

Step 3: $p^* = 0.60$ and $q^* = 0.40$

Step 4: $n = \{[z(\alpha/2)]^2 \cdot p^* \cdot q^*\}/E^2$
$= ((2.33^2)(0.6)(0.4))/(0.02^2) = 3257.34 = \underline{3258}$

9.169 $p' = x/n = 73/100 = 0.73$

$z = (p' - p)/\sqrt{pq/n}$

$z^* = (0.73 - 0.80)/\sqrt{(0.8)(0.2)/100} = -1.75$

$P = P(z < -1.75) = 0.5000 - 0.4599 = \underline{0.0401}$

9.171 a. Type A correct decision: The proportion who prefer the new crust is no more than 0.50 and it is decided that no more than 0.50 prefer the crust. Action – stay with current recipe

Type I error: The proportion who prefer the new crust is no more than 0.50 and it is decided that more than 0.50 prefer the crust. Action – change recipes and the majority of customers do not prefer the new recipe.

Type B correct decision: The proportion who prefer the new crust is more than 0.50 and it is decided that more than 0.50 prefer the crust. Action – change recipes and majority of customers are happy

Type II error – The proportion who prefer the new crust is more than 0.50 and it is decided that no more than 0.50 prefer the crust. Action – stay with current recipe when majority of customers prefer the new recipe

b. Type A correct decision: The proportion who prefer the new crust is at least 0.50 and it is decided that at least 0.50 prefer the crust. Action – change recipes and majority of customers are happy

Type I error: The proportion who prefer the new crust is at least 0.50 and it is decided that less than 0.50 prefer the crust. Action – stay with current recipe and the majority of customers are not happy.

Type B correct decision: The proportion who prefer the new crust is less than 0.50 and it is decided that less than 0.50 prefer the crust. Action – stay with current recipe and majority of customers are happy

Type II error – The proportion who prefer the new crust is less than 0.50 and it is decided that at least 0.50 prefer the crust. Action – change recipes and majority of customers will not be happy

c. If the position is "change only if p is significantly greater than 0.5" then use the alternative hypothesis in part (a).

9.173 a. $P = P[x = 9,10,11,\ldots,15 | B(n=15,p=0.5)]$

$\qquad = [0.153 + 0.092 + 0.042 + 0.014 + 0.003 + 2(0+)]$

$\qquad = \underline{0.304}$

$P > \alpha$; Fail to reject Ho   There is not significant evidence to conclude that there is a preference for the new crust.

b. $P = P[x = 120,121,122,\ldots,200 | B(n=200,p=0.5)]$

Binomial with n = 200 and p = 0.5

$$
\begin{array}{cc}
x & P(X <= x) \\
120 & 0.998183
\end{array}
$$

$P = 1.0 - 0.998183 = 0.001817$

$P < \alpha$; Reject Ho   There is significant evidence to conclude that there is a preference for the new crust.

c. Results using the binomial are the same as using the z-distribution.

9.175 a. The percentage of all people in the population who have a specific characteristic is a parameter; it is the binomial parameter p, P(success).

b. Step 1: The proportion of cell phone users ages 18-27 who have used text messaging within the past month

Step 2: a. The sample was randomly selected and each subject's response was independent of the others surveyed.

b. n = 1460, n > 20; np = (1460)(0.63) = 919.8, nq = (1460)(0.37) = 540.2, np and nq both > 5

c. $1 - \alpha = 0.95$

Step 3: n = 1460, p' = 0.63 (given)

Step 4: a. $z(\alpha/2) = z(0.025) = 1.96$

b. $E = z(\alpha/2) \cdot \sqrt{p'q'/n} = 1.96 \cdot \sqrt{(0.63)(0.37)/1460}$

$\qquad = (1.96)(0.013) = 0.025 = 0.03$

c. $p' \pm E = 0.63 \pm 0.03$

Step 5: $\underline{0.60 \text{ to } 0.66}$, the 0.95 interval for p =

$\qquad\qquad\qquad\qquad$ P(use text messaging)

c. The 63% is the point estimate, the ±3 percent is the
    maximum error, 63 ± 3 (60% to 66%) is the confidence
    interval and as shown, has the typical level of
    confidence, 95%.

9.177 Step 1: a. The standard deviation for the length of life for
                     60-watt light bulbs
                  b. H_o: $\sigma = 81$ [not larger] ($\leq$)
                     H_a: $\sigma > 81$ [larger]
         Step 2: a. normality indicated
                  b. χ^2 c. $\alpha = 0.05$
         Step 3: a. n = 101, s^2 = 8075
                  b. χ^{2*} = (n-1)s^2/σ^2 = (100)(8075)/81² = 123.1
         Step 4: -- using p-value approach --------------------
                  a. **P** = P(χ^2 > 123.1|df=100)
                     Using Table 8: 0.05 < **P** < 0.10
                     Using computer: **P** = 0.0584
                  b. **P** > α
                  -- using classical approach ------------------
                  a. χ^2(100, 0.05) = 124

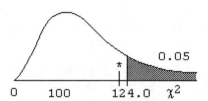

                  b. χ^{2*} falls in the noncritical region, see Step 4a
         ----
         Step 5: a. Fail to reject H_o
                  b. There is not sufficient reason to reject Bright-
                     lite's claim, at the 0.05 level of significance.

9.179 Step 1: a. The standard deviation of sales tickets at Julie's
ice cream restaurant franchise
b. H_o: σ = \$2.45   ($\leq$)
H_a: σ > \$2.45

Step 2: a. normality indicated
b. χ^2 c. α = 0.05

Step 3: a. n = 71,   s = 2.95
b. χ^{2*} = $(n-1)s^2/\sigma^2$ = $(70)(2.95^2)/2.45^2$ = 101.5

Step 4: -- using p-value approach -------------------
a. **P** = $P(\chi^2 > 101.5 | df=70)$
Using Table 8: 0.005 < **P** < 0.01
Using computer: **P** = 0.0082
b. **P** < α
-- using classical approach ------------------
a. $\chi^2(70, 0.05)$ = 90.5 $\chi^2 \geq$ 90.5

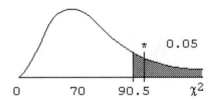

b. χ^{2*} falls in the critical region, see Step 4a

Step 5: a. Reject H_o
b. There is sufficient evidence that the variability
in sales at Julie's franchise is greater than the
variability for the company, at the 0.05 level of
significance.

9.181 a. The assumption of normality is reasonable in that the
sample size is 35, a size considered large by the CLT.

b. Step 1: a. The mean length of 2-inch nails
b. H_o: μ = 2.0
H_a: $\mu \neq$ 2.0
Step 2: a. Sample assumed approximately normal, CLT
satisfied
b. t   c. α = 0.05

Step 3: a. $n = 35$, $\overline{x} = 2.025$, $s = 0.048$

       b. $t = (\overline{x} - \mu)/(s/\sqrt{n})$

         $t* = (2.025 - 2.0)/(0.048/\sqrt{35}) = 3.08$

Step 4: -- using p-value approach --------------------

       a. $\mathbf{P} = 2P(t > 3.08|df = 34)$;

          Using Table 6, Appendix B, ES10-p813:

          $2(\mathbf{P} < 0.005) = P < 0.010$

          Using Table 7, Appendix B, ES10-p814:

          $\mathbf{P} = 2(0.002 < P < 0.003) = 0.004 < P < 0.006$

          Using computer: $\mathbf{P} = 0.0041$

       b. $\mathbf{P} < \alpha$

     -- using classical approach ------------------

       a. $\pm t(34, 0.025) = \pm 2.04$;

         $t \leq -2.04$ & $t \geq 2.04$ critical region

       b. $t*$ falls in the critical region

Step 5: a. Reject H_O

       b. There is sufficient evidence to reject the
idea that the nails have a mean length of 2
inches, at $\bullet = 0.05$.

c.

Step 1: a. The standard deviation, σ, of 2-inch nails

       b. H_O: $\sigma = 0.040$ $(\leq)$

         H_a: $\sigma > 0.040$

Step 2: a. normality assumed, based on $n = 35$ and CLT

       b. χ^2 c. $\alpha = 0.05$

Step 3: a. $n = 35$, $\overline{x} = 2.025$, $s = 0.048$

       b. $\chi^{2*} = (n-1)s^2/\sigma^2 = (34)(0.048^2)/0.04^2 = 48.96$

Step 4: -- using p-value approach --------------------

       a. $\mathbf{P} = P(\chi^2 > 48.96|df = 34)$

      Using Table 8: $0.01 < \mathbf{P} < 0.025$ using df = 30

      Using computer: $\mathbf{P} = 0.0466$

       b. $\mathbf{P} < \alpha$

      -- using classical approach ------------------

      a. $\chi^2(34, 0.05) = 43.8$

      b. χ^{2*} falls in the critical region

Step 5: a. Reject H_O

       b. There is sufficient reason to conclude that the
standard deviation is greater than 0.040 inches
on this production run, at the 0.05 level of
significance.

d. Answers will vary but will include that the length and variability did not meet specifications.

9.183 a.

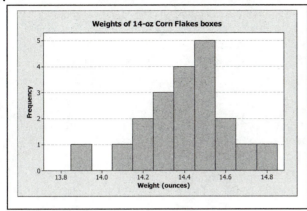

b. $\overline{x}$ = 14.386   s = 0.217

c.   1/20 = 0.05 = 5% below 14 oz.

d.

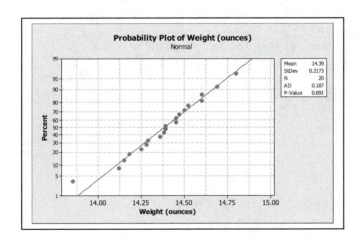

The data values follow closely to a straight line indicating a normal distribution. The P-value also indicates the normality of the data.

e. Step 1: The mean weight of Corn Flakes boxes
   Step 2: a. normality assumed, see part d above.
          b. t   c. $1-\alpha = 0.95$
   Step 3: n = 20, $\bar{x}$ = 14.386,   s = 0.217
   Step 4: a. $\alpha/2 = 0.05/2 = 0.025$; df = 19; t(19, 0.025) = 2.09
          b. E = $t(df,\alpha/2)\cdot(s/\sqrt{n})$ = $(2.09)(0.217/\sqrt{20})$
                = 0.101
          c. $\bar{x} \pm E$ = 14.386 $\pm$ 0.101
   Step 5: 14.285 to 14.487, the 0.95 estimate
          For μ

f.
Step 1: a. The standard deviation, σ, of fill for Corn Flakes
           boxes
        b. H_o: $\sigma = 0.2$ ($\leq$)
           H_a: $\sigma > 0.2$
Step 2: a. normality indicated
        b. χ^2 c. $\alpha = 0.01$
Step 3: a. n = 20, $\bar{x}$ = 14.386, s = 0.217
        b. χ^{2*} = $(n-1)s^2/\sigma^2$ = $(19)(0.217^2)/0.2^2$ = 22.37
Step 4: -- using p-value approach --------------------
        a. $P = P(\chi^2 > 22.37 | df = 19)$
           Using Table 8: $0.25 < P < 0.50$
           Using computer: $P = 0.2662$
        b. $P > \alpha$
        -- using classical approach ------------------
        a. $\chi^2(19, 0.01) = 36.2$
        b. χ^{2*} falls in the noncritical region

Step 5: a. Fail to reject H_o
        b. There is not sufficient reason to reject that the
           filling process is running with a standard
           deviation of no more than 0.2 oz, at the 0.01
           level of significance.

9.185 a. $P(x > 14.2) = P(z > (14.2 - 14.386)/0.217) = P(z > -0.86)$
$$= 0.3051 + 0.5000 = 0.8051$$

b. $P(x > 14.2) = P(z > (14.2 - 14.153)/0.0414) = P(z > 1.14)$
$$= 0.5000 - 0.3729 = 0.1271$$

c. $1000(14.386) = 14386$ oz; $14386/14.153 = 1016.46$ boxes

d. For every 1000 boxes at current fill-rate, there would be 1016.5 boxes with the new machine. That is an extra 16.5 boxes to sell, which is an increase of 1.65% in revenue at no extra cost of product, except for the initial cost of the machine.

CHAPTER 10 ∇ INFERENCES INVOLVING TWO POPULATIONS

Chapter Preview

In Chapters 8 and 9, the concepts of confidence intervals and hypothesis tests were introduced. Each of these was demonstrated with respect to a single mean, standard deviation, variance or proportion. In Chapter 10, these concepts will be extended to include two means, two proportions, two standard deviations, or two variances, thereby enabling us to compare two populations. Distinctions will have to be made with respect to dependent and independent samples, in order to select the appropriate testing procedure and test statistic.

A report published by Nellie Mae on the use of credit cards by college students is presented in the chapter's opening section, ''Students, Credit Cards, and Debt''.

SECTION 10.1 EXERCISES

10.1 a. College students

b. Freshmen: 97/200 = 0.485 or 48.5%; Nellie Mae = 54%
   Sophomore: 187/200 = 0.935 or 93.5%; Nellie Mae = 92%

c. mounded, but skewed with the right-hand tail being much longer

d.

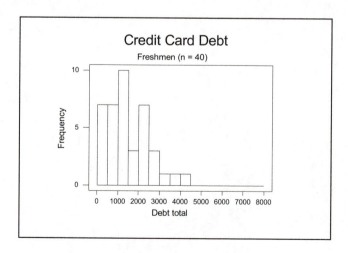

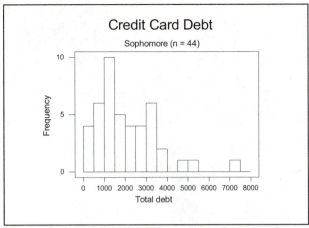

Both distributions are mounded and skewed to the
right. The distribution of debts for sophomores is
more dispersed.

e. Freshmen: mean = $1519, standard deviation = $1036
   Sophomore: mean = $2079, standard deviation = $1434

f. Both distributions are skewed to the right with the
   majority of debt between $0 and $3000. The maximum
   for freshman is around $4500, whereas the maximum for
   sophomores is around $7500.

SECTION 10.2 EXERCISES

INDEPENDENT SAMPLES - Two samples are independent if the selection
of one sample from a population has no effect on the selection of
the other sample from another population. (They do not have to be
different populations.)
ex.: the repair costs for two different brands of VCRs

DEPENDENT SAMPLES (paired samples) - two samples are dependent if
the objects or individuals selected for one sample from a
population are paired in some meaningful way with the objects or
individuals selected for the second sample from the same or another
population.
ex.: "before and after" experiments - change in weight for smokers
who became nonsmokers

10.3 Identical twins are so much alike that the information obtained from one would not be independent from the information obtained from the other twin.

10.5 Independent samples. The two samples are from two separate and different sets of students, males and females.

10.7 Dependent samples. The two sets of data were obtained from the same set of 20 people, each person providing one piece of data for each sample.

10.9 Independent samples. The gallon of paint serves as the population of many (probably millions) particles. Each set of 10 specimens forms a separate and independent samples.

10.11 a. Independent samples will result if the two sets are selected in such a way that there is no relationship between the two resulting sets.

b. Dependent samples will result if the 1,000 men and women were husband and wife or if they were brother and sister, or related in some way.

SECTION 10.3 EXERCISES

10.13 a.

Pairs	1	2	3	4	5
d=A-B	1	1	0	2	-1

b. $n = 5$, $\Sigma d = 3$, $\Sigma d^2 = 7$

$\bar{d} = \Sigma d/n = 3/5 = \underline{0.6}$

c. $s_d = \sqrt{(\Sigma d^2 - (\Sigma d)^2/n)/(n-1)} = \sqrt{(7 - (3)^2/5)/(4)}$

$= \sqrt{1.3} = \underline{1.14}$

```
┌─────────────────────────────────────────────────────────────────────────┐
│                                                                           │
│           Estimating μ_d - the population mean difference                 │
│                                                                           │
│                              Σ d                                          │
│  1. point estimate:   d̄ = ─────                                           │
│                               n                                           │
│                                                                           │
│  2. confidence interval:       d̄    ±    t(df, α/2)   ·   (s_d/√n)         │
│                                ↑             ↑              ↑              │
│                              point       confidence     estimated         │
│                              estimate     coefficient    standard error   │
│                                          └─────────────────────────┘      │
│                                                                           │
│                             maximum error of estimate                     │
│                                                                           │
│  Follow the steps outlined in "The Confidence Interval: A Five-Step       │
│  Procedure" in: ES10-p404, SSM-p256.                                      │
│                                                                           │
│  Computer and/or calculator commands to construct a confidence            │
│  interval for the mean difference can be found in ES10-p554.              │
│                                                                           │
└─────────────────────────────────────────────────────────────────────────┘
```

10.15   a. Step 1: The mean difference, μ_d
               Step 2: a. normality assumed
                       b. t   c. $1-\alpha = 0.95$
               Step 3: n = 26, $\bar{d}$ = 6.3, s_d = 5.1
               Step 4: a. $\alpha/2 = 0.05/2 = 0.025$; df = 25;
                          t(25, 0.025) = 2.06
                       b. E = $t(df,\alpha/2)\cdot(s_d/\sqrt{n})$ = (2.06)(5.1/$\sqrt{26}$)
                          = (2.06)(1.0002) = 2.06
                       c. $\bar{d} \pm E$ = 6.3 ± 2.06
               Step 5: <u>4.24 to 8.36</u>, the 0.95 confidence interval
                       for μ_d

        b. The same $\bar{d}$ and s_d values were used with a much larger n,
            resulting in a narrower confidence interval.

10.17   Data Summary: n = 10, Σd = 263, Σd^2 = 12319

        $\bar{d}$ = $\Sigma d/n$ = 263/10 = <u>26.3</u>

        s_d = $\sqrt{(\Sigma d^2 - (\Sigma d)^2/n)/(n-1)}$

             = $\sqrt{(12319 - (263^2 / 10))/9}$ = $\sqrt{600.2333}$ = <u>24.4997</u>

        SE Mean = $s_d/\sqrt{n}$ = 24.4997/$\sqrt{10}$ = <u>7.747</u>

        $t(df,\alpha/2)$ = t(9,0.025) = 2.26

$$E = t(df, \alpha/2) \cdot (s_d/\sqrt{n}) = (2.26)(7.747) = 17.51$$

$$\overline{d} \pm E = 26.3 \pm 17.51$$

<u>8.8 to 43.8</u>, the 0.95 confidence interval for μ_d

10.19   Data Summary: $n = 8$, $\Sigma d = 8$, $\Sigma d^2 = 48$

a. Point estimate = <u>1.0</u>

b. Step 1: The mean reduction in diastolic blood pressure
          following a two week salt-free diet (d = B - A)
   Step 2: a. normality indicated
          b. t   c. $1-\alpha = 0.98$
   Step 3: $n = 8$, $\overline{d} = 1$, $s_d = 2.39$
   Step 4: a. $\alpha/2 = 0.02/2 = 0.01$; df = 7;
          $t(7, 0.01) = 3.00$
          b. $E = t(df, \alpha/2) \cdot (s_d/\sqrt{n}) = (3.00)(2.39/\sqrt{8})$
           $= (3.00)(0.845) = 2.53$
          c. $\overline{d} \pm E = 1 \pm 2.53$
   Step 5: <u>-1.53 to 3.53</u>, the 0.98 confidence interval for

10.21   Sample statistics: d = I - II

     $n = 10$, $\overline{d} = 0.8$, $s_d = 1.32$

Step 1: The mean difference between two routes
Step 2: a. normality indicated
       b. t   c. $1-\alpha = 0.95$
Step 3: $n = 10$, $\overline{d} = 0.8$, $s_d = 1.32$
Step 4: a. $\alpha/2 = 0.05/2 = 0.025$; df = 9;
       $t(9, 0.025) = 2.26$
       b. $E = t(df, \alpha/2) \cdot (s_d/\sqrt{n}) = (2.26)(1.32/\sqrt{10})$
        $= (2.26)(0.4174) = 0.943$
       c. $\overline{d} \pm E = 0.8 \pm 0.943$
Step 5: <u>-0.143 to 1.743</u>, the 0.95 confidence interval for μ_d

```
┌─────────────────────────────────────────────────────────────────────┐
│           WRITING HYPOTHESES FOR TEST OF TWO DEPENDENT MEANS          │
│                                                                       │
│              μ_d = population mean difference                         │
│                                                                       │
│        null hypothesis -   H_O: μ_d = 0                               │
│                ("the mean difference equals 0, that is, there is      │
│                   no difference within the pairs of data")            │
│                                                                       │
│        possible alternative hypotheses -                              │
│                       H_a: μ_d > 0                                    │
│                       H_a: μ_d < 0                                    │
│                       H_a: μ_d ≠ 0,                                   │
│              ("the mean difference is significant, that is,           │
│                  there is a difference within the pairs of data")     │
└─────────────────────────────────────────────────────────────────────┘
```

10.23 a. H_O: $\mu_d = 0$ ($\leq$); H_a: $\mu_d > 0$; d = posttest - pretest

b. H_O: $\mu_d = 0$; H_a: $\mu_d \neq 0$; d = after - before

c. H_O: $\mu_d = 0$; H_a: $\mu_d \neq 0$; d = reading1 - reading2

d. H_O: $\mu_d = 0$ ($\leq$); H_a: $\mu_d > 0$; d = post score - pre score

10.25 a. $\mathbf{P} = P(t > 1.86 | df = 19)$
Using Table 6, Appendix B, ES10-p813:
    $0.025 < \mathbf{P} < 0.05$
Using Table 7, Appendix B, ES10-p814:
    $0.036 < \mathbf{P} < 0.044$
Using computer, $\mathbf{P} = 0.0392$

b. $P = 2P(t < -1.86 | df = 19) = 2P(t > 1.86 | df = 19)$
Using Table 6, Appendix B, ES10-p813:
    $2(0.025 < \mathbf{P} < 0.05) = 0.05 < P < 0.10$
Using Table 7, Appendix B, ES10-p814:
    $2(0.036 < \mathbf{P} < 0.044) = 0.072 < P < 0.088$
Using computer, $\mathbf{P} = 0.0784$

c. $P = P(t < -2.63 | df = 28) = P(t > 2.63 | df = 28)$
Using Table 6, Appendix B, ES10-p813:
    $0.005 < \mathbf{P} < 0.01$
Using Table 7, Appendix B, ES10-p814:
    $0.006 < \mathbf{P} < 0.008$
Using computer, $\mathbf{P} = 0.0069$

d. $P = P(t > 3.57 | df = 9)$
   Using Table 6, Appendix B, ES10-p813:
       P < 0.005
   Using Table 7, Appendix B, ES10-p814:
       0.002 < **P** < 0.004
   Using computer, **P** = 0.003

Reviewing how to determine the test criteria in: ES10-pp450-451,
SSM-p277, may be helpful. Remember the t-distribution uses Table 6
(Appendix B, ES10-p813), therefore α and degrees of freedom,
df = n - 1, are needed.

Hypothesis Tests for Two Dependent Means

In this form of hypothesis test, each data value of the first
sample is compared to its corresponding (or paired) data value in
the second sample. The differences between these paired data
values are calculated, thereby forming a sample of differences or d
values. It is these differences or d values that we wish to use to
test the difference between two dependent means.

Review the parts to a hypothesis test (p-value & classical) as
outlined in: ES10-pp427&444-445, SSM-pp265&275, if needed. Changes
will occur in:

   1) the calculated value of the test statistic, t;

$$t = \frac{\overline{d} - \mu_d}{s_d / \sqrt{n}} \quad , \quad \text{where } \overline{d} = \frac{\sum d}{n} , s_d = \sqrt{\frac{\sum d^2 - (\sum d)^2 / n}{n - 1}}$$

$$\text{and } n = \text{\# of paired differences}$$

   2)a. p-value approach
       Use Table 6 (Appendix B, ES10-p813) to <u>estimate</u> the
       p-value
          1) Locate df row.
          2) Locate the absolute value of the calculated t-
             value between two critical values in the df row.
          3) The p-value is in the interval between the two
             corresponding probabilities at the top of the
             columns; read the bounds from the one-tailed
             heading if Ha is one-tailed, or from the two-
             tailed headings if Ha is two-tailed.
                                                    . . .

OR
Use Table 7 (Appendix B, ES10-p814) to estimate or place bounds on the p-value
1) locate the absolute value of the calculated t-value along the left margin and the df along the top, then read the p-value directly from the table where the row and column intersect
OR
2) locate the absolute value of the calculated t-value and its df between appropriate bounds. From the box formed at the intersection of these row(s) and column(s), use the upper left and lower right values for the bounds on P.

b. classical approach
Use Table 6 (Appendix B, ES10-p813) with df = n - 1 and α to find the critical value

3) if Ho is rejected, a significant difference as stated in Ha is indicated; if Ho is not rejected, no significant difference is indicated

Computer and/or calculator commands to perform a hypothesis test for μd can be found in ES10-p483.

NOTE: To find $\bar{d}$ and sd, set up a table of corresponding pairs of data. Calculate d, the difference (be careful to subtract in the same direction each time). Calculate a d2 for each pair and find summations, Σd and $\Sigma(d^2)$.
The sample of paired differences are assumed to be selected from an approximately normally distributed population with a mean μ_d and a standard deviation σ_d. Since σ_d is unknown, the calculated t-statistic is found using an estimated standard error of $s_d/\sqrt{n}$.

Hint for writing the hypotheses for exercise 10.27
Look at the second and third sentences of the exercise; "The data
... where d is the amount of corrosion on the coated portion
subtracted from the amount of corrosion on the uncoated portion.
Does this sample provide sufficient reason to conclude that the
coating is beneficial?" For the coating to be beneficial, there
would have to be less corrosion on the coated portion. This
implies that the differences, "uncoated - coated," if beneficial,
would be positive, which in turn implies a greater than zero (> 0).
Therefore the alternative hypothesis is a greater than (>). The
negation is "NOT beneficial", which indicates < or =. Therefore
the null hypothesis should be written with the equal sign (=), but
include the less-than or equal-to sign (≤) in parentheses after it.

10.27   Sample statistics: $n = 40$, $\bar{d} = 220/40 = 5.5$, $s_d = 11.34$

Step 1: a. The mean difference between coated and uncoated
sections of steel pipe
b. H_o: $\mu_d = 0$
H_a: $\mu_d > 0$ (beneficial)
Step 2: a. normality assumed, CLT with $n = 40$.
b. t   c. $\alpha = 0.01$
Step 3: a. $n = 40$, $\bar{d} = 5.5$, $s_d = 11.34$
b. $t* = (\bar{d} - \mu_d)/(s_d/\sqrt{n})$
$= (5.5 - 0.0)/(11.34/\sqrt{40}) = 3.067$
Step 4: -- using p-value approach --------------------
a. **P** = $P(t > 3.067 | df = 39)$;
Using Table 6, Appendix B, ES10-p813:
P < 0.005
Using Table 7, Appendix B, ES10-p814:
P ≈ 0.002
Using computer, **P** $= 0.002$
b. $P < \alpha$

```
      -- using classical approach --------------------
   a. t(39, 0.01) ≈ t(35, 0.01) = 2.44, using Table 6
      t(39, 0.01) = 2.426, by computer
```

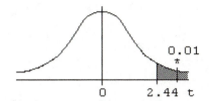

```
                                 0.01
                                  *

              0        2.44 t
```

```
   b. t* falls in the critical region, see * Step 4a
   ------------------------------------------------
Step 5: a. Reject H_o
        b. At the 0.01 level of significance, there is a
           significant benefit to coating the pipe.
```

Computer and/or calculator commands to perform a hypothesis test
for μ_d can be found in ES10-pp556-557.
The order of subtraction needs to match the "planned" approach as
determined by H_a.

10.29 Verify - answers given in exercise.

10.31 Data Summary: n = 5, Σd = 90, Σd^2 = 2700

```
        Step 1: a. The mean difference, μ_d
                b. H_o: μ_d = 0
                   H_a: μ_d > 0
        Step 2: a. normality indicated
                b. t              c. α = 0.05
        Step 3: a. n = 5, d̄ = 18,  s_d = 16.43
                b. t* = (d̄ - μ_d)/(s_d/√n) = (18 - 0)/(16.43/√5)
                      = 2.45
        Step 4: -- using p-value approach --------------------
                a. P = P(t > 2.45|df = 4);
                   Using Table 6, Appendix B, ES10-p813:
                   0.025 < P < 0.05
                   Using Table 7, Appendix B, ES10-p814:
                   0.033 < P < 0.037
                   Using computer: P = 0.0352
                b. P < α
```

a. $t(4, 0.05) = 2.13$

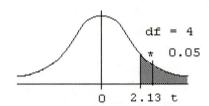

df = 4

* 0.05

0   2.13 t

b. $t*$ falls in the critical region, see Step 4a

Step 5: a. Reject H_o
   b. At the 0.05 level of significance, there is
      sufficient evidence that the mean difference is
      greater than zero.

10.33   Data Summary: $n = 5$, $\Sigma d = 12$, $\Sigma d^2 = 92$

Step 1: a. The mean difference, μ_d
   b. H_o: $\mu_d = 0$
      H_a: $\mu_d \neq 0$
Step 2: a. normality indicated
   b. t   c. $\alpha = 0.01$
Step 3: a. $n = 5$, $\bar{d} = 2.4$, $s_d = 3.97$
   b. $t* = (\bar{d} - \mu_d)/(s_d/\sqrt{n}) = (2.4 - 0)/(3.97/\sqrt{5})$
      $= 1.35$
Step 4: -- using p-value approach --------------------
   a. **P** $= 2P(t > 1.35 | df = 4)$;
      Using Table 6, Appendix B, ES10-p813:
      $0.10 < \frac{1}{2}P < 0.25$; $0.20 < P < 0.50$
      Using Table 7, Appendix B, ES10-p814:
      $0.117 < \frac{1}{2}P < 0.132$; $0.234 < P < 0.264$
      Using computer: **P** $= 0.2484$
   b. **P** $> \alpha$

-- using classical approach ------------------
a. $\pm t(4, 0.005) = \pm 4.60$

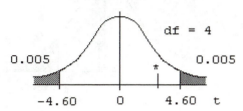

df = 4

0.005   0.005

-4.60   0   4.60   t

b. t* falls in the noncritical region, see Step 4a

Step 5: a. Fail to reject H_o
        b. At the 0.01 level of significance, there is not
           sufficient evidence that the mean difference is
           different than zero.

10.35 Step 1: a. The mean difference, μ_d, control group
           b. H_o: $\mu_d = 0$ (pre - post)
              H_a: $\mu_d > 0$ (improvement)
        Step 2: a. normality assumed
           b. t   c. $\alpha = 0.05$
        Step 3: a. n = 10, $\bar{d} = 0.80$, $s_d = 4.492$
           b. $t* = (\bar{d} - \mu_d)/(s_d/\sqrt{n})$
              $= (0.80 - 0)/(4.492/\sqrt{10}) = 0.56$
        Step 4: -- using p-value approach --------------------
           a. **P** = P(t > 0.56|df = 9);
              Using Table 6, Appendix B, ES10-p813:
              P > 0.25
              Using Table 7, Appendix B, ES10-p814:
              0.281 < **P** < 0.315
              Using computer: **P** = 0.2946
           b. **P** > α

       a. t(9, 0.05) = 1.83

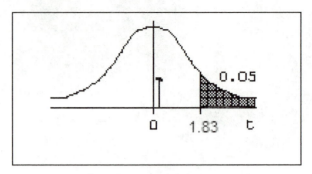

       b. t* falls in the noncritical region, see Step 4a

Step 5: a. Fail to reject H$_O$
        b. At the 0.05 level of significance, there is
           insufficient evidence to show that one's self
           esteem increases after participation in college
           courses.

10.37 a. The null hypothesis is, "the average difference is
             zero."
          b. The "t-calculated" and the "t-critical" values are
             being used to make the decision as in the classical
             approach.
          c. The test is two-tailed and the t-distribution is
             symmetric, making the number of multiples each
             statistic is from zero the only information needed.
             Further, the absence of negative numbers makes the
             table less confusing to most.
          d. The decision was to "fail to reject the null
             hypothesis" in 12 of them. Actually the calculated
             t (2.224) for the No. 4 sieve size is less than the
             critical value (2.228) and therefore leads to a
             "fail to reject" decision also. But for some
             reason, they viewed it as too close.
          e. The conclusion reached was, "the two methods of
             sampling are equivalent with respect to Gmb, Gmm,
             asphalt binder content and gradation."
          f. The recommended action was, "the revised Florida
             method for sampling (FM 1-T 168) be accepted and
             implemented statewide."

SECTION 10.4 EXERCISES

Estimating $(\mu_1 - \mu_2)$ - the difference between two population means, independent samples

1. Point Estimate: $\overline{x}_1 - \overline{x}_2$

2. Confidence Interval

$$(\overline{x}_1 - \overline{x}_2) \ \pm \ t(df, \ \alpha/2) \cdot \sqrt{(s_1^2 / n_1) + (s_2^2 / n_2)}$$

$$\uparrow \qquad\qquad \uparrow \qquad\qquad\qquad \uparrow$$

| point estimate | confidence coefficient | estimated standard error |

maximum error of estimate

estimate df by using the smaller value of df_1 or df_2

10.39 $\sqrt{(s_1^2 / n_1) + (s_2^2 / n_2)} = \sqrt{(190 / 12) + (150 / 18)} = \sqrt{24.1667} = \underline{4.92}$

10.41 Case I: df will be between 17 and 40
Case II; df = 17 (smaller df)

Case I will occur when completing the inference on a computer or calculator, where the df are calculated. Case II will occur when completing the test without software, using Table 6 and a conservative approach.

Review "The Confidence Interval: A Five-Step Procedure" in: ES10-p404, SSM-p256, if necessary.
Subtract sample means $(\overline{x}_1 - \overline{x}_2$ or $\overline{x}_2 - \overline{x}_1)$ in whichever order results in a positive difference.
Also, use appropriate subscripts to designate the source.

10.43 Step 1: The difference between two means, $\mu_1 - \mu_2$
Step 2: a. normality indicated
        b. t   c. $1-\alpha = 0.90$
Step 3: sample information given in exercise;
        $\overline{x}_1 - \overline{x}_2 = 35 - 30 = 5$

Step 4: a. $\alpha/2 = 0.10/2 = 0.05$; df $= 14$;
$\quad\quad\quad$ t$(14, 0.05) = 1.76$
$\quad\quad$ b. E $= t(df,\alpha/2) \cdot \sqrt{(s_1^2 / n_1) + (s_2^2 / n_2)}$
$\quad\quad\quad\quad = (1.76)\sqrt{(22^2/20) + (16^2/15)}$
$\quad\quad\quad\quad = (1.76)(6.42) = 11.3$
$\quad\quad$ c. $(\overline{x}_1 - \overline{x}_2) \pm E = 5 \pm 11.3$
Step 5: <u>-6.3 to 16.3</u>, the 0.90 confidence interval for $\mu_1 - \mu_2$

10.45$\quad\quad$ Confidence interval using t from Table 6:
Step 1: The difference between the average daily car rental
$\quad\quad\quad\quad$ rates in Boston and NYC, $\mu_{Bos} - \mu_{NYC}$
Step 2: a. normality indicated
$\quad\quad$ b. t $\quad\quad\quad\quad\quad\quad$ c. $1-\alpha = 0.95$
Step 3: sample information given in exercise;
$\quad\quad\quad$ $\overline{x}_{Bos} - \overline{x}_{NYC} = 128.25 - 116.60 = 11.65$
Step 4: a. $\alpha/2 = 0.05/2 = 0.025$; df $= 9$;
$\quad\quad\quad$ t$(9, 0.025) = 2.26$
$\quad\quad$ b. E $= t(df,\alpha/2) \cdot \sqrt{(s_{Bos}^2/n_{NYC}) + (s_{Bos}^2/n_{NYC})}$

$\quad\quad\quad\quad = (2.26)\sqrt{(7.50^2/10) + (8.90^2/15)}$
$\quad\quad\quad\quad = (2.26)(3.30237) = 7.463$
$\quad\quad$ c. $(\overline{x}_{Bos} - \overline{x}_{NYC}) \pm E = 11.65 \pm 7.46$
Step 5: <u>$\$4.19$ to $\$19.11$</u>, the 0.95 confidence interval for

OR$\quad$ Confidence interval using calculated df:
Difference = mu (Bos) - mu (NYC)
Estimate for difference: 11.6500
95% CI for difference: (4.7823, 18.5177)
DF = 21

Computer and/or calculator commands to construct a confidence
interval for the difference between two means can be found in ES10-
pp571-572.

10.47$\quad$ Verify - answer given in exercise

10.49$\quad$ Sample statistics:

N. Dakota: $\quad$ n = 11, $\quad$ $\overline{x}$ = 976.2, $\quad$ s = 255.7
S. Dakota: $\quad$ n = 14, $\quad$ $\overline{x}$ = 1370, $\quad$ s = 397

Confidence interval using t from Table 6:
Step 1: The difference between mean sunflower yields for
$\quad\quad\quad\quad$ North and South Dakota, $\quad \mu_1 - \mu_2$

Step 2: a. normality assumed.
        b. t   c. $1-\alpha = 0.95$
Step 3: sample information given above;
        $\overline{x}_S - \overline{x}_N = 1370 - 976.2 = 393.8$
Step 4: a. $\alpha/2 = 0.05/2 = 0.025$; df $= 10$;
        $t(10, 0.025) = 2.23$

        b. $E = t(df, \alpha/2) \cdot \sqrt{(s_S^2/n_S)+(s_N^2/n_N)}$

            $= (2.23) \sqrt{\left(397^2\right)+\left(255.7^2/11\right)}$

            $= (2.23)(131.155) = 292.476$

        c. $(\overline{x}_S - \overline{x}_N) \pm E = 393.8 \pm 292.5$
Step 5: <u>101.3 to 686.3</u>, the 0.95 confidence interval
        for $\mu_S - \mu_N$

OR   Confidence interval using calculated df:
Difference = mu (S. Dakota) - mu (N. Dakota)
Estimate for difference: 393.818
95% CI for difference: (121.761, 665.875)
DF = 22

WRITING HYPOTHESES FOR THE DIFFERENCE BETWEEN TWO MEANS

null hypothesis:

      $H_O: \mu_1 = \mu_2$ <u>or</u> $H_O: \mu_1 - \mu_2 = 0$ <u>or</u> $H_O: \mu_1 - \mu_2 = \#$

possible alternative hypotheses:

      $H_a: \mu_1 > \mu_2$ <u>or</u> $H_a: \mu_1 - \mu_2 > 0$ <u>or</u> $H_a: \mu_1 - \mu_2 > \#$
      $H_a: \mu_1 < \mu_2$ <u>or</u> $H_a: \mu_1 - \mu_2 < 0$ <u>or</u> $H_a: \mu_1 - \mu_2 < \#$
      $H_a: \mu_1 \neq \mu_2$ <u>or</u> $H_a: \mu_1 - \mu_2 \neq 0$ <u>or</u> $H_a: \mu_1 - \mu_2 \neq \#$

10.51   a. $H_O: \mu_1 - \mu_2 = 0$ vs. $H_a: \mu_1 - \mu_2 \neq 0$

     b. $H_O: \mu_1 - \mu_2 = 0 \ (\leq)$ vs. $H_a: \mu_1 - \mu_2 > 0$

     c. $H_O: \mu_N - \mu_S = 0 \ (\bullet)$ vs. $H_a: \mu_N - \mu_S < 0$
            or equivalently
        $H_O: \mu_S - \mu_N = 0 \ (\leq)$ vs. $H_a: \mu_S - \mu_N > 0$

     d. $H_O: \mu_M - \mu_F = 0$ vs. $H_a: \mu_M - \mu_F \neq 0$

10.53 st.err. $= \sqrt{(s_1^2 / n_1) + (s_2^2 / n_2)}$

 a. $\sqrt{(12/16) + (15/21)} = \underline{1.21}$

 b. $\sqrt{(0.054/8) + (0.087/10)} = \underline{0.1243}$

 c. $\sqrt{(2.8^2/16) + (6.4^2/21)} = \sqrt{2.44} = \underline{1.56}$

10.55 $t* = [(\overline{x}_2 - \overline{x}_1) - (\mu_2 - \mu_1)] / \sqrt{(s_2^2 / n_2) + (s_1^2 / n_1)}$

 $= [(1.66 - 1.43) - 0] / \sqrt{(0.29^2/21) + (0.18^2/9)} = \underline{2.64}$

Review the rules for calculating the p-value in: ES10-p433, SSM-p270, if necessary. Remember to use the t-distribution, therefore either Table 6, Table 7 or the computer/calculator will be used to find probabilities.
Review of the use of the tables can be found in: ES10-pp483-484, SSM-p305-306.

Use Table 6 (Appendix B, ES10-p813) with the smaller of df_1 or df_2 and the given α to find the critical value(s). Reviewing how to determine the test criteria in: ES10-p477-478, SSM-pp300, as it is applied to the t-distribution may be helpful.

10.57 a. $\alpha = 0.05$ df=15   b. $\alpha = 0.01$ df=26

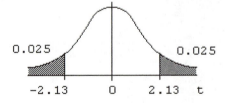

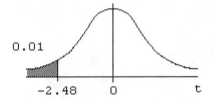

c. $\alpha = 0.10$ df=7   d. $\alpha = 0.05$ df=13

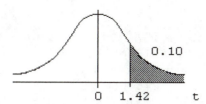

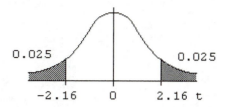

10.59 With the smaller degrees of freedom, df = 9, a higher calculated value is needed making it more difficult to reject H_O. This is due to the lack of reliability with a small sample.

Hypothesis Test for the Difference Between Two Means,
Independent Samples

Review the parts to a hypothesis test as outlined in: ES10-pp427&444-445, SSM-pp265&275, if needed. Slight changes will occur in:

  1. **the hypotheses**:   (see box before exercise 10.51)

  2. **the calculated test statistic**
  $$t = \frac{(\overline{x}_1 - \overline{x}_2) - (\mu_1 - \mu_2)}{\sqrt{(s_1^2 / n_1) + (s_2^2 / n_2)}}$$, using df = smaller of df_1 or df_2

  3. If H_O is rejected, a significant difference between the means is indicated.
      If H_O is not rejected, no significant difference between the means is indicated.

Any subscripts may be used on the hypotheses. Try to use letters that indicate the source. The form H_O: $\mu_2 - \mu_1 = 0$ (versus H_O: $\mu_1 = \mu_2$) is the preferred form since it establishes the order for subtraction that will be needed when calculating the test statistic.

Computer and/or calculator commands to perform a hypothesis test for the difference between two means can be found in ES10-pp571-572.

10.61 $t = (1.2 - 1.05)/\sqrt{((0.32^2/16) + (0.25^2/14))}$

$= 0.15/\sqrt{(0.0064 + 0.00446)} = 0.15/0.104 = 1.44$

p-value $= 2 \cdot P(t > 1.44|df=13)$
  Using computer/calculator, **P** = 0.1735
   Using Table 6, Appendix B, ES10-p813:
        $2(0.05 < \mathbf{P} < 0.10) = 0.10 < \mathbf{P} < 0.20$
    Using Table 7, Appendix B, ES10-p814:
        $[0.077 < \frac{1}{2}\mathbf{P} < 0.093]; \; 0.154 < \mathbf{P} < 0.186$
OR P $= 2P(t > 1.44|df = 27)$;
   Using computer/calculator, **P** = 0.1614

Hint for writing the hypotheses for exercise 10.63
Look at the second to last sentence of the exercise; "Do these results show … with an ESG to help them is significantly <u>greater than</u> those not using an ESG?''. The words "greater than'' indicates a (>), which can only go in the alternative hypothesis. The negation becomes "not greater than" and the null hypothesis would be written with an equality sign (=) and (•) after it. The direction of subtraction,
$(\mu_1 - \mu_2)$ will give a positive differenc.

10.63   Step 1: a. The difference between the mean scores of students   using an electronic study guide to help them learn accounting principles and those not using one
           b. $H_o: \mu_1 - \mu_2 = 0$
              $H_a: \mu_1 - \mu_2 > 0$
      Step 2: a. normality assumed, CLT with $n_1 = 38$ and $n_2 = 36$.
           b. t   c. $\alpha = 0.01$
      Step 3: a. sample information given in exercise
           b. $t* = [(\bar{x}_1 - \bar{x}_2) - (\mu_1 - \mu_2)]/\sqrt{(s_E^2/n_E) + (s_C^2/n_C)}$

              $= [(79.6 - 72.8)-0]/[\sqrt{(6.9^2/38) + (7.6^2/36)}]$

              $= 4.02$
      Step 4: -- using p-value approach --------------------
           a. **P** $= P(t > 4.02|df =35)$;
              Using computer/calculator, **P** = 0.0001
               Using Table 6, Appendix B, ES10-p813:
              P < 0.005
               Using Table 7, Appendix B, ES10-p814:
              P = 0+

                    OR P = P(t > 4.02|df = 70);
                          Using computer/calculator, **P** = 0.0001
                    b. **P** < α
                 -- using classical approach ------------------
                    a. critical region: t ≥ 2.44
                    b. t* is in the critical region

        Step 5: a. Reject H$_O$
                b. At the 0.01 level of significance, there is
                   sufficient evidence to conclude the ESG students
                   are doing better in accounting principles than
                   those not using an ESG.

10.65 No, the brunettes overall grade average was lower than the
        overall grade average for the blondes. Lauren could say
        that there is no difference between blondes and brunettes
        intelligence and do a hypothesis test and hope the means
        are not significantly different.

10.67 a. $\sqrt{[((1.589)^2 / 15) + ((3.023)^2 / 13)]}$ = $\sqrt{[0.168328 + 0.702964]}$
               = $\sqrt{0.871292}$ = 0.9334 = 0.933

        b. (9.6667 - 3.1538)/0.933 = 6.5129/0.933 = 6.9806

        c. 12 to 26
        d. It was calculated using a computer program.
        e. 12, the number of degrees of freedom associated
            with the smaller sample
        f. t* = 6.98 is significant for a very small level of
            significance, p-value is less than 0.0005.

10.69 a. Verify - answers given in exercise.

        b. **P** = 2P(t > 0.59|df = 20);   (Table 7)
              2(0.277 < ½**P** < 0.312),   0.554 < **P** < 0.624

        c. Using Table 7: **P** = 2P(t > 0.59|df = 12);
              2(0.280 < ½**P** < 0.313),   0.560 < **P** < 0.626

10.71   Step 1: a. The difference between mean weight gained on two
                   diets, A and B, $\mu_B - \mu_A$
                b. H_o: $\mu_B - \mu_A = 0$
                   H_a: $\mu_B - \mu_A > 0$
          Step 2: a. normality assumed
                b. t   c. $\alpha = 0.05$
          Step 3: a. $n_A = 10$, $\overline{x}_A = 10.0$, $s_A^2 = 10.44$
                   $n_B = 10$, $\overline{x}_B = 14.7$, $s_B^2 = 46.01$

                b. $t\overset{*}{.} = [(\overline{x}_B - \overline{x}_A) - (\mu_B - \mu_A)] / \sqrt{(s_B^2 / n_B) + (s_A^2 / n_A)}$
                   $= [(14.7 - 10.0) - 0] / [\sqrt{(10.44/10) + (46.01/10)}]$
                   $= 1.98$
          Step 4: -- using p-value approach --------------------
                a. $\mathbf{P} = P(t > 1.98 | df = 9)$;
                   Using computer/calculator, $\mathbf{P} = 0.0395$
                   Using Table 6, Appendix B, ES10-p813:
                   $0.025 < \mathbf{P} < 0.05$
                   Using Table 7, Appendix B, ES10-p814:
                   $0.037 < \mathbf{P} < 0.047$
                OR $\mathbf{P} = P(t > 1.98 | df = 12)$;
                   Using computer/calculator, $\mathbf{P} = 0.036$
                b. $\mathbf{P} < \alpha$
                -- using classical approach ------------------
                a. critical region: $t \geq 1.83$
                b. $t*$ is in the critical region

          Step 5: a. Reject H_o
                b. At the 0.05 level of significance, there is
                   sufficient evidence to show that Diet B had a
                   significantly higher weight gain.

10.73   Step 1: a. The difference between mean real estate
                   transaction in Penfield and Perinton, μ_{PN} -
                   μ_{PR}
                b. H_o: $\mu_{PN} - \mu_{PR} = 0$
                   H_a: $\mu_{PN} - \mu_{PR} > 0$
          Step 2: a. normality assumed
                b. t   c. $\alpha = 0.10$
          Step 3: a. $n_{PR} = 10$, $\overline{x}_{PR} = 158940$, $s_{PR} = 67421$
                   $n_{PN} = 5$, $\overline{x}_{PN} = 206680$, $s_{PN} = 62332$

                b. $t* = [(\overline{x}_{PN} - \overline{x}_{PR}) - (\mu_{PN} - \mu_{PR})] / \sqrt{(s_{PN}^2/n_{PN}) + (s_{PR}^2/n_{PR})}$

                   $= [(206680 - 158940) - 0] / [\sqrt{(62332^2 / 5) + (67421^2 / 10)}]$
                   $= 1.36$

Step 4: -- using p-value approach --------------------
a. $P = P(t > 1.36 | df = 4)$;
Using computer/calculator, $P = 0.1227$
Using Table 6, Appendix B, ES10-p813:
$0.10 < P < 0.25$
Using Table 7, Appendix B, ES10-p814:
$0.117 < P < 0.132$
OR $P = P(t > 1.36 | df = 8)$;
Using computer/calculator, $P = 0.105$
b. $P > \alpha$
-- using classical approach ------------------
a. critical region: $t \geq 1.53$
b. $t*$ is in the noncritical region

Step 5: a. Fail to reject H_o
b. At the 0.10 level of significance, there is not
sufficient evidence to show that the level of
housing in Penfield is greater than in Perinton.

10.75 Everybody will get different results, but they can all
be expected to look very similar to the following.
a.   Minitab commands:
Choose: Calc > Random > Normal to generate both
distributions into C1 and C2
Choose: Stat > Basic Statistics > Display
Descriptive Statistics for the means and standard
deviations.
Choose: Graph > Histogram
Use cutpoints as noted for both distributions.
Excel commands:
Choose: Tools > Data Analysis > Random Number
Generation > Normal to generate both
distributions in columns A and B.
Choose: Tools > Data Analysis > Descriptive
Statistics for the means and standard deviations.
Choose: Tools > Data Analysis > Histogram
Use cutpoints as noted for both distributions.

N(100,20)

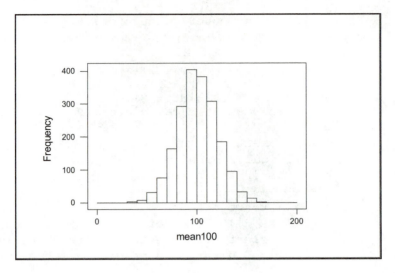

N(120,20)

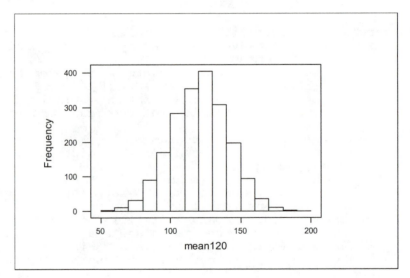

b. The sampling distribution is expected to be normal
    in shape with a mean of 20 (120-100) and have a
    standard error of $\sqrt{\dfrac{20^2}{8}+\dfrac{20^2}{8}}$ or 10.

c.   <u>Minitab commands</u>:

Choose: Calc > Random > Normal
Generate 100 rows into C3-C10 with mean = 100 and
standard deviation = 20

Choose: Calc > Row Statistics
Select Mean for C3-C10 and store in C11.
Repeat above for 100 rows into C12-C19 with mean =
120
and standard deviation = 20. Also select Mean for
C12-C19 and store in C20.
Choose: Calc > Calculator
Store in C21 the expression: C20 - C11

<u>Excel commands</u>:

Choose: Tools > Data Analysis > Random Number
Generation > Normal
Generate 100 rows into columns C through J with
mean 100 and standard deviation = 20.

Choose: Insert function > All > Average
Find the average for C1-J1 and store in K1. Drag
down for other averages.

Repeat above for 100 rows into L through S with
mean 120 and standard deviation = 20. Also
calculate means into column T.

Choose: Edit formula (=)
Find T1 - K1 and store in U1. Drag down for other
subtractions.

d.   <u>Minitab commands</u>:

Choose: Stat > Basic Statistics > Display
Descriptive Statistics for C21

Choose: Graph > Histogram
Use cutpoints -20:60/10 for C21

<u>Excel commands</u>:

Choose: Tools > Data Analysis > Descriptive
Statistics for column U.

Choose: Tools > Data Analysis > Histogram
Use classes from -20 to 60 in increments of 10 for
column U.

100 values for the difference between two sample means:

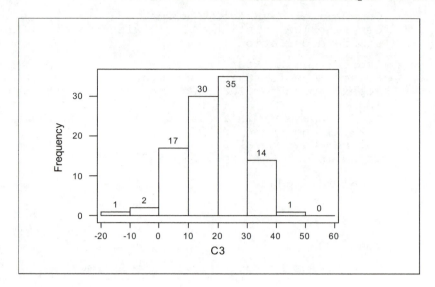

e. For the empirical sampling distribution, the mean
   is 19.51 and the standard error is 10.71. There is
   65%, 96% and 99% of the values within one, two and
   three standard errors of the expected mean of 20.
   This seems to agree closely with the empirical
   rule, thus suggesting a normal distribution
   occurred.

f. You can expect very similar results to occur on
   repeated trials.

10.77 Everybody will get different results, but they can all be
expected to look very similar to the results found in
exercise 10.75. It turns out that the t* statistic is very
"robust", meaning ''it works quite well even when the
assumptions are not met.'' This it one of the reasons the
t-test for the mean and the t-test for the difference
between to means are such important tests.

Estimating $(p_1 - p_2)$ - the difference between two population
proportions - independent samples
(large samples)

1. Point estimate: $p_1' - p_2'$

2. Confidence interval:

$$(p_1' - p_2') \pm z(\alpha/2) \cdot \sqrt{(p_1' \cdot q_1' / n_1) + (p_2' \cdot q_2' / n_2)}$$

$\uparrow$	$\uparrow$	$\uparrow$
point estimate	confidence coefficient	estimated standard error

Maximum error of the estimate

Computer and/or calculator commands to construct a confidence
interval for the difference between two proportions can be found in
ES10-p584.

10.79 $x = \underline{75}$, $n = \underline{250}$,
$p' = x/n = 75/250 = \underline{0.30}$, $q' = 1 - p' = 1 - 0.30 = \underline{0.70}$

10.81   a. $\sqrt{(p_1' \cdot q_1' / n_1) + (p_2' \cdot q_2' / n_2)} =$

$\sqrt{(0.8 \cdot 0.2 / 40) + (0.8 \cdot 0.2 / 50)} = \sqrt{0.0072} = 0.085$

b. $\sqrt{(p_1' \cdot q_1' / n_1) + (p_2' \cdot q_2' / n_2)} =$

$\sqrt{(0.6 \cdot 0.4 / 33) + (0.65 \cdot 0.35 / 38)} = \sqrt{0.013259} = 0.115$

10.83   Step 1: The difference in proportions of nurses who
experienced a change in position based on their
participation in a program, $p_w - p_n$
Step 2: a. n's > 20,   np's and nq's all > 5
b. z   c. $1 - \alpha = 0.99$
Step 3: $n_w = 341$, $x_w = 87$, $p_w' = 87/341 = 0.255$,

$q_w' = 1 - 0.255 = 0.745$

$n_n = 40$, $x_n = 9$, $p_n' = 9/40 = 0.225$,

$q_n' = 1 - 0.225 = 0.775$

$p_w' - p_n' = 0.255 - 0.225 = 0.03$

Step 4: a. $\alpha/2 = 0.01/2 = 0.005$; $z(0.005) = 2.58$

  b. $E = z(\alpha/2) \cdot \sqrt{(p'_w \cdot q'_w / n_w) \ + \ (p'_n \cdot q'_n / n_n)}$

   $= 2.58 \cdot \sqrt{(0.255)(0.745)/341 + (0.225)(0.775)/40}$

   $= (2.58)(0.07) = 0.18$

  c. $(p'_w - p'_n) \pm E = 0.03 \pm 0.18$

Step 5: <u>−0.15 to 0.21</u>, the 0.99 interval for $p_W - p_n$

10.85 Step 1: The difference between proportions, $p_m - p_f$
Step 2: a. n's > 20, np's and nq's all > 5
  b. z   c. $1-\alpha = 0.95$
Step 3: sample information given in exercise

   $p'_m = 0.66$, $q'_m = 1 - 0.66 = 0.34$

   $p'_f = 0.37$, $q'_f = 1 - 0.37 = 0.63$

   $\mathbf{p'_m - p'_f} = 0.66 - 0.37 = 0.29$

Step 4: a. $\alpha/2 = 0.05/2 = 0.025$; $z(0.025) = 1.96$

  b. $E = z(\alpha/2) \cdot \sqrt{(p'_m \cdot q'_m / n_m) + (p'_f \cdot q'_f / n_f)}$

   $= (1.96) \cdot \sqrt{(0.66)(0.34)/200 + (0.37)(0.63)/199}$

   $= (1.96)(0.04789) = 0.094$

  c. $(p'_A - p'_B) \pm E = 0.29 \pm 0.094$

Step 5: <u>0.196 to 0.384</u>, the 0.95 confidence interval
  for $p_m - p_f$

The confidence interval constructed above for the difference
between two proportions does not include zero. This
indicates that the two proportions are not equal to each
other, at the 0.95 confidence level. Being a positive
difference indicates that the male teenagers' proportion is
significantly greater than the female teenagers' proportion.
In exercise 9.74, it was written that: The proportion
of males that 'have ever gambled' is greater than the
proportion of females. The entire confidence interval for
females (0.303 to 0.437) is below the lowest confidence
limit for the males (0.594 to 0.726).

10.87 Step 1: The difference in proportions of defectives parts
     produced by two machines, $p_1 - p_2$
Step 2: a. n's > 20, np's and nq's all > 5
  b. z   c. $1-\alpha = 0.90$

Step 3: $n_1 = 150$, $x_1 = 12$, $p_1' = 12/150 = 0.08$,

$\quad q_1' = 1 - 0.08 = 0.92$

$\quad n_2 = 150$, $x_2 = 6$, $p_2' = 6/150 = 0.04$,

$\quad q_2' = 1 - 0.04 = 0.96$

$\quad p_1' - p_2' = 0.08 - 0.04 = 0.04$

Step 4: a. $\alpha/2 = 0.10/2 = 0.05$; $z(0.05) = 1.65$

$\quad$ b. $E = z(\alpha/2) \cdot \sqrt{(p_1' \cdot q_1' / n_1) + (p_2' \cdot q_2' / n_2)}$

$\quad\quad = 1.65 \sqrt{(0.08 \cdot 0.92/150) + (0.04 \cdot 0.96/150)}$

$\quad\quad = (1.65)(0.027) = 0.04$

$\quad$ c. $(p_1' - p_2') \pm E = 0.04 \pm 0.04$

Step 5: <u>0.000 to 0.080</u>, the 0.90 interval for $p_1 - p_2$

WRITING HYPOTHESES FOR THE DIFFERENCE BETWEEN TWO PROPORTIONS

$\quad$ a) null hypothesis:

$$H_o: p_1 = p_2 \quad \underline{or} \quad p_1 - p_2 = 0$$

$\quad$ b) possible alternative hypotheses:

$$H_a: p_1 > p_2 \quad \underline{or} \quad H_a: p_1 - p_2 > 0$$
$$H_a: p_1 < p_2 \quad \underline{or} \quad H_a: p_1 - p_2 < 0$$
$$H_a: p_1 \neq p_2 \quad \underline{or} \quad H_a: p_1 - p_2 \neq 0$$

10.89 a. $H_o: p_m - p_w = 0$ vs. $H_a: p_m - p_w \neq 0$

$\quad$ b. $H_o: p_b - p_g = 0$ ($\leq$) vs. $H_a: p_b - p_g > 0$

$\quad$ c. $H_o: p_c - p_{nc} = 0$ ($\leq$) vs. $H_a: p_c - p_{nc} > 0$

Hypothesis Test for the Difference Between Two Proportions,
Independent Samples (Large Samples)

Review parts to a hypothesis test as outlined in: ES10-pp427&444-445, SSM-pp265&275, if needed. Changes will occur in:

$\quad$ 1. **the hypotheses**: (see box before exercise 10.89) . . .

2. the calculated test statistic

$$z = \frac{(p_1' - p_2') - (p_1 - p_2)}{\sqrt{p_p'q_p'(\dfrac{1}{n_1} + \dfrac{1}{n_2})}} \quad , \text{ where } p_1' = \frac{x_1}{n_1} \quad , \quad p_2' = \frac{x_2}{n_2}$$

$$p_p' = \frac{x_1 + x_2}{n_1 + n_2} \quad \text{and} \quad q_p' = 1 - p_p'$$

3. If H_O is rejected, a significant difference in proportions is indicated. If H_O is not rejected, no significant difference is indicated.

NOTE: The sampling distribution of $p_1' - p_2'$ is approximately normally distributed with a mean $(p_1 - p_2)$ and a standard error of $\sqrt{p_1 q_1 / n_1 + p_2 q_2 / n_2}$, if n_1 and n_2 are sufficiently large. Since H_O is assumed to be true, $p_1 - p_2$ is considered equal to 0. Since p_1 and p_2 are also unknown, the best estimate for $p(p = p_1 = p_2)$ is a pooled estimate p_p'.

Computer and/or calculator commands to perform a hypothesis test for the difference between two proportions can be found in ES10-p587.

10.91 $\quad p_p' = (x_E + x_R)/(n_E + n_R) = (15 + 25)/(250 + 275)$
$$= 40/525 = \underline{0.076}$$
$\quad q_p' = 1 - p_p' = 1 - 0.076 = \underline{0.924}$

10.93 Rewrite the alternative hypothesis for easier understanding:
$\quad H_a: p_R - p_E > 0$

$\quad p_R' = 25/275 = 0.091, \quad p_E' = 15/250 = 0.06$

$\quad z^* = (p_R' - p_E')/\sqrt{(p_p')(q_p')[(1/n_R)+(1/n_E)]}$
$$= (0.091 - 0.06)/\sqrt{(0.076)(0.924)[(1/275)+(1/250)]}$$
$$= 0.031/0.0232 = 1.34$$

$\quad \mathbf{P} = P(z > 1.34) = (0.5000 - 0.4099) = \underline{0.0901}$

10.95   a. α = 0.05   b. α = 0.05

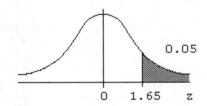   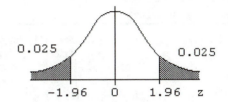

c. α = 0.04   d. α = 0.01

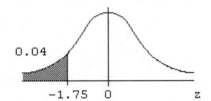   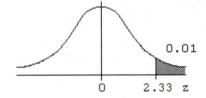

Hint for writing the hypotheses for exercise 10.97

Look at the last sentence of the exercise; "Is there sufficient evidence to show a <u>difference</u> in the effectiveness of the <u>2 image campaigns</u>...?" The words "2 image campaigns" imply 2 populations, namely the citizens exposed to a conservative campaign and citizens exposed to a moderate campaign. The results are given in the form of proportions, therefore a difference between 2 proportions is suggested. The word "difference" indicates a <u>not equal to</u>. Therefore, the alternative is <u>not equal to</u> ($\neq$). The negation becomes "equal to" and the null hypothesis would be written with an equality sign (=).

10.97 Step 1: a. The difference in the proportions for the effectiveness of two campaign images, $p_m - p_c$

b. H_o: $p_m - p_c = 0$

H_a: $p_m - p_c \neq 0$

Step 2: a. n's > 20, np's and nq's all > 5

b. z   c. $\alpha = 0.05$

Step 3: a. $n_m = 100$, $p'_m = 0.50$, $n_c = 100$, $p'_c = 0.40$

$p'_p = (x_m + x_c)/(n_m + n_c) = (50+40)/(100+100)$

$= 0.45$

$q'_p = 1 - p'_p = 1.000 - 0.45 = 0.55$

b. $z = [(p'_m - p'_c) - (p_m - p_c)] / \sqrt{(p'_p)(q'_p)[(1/n_m) + (1/n_c)]}$

$z^* = (0.50 - 0.40)/\sqrt{(0.45)(0.55)[(1/100) + (1/100)]}$

$= 0.10/0.0704 = 1.42$

Step 4: -- using p-value approach --------------------

a. **P** = 2P(z > 1.42);

Using Table 3, Appendix B, ES10-p810:

P = 2(0.5000 - 0.4222) = 2(0.0778) = 0.1556

Using Table 5, Appendix B, ES10-p812:

2(0.0735 < ½**P** < 0.0808);   0.1470 < **P** < 0.1616

a. **P** > α

-- using classical approach ------------------

a. Critical region: z ≤ -1.96 and z ≥ 1.96

b. z* falls in the noncritical region

Step 5: a. Fail to reject H_o

b. There is not sufficient evidence to show a difference, at the 0.05 level.

10.99 a. Answers may vary but it there does seem to be a difference in the proportions of men and women who say 'Executives are paid too much'. 60% is a well over a majority and the 50% is not quite a majority.

b. Step 1: a. The difference in the proportion of men and women who say that 'Executives are paid too much', $p_w - p_m$

   b. H_o: $p_w - p_m = 0$

   H_a: $p_w - p_m \neq 0$

Step 2: a. np's and nq's all > 5

   b. z   c. $\alpha = 0.05$

Step 3: a. $n_w = 20$, $p'_w = 0.60$; x = (20)(0.60) = 12

   $n_m = 20$, $p'_m = 0.50$; x = (20)(0.50) = 10

   $p'_p = (x_w + x_m)/(n_w + n_m) = (12+10)/(20+20) = 0.55$

   $q'_p = 1 - p'_p = 1.000 - 0.55 = 0.45$

   b. $z = [(p'_w - p'_m) - (p_w - p_m)]/\sqrt{(p'_p)(q'_p)[(1/n_w) + (1/n_m)]}$

   $z* = (0.60 - 0.50)/\sqrt{(0.55)(0.45)[(1/20) + (1/20)]}$

   $= 0.10/0.1573 = 0.64$

Step 4: -- using p-value approach --------------------

   a. $P = 2P(z* > 0.64)$;

   Using Table 3, Appendix B, ES10-p810:

   $P = 2(0.5000 - 0.2389) = 2(0.2611) = 0.5222$

   Using Table 5, Appendix B, ES10-p812:

   $(0.2578 < \frac{1}{2}P < 0.2743)$; $0.5156 < P < 0.5486$

   a. $P > \alpha$

   -- using classical approach ------------------

   b. Critical region: $z \leq -1.96$ and $z \geq 1.96$

   b. z* falls in the noncritical region

Step 5: a. Fail to reject H_o

   b. There is not sufficient evidence to show a difference, at the 0.05 level of significance.

c. Step 1: a. The difference in the proportion of men and women who say that 'Executives are paid too much', $p_w - p_m$

   b. H_o: $p_w - p_m = 0$

   H_a: $p_w - p_m \neq 0$

Step 2: a. n's > 20, np's and nq's all > 5

   b. z   c. $\alpha = 0.05$

Step 3: a. $n_w = 500$, $p'_w = 0.60$; $x = (500)(0.60) = 300$

$n_m = 500$, $p'_m = 0.50$; $x = (500)(0.50) = 250$

$p'_p = (x_w + x_m)/(n_w + n_m) = (300+250)/(500+500) = 0.55$

$q'_p = 1 - p'_p = 1.000 - 0.55 = 0.45$

b. $z = [(p'_w - p'_m) - (p_w - p_m)]/\sqrt{(p'_p)(q'_p)[(1/n_w) + (1/n_m)]}$

$z* = (0.60 - 0.50)/\sqrt{(0.55)(0.45)[(1/500) + (1/500)]}$

$= 0.10/0.03146 = 3.18$

Step 4: -- using p-value approach --------------------

a. $P = 2P(z* > 3.18)$;

Using Table 3, Appendix B, ES10-p810:

$P = 2(0.5000 - 0.4993) = 2(0.0007) = 0.0014$

Using Table 5, Appendix B, ES10-p812:

$(0.0007 < ½P < 0.0008)$; $0.0014 < P < 0.0016$

b. $P < \alpha$

-- using classical approach ------------------

c. Critical region: $z \leq -1.96$ and $z \geq 1.96$

b. $z*$ falls in the critical region

Step 5: a. Reject H_O

b. There is sufficient evidence to show a
difference, at the 0.05 level of significance.

d. Even with a 10% difference, it takes a reasonablely large
sample size to show significance.

10.101 Step 1: a. The difference in the proportion of reject
product between two methods, $p_2 - p_1$

b. H_O: $p_2 - p_1 = 0$

H_a: $p_2 - p_1 \neq 0$

Step 2: a. n's > 20,   np's and nq's all > 5

b. z c. $\alpha = 0.05$

Step 3: a. $n_2 = 992$, $p'_2 = 26/992 = 0.0262$,

$n_1 = 320$, $p'_1 = 4/320 = 0.0125$

$p'_p = (x_2 + x_1)/(n_2 + n_1) = (26+4)/(992+320)$

$= 0.0229$

$q'_p = 1 - p'_p = 1.000 - 0.0229 = 0.9771$

b. $z = [(p'_2 - p'_1) - (p_2 - p_1)]/\sqrt{(p'_p)(q'_p)[(1/n_2) + (1/n_1)]}$

$z* = (0.0262 - 0.0125)/\sqrt{(0.0229)(0.9771)[(1/992) + (1/320)]}$

$= 0.0137/0.0096 = 1.43$

Step 4: -- using p-value approach --------------------
    a. $P = 2P(z > 1.43)$;
       Using Table 3, Appendix B, ES10-p810:
       $P = 2(0.5000 - 0.4236) = 2(0.0764) = 0.1528$
       Using Table 5, Appendix B, ES10-p812:
       $(0.0735 < \frac{1}{2}P < 0.0808)$; $0.1470 < P < 0.1616$
    b. $P > \alpha$
  -- using classical approach ------------------
  a. Critical region: $z \le -1.96$ and $z \ge 1.96$
  b. $z*$ falls in the noncritical region

Step 5: a. Fail to reject H_O
    b. There is not sufficient evidence to show a
       difference, at the 0.05 level of significance.

10.103 Step 1: a. The difference in the proportion of men and women
    who think it is OK for women to make marriage
    proposals to men, $p_M - p_W$
  b. H_O: $p_M - p_W = 0$
    H_a: $p_M - p_W \ne 0$
Step 2: a. n's > 20, np's and nq's all > 5
  b. z   c. $\alpha = 0.05$
Step 3: a. $n_M = 250$, $p'_M = 0.63$
      $n_W = 250$, $p'_W = 0.55$
  $p'_p = (x_M + x_W)/(n_M + n_W) = (158+138)/(250+250) = 0.592$
      $q'_p = 1 - p'_p = 1.000 - 0.592 = 0.408$
  b. $z = [(p'_M - p'_W) - (p_M - p_W)]/\sqrt{(p'_p)(q'_p)[(1/n_M) + (1/n_W)]}$
  $z* = (0.63 - 0.55)/\sqrt{(0.592)(0.408)[(1/250) + (1/250)]}$
      $= 0.08/0.0439578 = 1.82$
Step 4: -- using p-value approach --------------------
    a. $P = 2P(z* > 1.82)$;
       Using Table 3, Appendix B, ES10-p810:
       $P = 2(0.5000 - 0.4656) = 2(0.0344) = 0.0688$
       Using Table 5, Appendix B, ES10-p812:
       $(0.0322 < \frac{1}{2}P < 0.0359)$; $0.0644 < P < 0.0718$
    b. $P > \alpha$
  -- using classical approach ------------------
  a. Critical region: $z \le -1.96$ and $z \ge 1.96$
  b. $z*$ falls in the noncritical region

Step 5: a. Fail to reject H_O
b. There is not sufficient evidence to show a difference, at the 0.05 level.

b. Steps 1 & 2 are same as in part a.
   Step 3: a. $n_M = 500$, $p'_M = 0.63$

   $$n_W = 500, \ p'_W = 0.55$$

   $$p'_p = (x_M + x_W)/(n_M + n_W) = (315+275)/(500+500) = 0.59$$

   $$q'_p = 1 - p'_p = 1.000 - 0.59 = 0.41$$

   b. $z = [(p'_M - p'_W)-(p_M-p_W)]/\sqrt{(p'_p)(q'_p)[(1 / n_M) \ + \ (1 / n_W)]}$

   $$z* = (0.63 - 0.55)/\sqrt{(0.59)(0.41)[(1 / 500) + (1 / 500)]}$$

   $$= 0.08/0.0311 = 2.57$$

   Step 4: -- using p-value approach --------------------
   a. **P** $= 2P(z* > 2.57)$;
      Using Table 3, Appendix B, ES10-p810:
      P $= 2(0.5000 - 0.4949) = 2(0.0051) = 0.0102$
      Using Table 5, Appendix B, ES10-p812:
      $(0.0047 < \frac{1}{2}\textbf{P} < 0.0054)$; $0.0094 < \textbf{P} < 0.0108$
   b. **P** $< \alpha$
   -- using classical approach ------------------
   a. Critical region: $z \leq -1.96$ and $z \geq 1.96$
   b. $z*$ falls in the critical region

   Step 5: a. Reject H_O
   b. There is sufficient evidence to show a difference, at the 0.05 level.

c. P = 0.05 for a 2-tailed test - 0.025 on each tail
   $z(0.025) = 1.96$
   $1.96 = (0.63 - 0.55)/\sqrt{(0.59)(0.41)[(1 / n) + (1 / n)]}$
   $1.96 = (0.08)/\sqrt{(0.4838) / n}$
   $1.96(\sqrt{(0.4838) / n}) = 0.08$

   $\sqrt{(0.4838) / n} = 0.0408$

   $0.4838/n = 0.00166$

   $0.4838 = 0.00166n$

   $0.4838/0.00166 = n$

   $290.4 = n = \underline{291}$

WRITING HYPOTHESES FOR THE RATIO BETWEEN
TWO STANDARD DEVIATIONS OR VARIANCES

null hypothesis:

$$H_O: \sigma_1 = \sigma_2 \quad \underline{or} \quad H_O: \sigma_1^2 = \sigma_2^2 \quad \underline{or} \quad H_O: \sigma_1^2 / \sigma_2^2 = 1$$

possible alternative hypotheses:

$$H_a: \sigma_1 > \sigma_2 \quad \underline{or} \quad H_O: \sigma_1^2 > \sigma_2^2 \quad \underline{or} \quad H_O: \sigma_1^2 / \sigma_2^2 > 1$$

$$H_O: \sigma_1 \neq \sigma_2 \quad \underline{or} \quad H_O: \sigma_1^2 \neq \sigma_2^2 \quad \underline{or} \quad H_O: \sigma_1^2 / \sigma_2^2 \neq 1$$

$$H_O: \sigma_2 > \sigma_1 \quad \underline{or} \quad H_O: \sigma_2^2 > \sigma_1^2 \quad \underline{or} \quad H_O: \sigma_2^2 / \sigma_1^2 > 1 \quad **$$

**Note change for "less than", reverse order.

10.105 a. $H_O: \sigma_A^2 = \sigma_B^2 \qquad$ vs. $\qquad H_a: \sigma_A^2 \neq \sigma_B^2$

b. $H_O: \sigma_I / \sigma_{II} = 1 \quad$ vs. $\quad H_a: \sigma_I / \sigma_{II} > 1$

c. $H_O: \sigma_A^2 / \sigma_B^2 = 1 \quad$ vs. $\quad H_a: \sigma_A^2 / \sigma_B^2 \neq 1$

d. $H_O: \sigma_C^2 / \sigma_D^2 = 1 \quad$ vs. $\quad H_a: \sigma_C^2 / \sigma_D^2 < 1$
or equivalently,
$H_O: \sigma_D^2 / \sigma_C^2 = 1 \quad$ vs. $\quad H_a: \sigma_D^2 / \sigma_C^2 > 1$

10.107 Divide both sides of the original inequality by σ_p^2

F-DISTRIBUTION

Key facts about the F-distribution:
1. The total area under the F-distribution is 1.
2. It is zero or positively valued.
3. The shape is skewed right (much like χ^2).
4. A different curve exists for each pair of sample sizes.
5. Critical values are determined based on α and degrees of freedom in the numerator (df_n) and degrees of freedom in the denominator (df_d).
6. Degrees of freedom = df = n - 1.   . . .

Notation: $F(df_n, df_d, \alpha) =$

$\qquad F(df$ for numerator, df for denominator, area to the right$)$
$\qquad\quad \uparrow \qquad\qquad \uparrow \qquad\qquad\qquad \uparrow \qquad\qquad\qquad\quad \uparrow$
$\qquad$ Table 9 $\qquad$ column id # $\qquad$ row id # $\qquad\qquad$ Table 9a,b,c

$\quad$ ex: $F(10,12,0.05)$ means $df_n = 10$(column), $df_d = 12$(row) and
$\qquad\qquad\qquad\qquad \alpha = 0.05$ (Table 9A)
$\qquad$ Using Table 9A, $F(10,12,0.05) = 2.75 \qquad$ (df = n-1 for each
$\qquad\qquad$ sample)

Explore the F- distribution using the Chapter 10 Skillbuilder
Applets 'Properties of F-distribution' and 'F-distribution
Probabilities'.

10.109 a. $F(9,11,0.025)$ $\qquad$ b. $F(24,19,0.01)$
$\qquad$ c. $F(8,15,0.01)$ $\qquad$ d. $F(15,9,0.05)$

10.111 a. 2.51 $\qquad$ b. 2.20 $\qquad$ c. 2.91 $\qquad$ d. 4.10
$\qquad$ e. 2.67 $\qquad$ f. 3.77 $\qquad$ g. 1.79 $\qquad$ h. 2.99

Estimate the p-value using Tables 9A, B or C, (Appendix B, ES10-pp-816-821).

$\qquad P = P(F > F^*|df_n, df_d)$

Locate critical values for the given df_n and df_d on each of the
Tables. Compare F^* to each and give an interval estimate of P
using one or two of the following values: 0.05, 0.025, 0.01.
The p-value can be calculated using the cumulative probability
commands in ES10-pp594-595.

Draw a diagram of a F-curve and label the regions with the given
information.
For a <u>one-sided test</u>, be sure that the alternative hypothesis has a
"greater than", otherwise adjust the order of the numerator and
denominator (i.e. $\sigma_1 < \sigma_2 \Rightarrow \sigma_2 > \sigma_1$).
For a <u>two-sided test</u>, divide α by 2 and use the largest sample
variance in the numerator for the calculated test statistic. Use
Table 9A, B or C, based on $\alpha/2$.

10.113 $F(6,9,0.05) = \underline{3.37}$

10.115 $F^* = s_1^2 / s_2^2 = (3.2)^2/(2.6)^2 = 1.515 = \underline{1.52}$

10.117 $F^* = s_1^2 / s_2^2 = 14.44/29.16 = \underline{0.495}$
The smaller variance is in the numerator.

Hypothesis Test for the Ratio Between Two Standard Deviations
(or Variances), Independent Samples

Review the parts to a hypothesis test as outlined in: ES10-pp427&444-445, SSM-pp265&275, if needed. Slight changes will occur in:

1. **the hypotheses**: (see box before exercise 10.105)

2. **the calculated test statistic**
   $F = s_1^2 / s_2^2$, $df_1 = df_n = n_1 - 1$, $df_2 = df_d = n_2 - 1$

3. If H_O is rejected, a significant difference between the standard deviations (variances) is indicated.
   If H_O is not rejected, no significant difference between the standard deviations (variances) is indicated.

Use subscripts on the sample (or population) variables that identify the source.

Estimate the p-value using Tables 9A, B or C, (Appendix B, ES10-pp-816-821).
   $P = P(F > F^* | df_n, df_d)$

Locate critical values for the given df_n and df_d on each of the Tables. Compare F^* to each and give an interval estimate of P using one or two of the following values: 0.05, 0.025, 0.01.

The p-value can be calculated using the cumulative probability commands in ES10-pp594-595.

10.119 Step 1: a. The ratio of the variances for two ovens
   b. H_O: $\sigma_k^2 = \sigma_m^2$
      H_a: $\sigma_k^2 \neq \sigma_m^2$
Step 2: a. normality indicated, independence exists
   b. F   c. $\alpha = 0.02$
Step 3: a. $n_m = 16$, $s_m^2 = 2.4$, $n_k = 12$, $s_k^2 = 3.2$
   b. $F^* = s_k^2/s_m^2 = 3.2/2.4 = 1.33$

Step 4: -- using p-value approach --------------------
    a. $P = 2P(F > 1.33 | df = 11,15)$;
      Using Tables 9, Appendix B, ES10-pp-816-821:
       $P > 2(0.05)$; $P > 0.10$
      Using computer: 0.298
    b. $P > \alpha$
-- using classical approach -----------------
  a. $F(11, 15, 0.01) = 3.73$

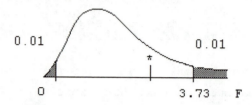

    b. F* falls in the noncritical region, see Step 4a

Step 5: a. Fail to reject H_O
    b. At the 0.02 level of significance, there is not
      sufficient evidence to conclude that a
      difference in variances exist for the two ovens.

Computer and/or calculator commands to perform a hypothesis test
between two standard deviations or two variances can be found in
ES10-p599.

The p-value can be calculated using the cumulative probability
commands in ES10-pp594-595.

10.121 Step 1: a. The ratio of the variances for sugar beet and
        sugar can cane sucrose percentages
      b. $H_O: \sigma_{sc} = \sigma_{sb}$

      $H_a: \sigma_{sc} \neq \sigma_{sb}$
Step 2: a. normality indicated, independence exists
    b. F   c. $\alpha = 0.05$
Step 3: a. $n_{sc} = 12$, $s_{sc} = 0.912$, $n_{sb} = 15$, $s_{sb} = 0.862$
    b. $F* = s_{sc}^2 / s_{sb}^2 = (0.912)^2 / (0.862)^2 = 1.12$

Step 4: -- using p-value approach --------------------
    a. $P = 2P(F > 1.12|df = 11,14)$;
        Using Tables 9, Appendix B, ES10-pp-816-821:
        $P > 2(0.05)$; $P > 0.10$
        Using a computer: $P = 0.830$
    b. $P > \alpha$
      -- using classical approach ------------------
    a. $F(11, 14, 0.025) = 3.10$

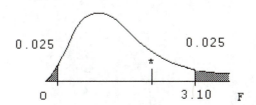

    b. F* falls in the noncritical region, see Step 4a

Step 5: a. Fail to reject H_o
        b. At the 0.05 level of significance, there is not
           sufficient evidence to show a significant
           difference between the standard deviations.

10.123 Multiply it by 2 to cover both sides of the distribution.

10.125 $F^* = s_a^2 / s_b^2 = (4.43)^2/(3.50)^2 = \underline{1.60}$

10.127 a. $n_1 = 8$, $\overline{x}_1 = 0.01525$, $s_1 = 0.00547$

        $n_2 = 25$, $\overline{x}_2 = 0.02856$, $s_2 = 0.00680$
    b. Step 1: a. The ratio of the variances for the two lots
            b. H_o: $\sigma_1^2 = \sigma_2^2$

            H_a: $\sigma_1^2 \neq \sigma_2^2$

    Step 2: a. normality indicated, independence exists
            b. F   c. $\alpha = 0.05$
    Step 3: a. $n_1 = 8$, $s_1 = 0.00547$; $n_2 = 25$, $s_2 = 0.00680$

            b. $F^* = s_2^2/s_1^2 = (0.00680)^2/(0.00547)^2 = 1.545 = 1.55$

Step 4: -- using p-value approach --------------------
       a. $P = 2P(F > 1.55 | df = 24,7)$;
           Using Tables 9, Appendix B, ES10-pp-816-821:
           $P > 2(0.05)$; $P > 0.10$
           Using a computer: $P = 0.575$
       b. $P > \alpha$
       -- using classical approach ------------------
         a. $F(24, 7, 0.025) = 4.42$

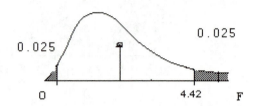

           b. F* falls in the noncritical region, see Step 4a
       ----
  Step 5: a. Fail to reject H_O
          b. At the 0.05 level of significance, there is no
             sufficient evidence to conclude that a
             difference in variances exist for the two lots.
c.
  Step 1: a. The mean levels of Critical Feature A in two
             lots, $\mu_2 - \mu_1$

          b. $H_O: \mu_2 - \mu_1 = 0$
             $H_a: \mu_2 - \mu_1 \cdot 0$
  Step 2: a. normality assumed
          b. t   c. $\alpha = 0.05$
  Step 3: a. $n_1 = 8$, $\bar{x}_1 = 0.01525$, $s_1 = 0.00547$
             $n_2 = 25$, $\bar{x}_2 = 0.02856$, $s_2 = 0.00680$
          b. $t^* = (0.02856 - 0.01525) / \sqrt{((0.00680^2 / 25) + (0.00547^2 / 8))}$
             $= 0.01331/0.00236 = 5.64$
  Step 4: -- using p-value approach --------------------
       a. $P = 2P(t > 5.64 | df = 7)$;
           Using Table 6, Appendix B, ES10-p813:
           $P < 2(0.005)$; $P < 0.01$
           Using Table 7, Appendix B, ES10-p814:
           $P < 2(0.003)$; $P < 0.006$
           Using computer: $P = 0.0008$
       b. $P < \alpha$

a. critical region: t ≤ - 2.36, t ≥ 2.36
b. t* falls in the critical region

Step 5: a. Reject H_O
b. At the 0.05 level of significance, there is
sufficient evidence to show that the meal levels
of Critical Feature A is different between the
two lots.

10.129 a. Men: $\overline{x}_m$ = \$68.14, s_m = \$47.95

Women: $\overline{x}_w$ = \$85.90, s_w = \$63.50

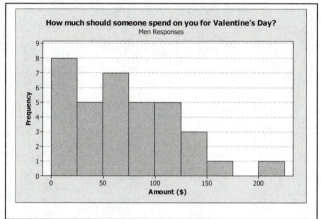

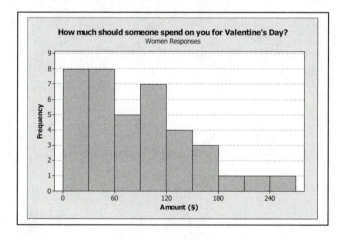

b.   The two distributions appear to be very similar with
regards to shape, center and spread.

c. Both samples appear to be skewed right distributions but the normality tests conducted below indicate that the samples might very well have come from a population with a normal distributed. The p-values are 0.216 and 0.052.

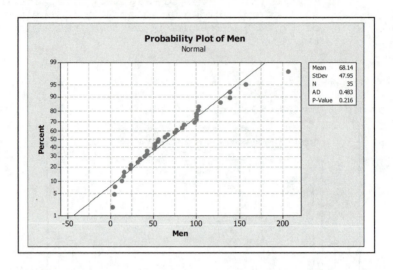

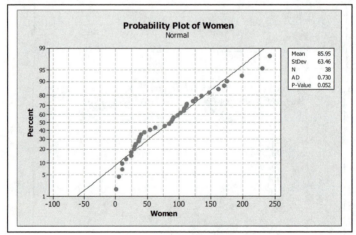

d. Step 1: a. The mean amount spent on Valentine's Day for men and women, $\mu_w - \mu_m$

b. $H_O: \mu_w - \mu_m = 0$
$H_a: \mu_w - \mu_m > 0$

Step 2: a. normality assumed

b. t   c. $\alpha = 0.05$

Step 3: a. $n_m = 35$, $\bar{x}_m = 68.14$, $s_m = 47.95$

$n_w = 38$, $\bar{x}_w = 85.90$, $s_w = 63.50$

b. $t* = (85.90 - 68.14) / \sqrt{(63.50^2 / 38) + (47.95^2 / 35)}$

$= 17.76/13.107 = 1.36$

Step 4: -- using p-value approach --------------------

a. $P = P(t > 1.36 | df = 34)$;

Using Table 6, Appendix B, ES10-p813:

$0.05 < P < 0.10$

Using Table 7, Appendix B, ES10-p814:

$0.085 < P < 0.102$

b. $P > \alpha$

-- using classical approach ------------------

a. critical region: $t \geq 1.70$

b. $t*$ falls in the noncritical region

Step 5: a. Fail to reject H_O

b. At the 0.05 level of significance, there is insufficient evidence to show that the mean amount that should be spent on Valentine's Day is greater for women than men.

e. Step 1: a. The standard deviations for amounts that should be spent for women and men on Valentine's Day

b. H_O: $\sigma_w = \sigma_m$

H_a: $\sigma_w \bullet \sigma_m$

Step 2: a. normality assumed, independence exists

b. F   c. $\alpha = 0.05$

Step 3: a. $n_m = 35$, $s_m = 47.95$; $n_w = 38$, $s_w = 63.50$

b. $F* = s_w^2 / s_m^2 = (63.50)^2/(47.95)^2 = 1.75$

Step 4: -- using p-value approach --------------------

a. $P = 2P(F > 1.75 | df = 37,34)$;

Using Tables 9, Appendix B, ES10-pp-816-821

$P \approx 2(0.05)$; $P \approx 0.10$

Using a computer: $P = 0.1026$

b. $P > \alpha$

-- using classical approach ------------------

a. critical region: $F \geq 2.72$

b. $F*$ falls in the noncritical region

Step 5: a. Fail to reject H_o
b. At the 0.05 level of significance, there
is insufficient evidence that the
standard deviations of amounts that
should be spent for Valentine's Day is
different between men and women.

f. The results of the hypothesis tests in parts (d) &
(e) show that even though there was a difference in
the sample statistics, the different was not
significant.difference

10.131   Everybody will get different results, but do not
expect the F* distribution to look as much like
the theoretical F distribution. There will be
similarities, but the F* distribution will be much
more wide spread. Compare the following empirical
distribution of 100 F*s to the theoretical
distribution in 10.130. You need to only compare
the proportions of the distributions that are
above 2.0 to see a huge difference.

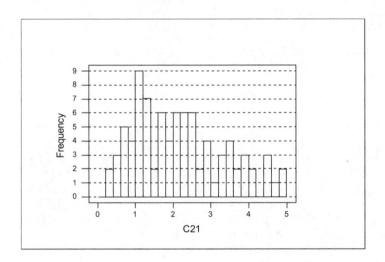

10.133 Step 1: The mean difference in IQ scores for oldest and youngest members of a family (d = O - Y)
Step 2: a. normality indicated
b. t   c. $1-\alpha = 0.95$
Step 3: n = 12, $\overline{d} = 3.583$, $s_d = 19.58$
Step 4: a. $\alpha/2 = 0.05/2 = 0.025$; df = 11; t(11, 0.025) = 2.20
b. E = $t(df, \alpha/2) \cdot (s_d/\sqrt{n})$ = (2.20)(19.58/$\sqrt{12}$) = (2.20)(5.65) = 12.435
c. $\overline{d} \pm E$ = 3.583 ± 12.435
Step 5: <u>-8.85 to 16.02</u>, the 0.95 interval for μ_d

10.135 Step 1: The mean difference in readings for filter 1 and filter 2 (d = $F_1 - F_2$)
Step 2: a. normality indicated
b. t   c. $1-\alpha = 0.90$
Step 3: n = 20, $\overline{d} = 2.00$, $s_d = 2.714$
Step 4: a. $\alpha/2 = 0.10/2 = 0.05$; df = 19; t(19, 0.05) = 1.73
b. E = $t(df, \alpha/2) \cdot (s_d/\sqrt{n})$ = (1.73)(2.714/$\sqrt{20}$) = (1.73)(0.607) = 1.05
c. $\overline{d} \pm E$ = 2.00 ± 1.05
Step 5: <u>0.95 to 3.05</u>, the 0.90 interval for μ_d

10.137 Sample statistics: d = LD - HD

$$n = 20, \quad \overline{d} = 1.35, \quad s_d = 1.631$$

Step 1: a. The mean difference in the level of pain and discomfort experienced by patients receiving heavier dosages of anesthetic prior to eye surgery

b. H_o: $\mu_d = 0$ (no difference)
H_a: $\mu_d > 0$ (less pain and discomfort)

Step 2: a. normality assumed

b. t   c. $\alpha = 0.01$

Step 3: a. $n = 20$, $\overline{d} = 1.35$, $s_d = 1.631$

b. $t* = (\overline{d} - \mu_d)/(s_d/\sqrt{n})$

$$= (1.35 - 0.0)/(1.631/\sqrt{20}) = 3.70$$

Step 4: -- using p-value approach --------------------

a. **P** = P(t > 3.70|df=19);
Using Table 6, Appendix B, ES10-p813:
P < 0.005
Using Table 7, Appendix B, ES10-p814:
P = 0.001
Using computer: P = 0.0008

b. **P** < α

-- using classical approach ------------------

a. critical region: t ≥ 2.54

a. t* falls in the critical region

Step 5: a. Reject H_o

b. At the 0.01 level of significance, there is sufficient evidence that there is less pain and discomfort.

10.139 Step 1: The difference between the mean anxiety scores for males and females, $\mu_f - \mu_m$

Step 2: a. normality assumed, CLT with $n_f = 50$ and $n_m = 50$.

b. t   c. $1-\alpha = 0.95$

Step 3: $n_f = 50$, $\overline{x}_f = 75.7$, $s_f = 13.6$
$n_m = 50$, $\overline{x}_m = 70.5$, $s_m = 13.2$
$\overline{x}_f - \overline{x}_m = 75.7 - 70.5 = 5.2$

Step 4: a. $\alpha/2 = 0.05/2 = 0.0025$; df = 49;
$\qquad$ $t(49, 0.025) = 2.02$

$\qquad$ b. $E = t(df, \alpha/2) \cdot \sqrt{(s_f^2 / n_f) + (s_m^2 / n_m)}$

$\qquad\qquad$ $= (2.02)\sqrt{(13.6^2/50) + (13.2^2/50)}$
$\qquad\qquad$ $= (2.02)(2.68) = 5.41$

$\qquad$ c. $(\overline{x}_f - \overline{x}_m) \pm E = 5.2 \pm 5.41$

Step 5: $\underline{-0.21 \text{ to } 10.61}$, the 0.95 interval for $\mu_f - \mu_m$

10.141 Step 1: The difference between mean rifle-shooting scores
$\qquad\qquad$ for two companies, $\mu_B - \mu_A$

$\qquad$ Step 2: a. normality assumed
$\qquad\qquad$ b. t $\qquad\qquad$ c. $1-\alpha = 0.95$

$\qquad$ Step 3: $n_A = 10$, $\overline{x}_A = 57.0$, $s_A^2 = 209.111$, $s_A = 14.46$
$\qquad\qquad$ $n_B = 10$, $\overline{x}_B = 62.2$, $s_B^2 = 193.289$, $s_B = 13.90$
$\qquad\qquad$ $\overline{x}_B - \overline{x}_A = 62.2 - 57.0 = 5.2$

$\qquad$ Step 4: a. $\alpha/2 = 0.05/2 = 0.025$; df = 9;
$\qquad\qquad$ $t(9, 0.025) = 2.26$

$\qquad\qquad$ b. $E = t(df, \alpha/2) \cdot \sqrt{(s_B^2 / n_B) + (s_A^2 / n_A)}$

$\qquad\qquad\qquad$ $= (2.26)\sqrt{(193.289/10) + (209.111/10)}$
$\qquad\qquad\qquad$ $= (2.26)(6.3435) = 14.34$

$\qquad\qquad$ c. $(\overline{x}_B - \overline{x}_A) \pm E = 5.2 \pm 14.34$

$\qquad$ Step 5: $\underline{-9.14 \text{ to } 19.54}$, the 0.95 confidence interval
$\qquad\qquad$ for $\mu_B - \mu_A$

10.143 Step 1: The difference between means of two methods used in
$\qquad\qquad$ ice fusion, $\mu_A - \mu_B$

$\qquad$ Step 2: a. normality assumed
$\qquad\qquad$ b. t $\qquad\qquad$ c. $1-\alpha = 0.95$

$\qquad$ Step 3: $n_A = 13$, $\overline{x}_A = 80.021$, $s_A^2 = 0.0005744$, $s_A = 0.02397$
$\qquad\qquad$ $n_B = 8$, $\overline{x}_B = 79.979$, $s_B^2 = 0.0009839$, $s_B = 0.03137$
$\qquad\qquad$ $\overline{x}_A - \overline{x}_B = 80.021 - 79.979 = 0.042$

$\qquad$ Step 4: a. $\alpha/2 = 0.05/2 = 0.025$; df = 7;
$\qquad\qquad$ $t(7, 0.025) = 2.36$

$\qquad\qquad$ b. $E = t(df, \alpha/2) \cdot \sqrt{(s_A^2 / n_A) + (s_B^2 / n_B)}$

$\qquad\qquad\qquad$ $= (2.36)\sqrt{(0.0005744/13) + (0.0009839/8)}$
$\qquad\qquad\qquad$ $= (2.36)(0.0129) = 0.030$

$\qquad\qquad$ c. $(\overline{x}_A - \overline{x}_B) \pm E = 0.042 \pm 0.030$

$\qquad$ Step 5: $\underline{0.012 \text{ to } 0.072}$, the 0.95 confidence interval
$\qquad\qquad$ for $\mu_A - \mu_B$

10.145 Step 1: a. The difference between the mean 40-yard sprint time recorded by football players on artificial turf and grass, $\mu_2 - \mu_1$

b. H_o: $\mu_2 - \mu_1 = 0$ (no difference)

H_a: $\mu_2 - \mu_1 > 0$ (artif. turf yields a lower time)

Step 2: a. normality assumed, CLT with $n_c = 22$ and $n_h = 22$.

b. t   c. $\alpha = 0.05$

Step 3: a. sample information given in exercise

b. $t = [(\overline{x}_2 - \overline{x}_1) - (\mu_2 - \mu_1)] / \sqrt{(s_2^2/n_2) + (s_1^2/n_1)}$

$t* = [(4.96 - 4.85) - 0] / [\sqrt{(0.42^2/22) + (0.31^2/22)}]$

$= 0.988$

Step 4: -- using p-value approach --------------------

a. $\mathbf{P} = P(t > 0.988 | df = 21)$;

Using Table 6, Appendix B, ES10-p813:

$0.10 < \mathbf{P} < 0.25$

Using Table 7, Appendix B, ES10-p814:

$0.164 < \mathbf{P} < 0.189$

Using computer: $\mathbf{P} = 0.1672$

b. $\mathbf{P} > \alpha$

-- using classical approach ------------------

a. critical region: $t \geq 1.72$

b. t* falls in the noncritical region

Step 5: a. Fail to reject H_o

b. At the 0.05 level of significance, there is insufficient evidence to show that the artificial turf had faster mean sprint times than that of grass.

10.147 Step 1: a. The difference between the mean number of days missed at work by people receiving CSM treatment and those undergoing physical therapy, CSM(1) and Therapy(2), $\mu_2 - \mu_1$

b. H_o: $\mu_2 - \mu_1 = 0$ (no difference)

H_a: $\mu_2 - \mu_1 > 0$ (physical therapy less effective)

Step 2: a. normality assumed

b. t   c. $\alpha = 0.01$

Step 3: a. $n_1 = 32$, $\overline{x}_1 = 10.6$, $s_1 = 4.8$

$n_2 = 28$, $\overline{x}_2 = 12.5$, $s_2 = 6.3$

b. $t* = [(\overline{x}_2 - \overline{x}_1) - (\mu_2 - \mu_1)] / \sqrt{(s_2^2/n_2) + (s_1^2/n_1)}$

$= [(12.5 - 10.6) - 0] / [\sqrt{(6.3^2 / 28) + (4.8^2 / 32)}]$

$= 1.30$

Step 4: -- using p-value approach --------------------

a. $P = P(t > 1.30 | df = 27)$;

Using Table 6, Appendix B, ES10-p813:

$0.10 < P < 0.25$

Using Table 7, Appendix B, ES10-p814:

$0.102 < P < 0.103$

Using computer: $P = 0.1023$

b. $P > \alpha$

-- using classical approach ------------------

a. critical region: $t \geq 2.47$

b. $t*$ is in the noncritical region

Step 5: a. Fail to reject H_O

b. At the 0.01 level of significance, there is insufficient evidence to show that people treated by chiropractors using CSM miss fewer days of work due to acute lower back pain than people undergoing physical therapy.

10.149 a. $n_A = 15$, $\overline{x}_A = \underline{15.53}$, $s_A^2 = \underline{1.98}$, $s_A = \underline{1.41}$

b. $n_B = 15$, $\overline{x}_B = \underline{12.53}$, $s_B^2 = \underline{1.98}$, $s_B = \underline{1.41}$

c. Step 1: a. The difference between mean torques required to remove screws from two different materials, $\mu_A - \mu_B$

b. H_O: $\mu_A - \mu_B = 0$

H_a: $\mu_A - \mu_B \neq 0$

Step 2: a. normality indicated

b. t   c. $\alpha = 0.01$

Step 3: a. sample information given above

b. $t* = [(\overline{x}_A - \overline{x}_B) - (\mu_A - \mu_B)] / \sqrt{(s_A^2 / n_A) + (s_B^2 / n_B)}$

$= [(15.53 - 12.53) - 0] / [\sqrt{(1.98/15) + (1.98/15)}]$

$= 5.84$

Step 4: a. $P = 2P(t* > 5.84 | df = 14)$;
Using Table 6, Appendix B, ES10-p813:
$P < 0.01$
Using Table 7, Appendix B, ES10-p814:
$2(\frac{1}{2}P < 0.001)$, $P < 0.002$
Using computer: $P = 0.00004$
b. $P < \alpha$
-- using classical approach ------------------
a. critical region: $t \leq -2.98$, $t \geq 2.98$
b. t* falls in the critical region

Step 5: a. Reject H_O
b. At the 0.01 level of significance, the mean
torques are not the same for Material A and
Material B.

10.151 a. M: $\overline{x} = 74.69$, $s = 10.19$, F: $\overline{x} = 79.83$, $s = 8.80$

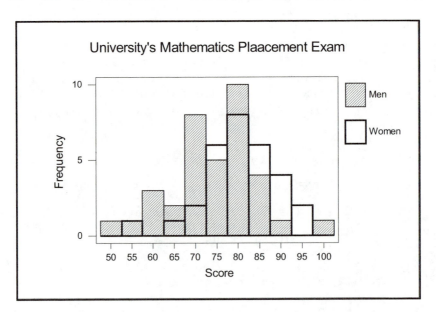

b.
Step 1: a. The mean mathematics placement score for all
men
b. H_O: $\mu = 77$
H_a: $\mu \neq 77$
Step 2: a. normality indicated
b. t   c. $\alpha = 0.05$

Step 3: a. $n = 36$, $\bar{x} = 74.69$, $s = 10.19$

       b. $t = (\bar{x} - \mu)/(s/\sqrt{n})$

          $t^* = (74.69 - 77.0)/(10.19/\sqrt{36}) = -1.36$

Step 4: -- using p-value approach --------------------

       a. **P** $= 2P(t > 1.36 | df = 35)$;

          Using Table 6, Appendix B, ES10-p813:

          $0.10 <$ **P** < 0.20

          Using Table 7, Appendix B, ES10-p814:

          $0.085 <$ ½**P** < 0.101; $0.170 <$ **P** < 0.202

          Using a computer: **P** $= 0.183$

       b. **P** $> \alpha$

       -- using classical approach ------------------

       a. $t \le -2.03$, $t \ge 2.03$

       b. t^* falls in the noncritical region, see Step 4a

Step 5: a. Fail to reject H_o

       b. The sample does provide sufficient evidence that the mean score for men is 77, at the 0.05 level of significance.

Step 1: a. The mean mathematics placement score for all women

       b. H_o: $\mu = 77$

          H_a: $\mu \ne 77$

Step 2: a. normality indicated

       b. t   c. $\alpha = 0.05$

Step 3: a. $n = 30$, $\bar{x} = 79.83$, $s = 8.80$

       b. $t = (\bar{x} - \mu)/(s/\sqrt{n})$

          $t^* = 79.83 - 77.0/(8.80/\sqrt{30} = 1.76$

Step 4: -- using p-value approach --------------------

       a. **P** $= 2P(t > 1.76 | df = 29)$;

          Using Table 6, Appendix B, ES10-p813:

          $0.05 <$ **P** < 0.10

          Using Table 7, Appendix B, ES10-p814:

          $0.041 <$ ½**P** < 0.050; $0.082 <$ **P** < 0.100

          Using a computer: **P** $= 0.089$

       b. **P** $> \alpha$

       -- using classical approach ------------------

       a. $t \le -2.05$, $t \ge 2.05$

       b. t^* falls in the noncritical region, see Step 4a

Step 5:   a. Fail to reject H_O

           b. The sample does provide sufficient
               evidence that the mean score for women is
               77, at the 0.05 level of significance.

c.   They're both not significantly different than 77.
     It is easy to jump to the wrong conclusion.

d.   Step 1: a. The difference between mean scores for male
                 and female students, $\mu_F - \mu_M$
            b. H_O: $\mu_F - \mu_M = 0$
              H_a: $\mu_F - \mu_M \neq 0$
      Step 2: a. normality indicated
            b. t   c. $\alpha = 0.05$
      Step 3: a. sample information given above

            b. $t* = (79.83 - 74.7)/\sqrt{(8.80^2 / 30) + (10.2^2 / 36)}$
               $= 2.19$
      Step 4: a. $\mathbf{P} = 2P(t* > 2.19 | df = 29)$;
               Using Table 6, Appendix B, ES10-p813:
               $0.02 < \mathbf{P} < 0.05$
               Using Table 7, Appendix B, ES10-p814:
               $2(0.018 < \tfrac{1}{2}\mathbf{P} < 0.022)$, $0.036 < \mathbf{P} < 0.044$
               Using computer: $P = 0.0368$
            b. $\mathbf{P} < \alpha$
           -- using classical approach ------------------
           a. critical region: $t \leq -2.05$, $t \geq 2.05$
           b. t* falls in the critical region

      Step 5: a. Reject H_O
            b. At the 0.05 level of significance, the mean
               scores for men and women are not equal.

e. & f. No, a significant difference was found. The
        questions of (b) and (d) are asking different
        questions. In (b), individually two hypothesis tests
        are testing if the means are different than 77. In
        this case, the two sample means are on opposite
        sides of 77, but not significantly far from 77. Yet
        the two sample means are themselves far enough apart
        to be significantly different.

10.153 Step 1: The difference in proportions requiring service from
         two manufacturers, $p_1 - p_2$
     Step 2: a. n's > 20, np's and nq's all > 5
         b. z   c. $1 - \alpha = 0.95$

Step 3: sample information given in exercise
$$p_1' - p_2' = 0.15 - 0.09 = 0.060$$
Step 4: a. $\alpha/2 = 0.05/2 = 0.025$; $z(0.025) = 1.96$
b. $E = z(\alpha/2) \cdot \sqrt{(p_1' \cdot q_1' / n_1) + (p_2' \cdot q_2' / n_2)}$
$$= 1.96 \sqrt{(0.15 \cdot 0.85/75) + (0.09 \cdot 0.91/75)}$$
$$= (1.96)(0.0528) = 0.104$$
c. $(p_1' - p_2') \pm E = 0.060 \pm 0.104$
Step 5: <u>−0.044 to 0.164</u>, the 0.95 interval for $p_1 - p_2$

10.155   Find the p-value for each sample size for the situation:
$$H_0: p_w - p_m = 0$$
$$H_a: pw - pm \neq 0$$
a. n = 100
If samples are of same size, then
$$p_p' = (p_w + p_m)/2 = (0.88+0.75)/2 = 0.815$$
$$q_p' = 1 - p_p' = 1.000 - 0.815 = 0.185$$
$$z = [(p_w' - p_m') - (p_w - p_m)] / \sqrt{(p_p')(q_p')[(1/n_w)+(1/n_m)]}$$
$$z^* = (0.88 - 0.75) / \sqrt{(0.815)(0.185)[(1/100)+(1/100)]}$$
$$= 0.13/0.0549 = 2.37$$
$\mathbf{P} = 2P(z > 2.37)$;
Using Table 3, Appendix B, ES10-p810:
$\mathbf{P} = 2(0.5000 - 0.4911) = 2(0.0089) = 0.0178$
Using Table 5, Appendix B, ES10-p812:
$2(0.0082 < \mathbf{P} < 0.0094)$; $0.0164 < P < 0.0188$
Using a computer: p-value = 0.0178
There is a significant difference for $\alpha \geq 0.02$.

b. n = 150
If samples are of same size, then
$$p_p' = (p_w + p_m)/2 = (0.88+0.75)/2 = 0.815$$
$$q_p' = 1 - p_p' = 1.000 - 0.815 = 0.185$$
$$z = [(p_w' - p_m') - (p_w - p_m)] / \sqrt{(p_p')(q_p')[(1/n_w)+(1/n_m)]}$$
$$z^* = (0.88 - 0.75) / \sqrt{(0.815)(0.185)[(1/150) + (1/150)]}$$
$$= 0.13/0.0448 = 2.90$$
$\mathbf{P} = 2P(z > 2.90)$;
Using Table 3, Appendix B, ES10-p810:
$\mathbf{P} = 2(0.5000 - 0.4981) = 2(0.0019) = 0.0038$
Using Table 5, Appendix B, ES10-p812:
$\tfrac{1}{2}\mathbf{P} = 0.0019$; $\mathbf{P} = 0.0038$
Using a computer: p-value = 0.0036
There is a significant difference for $\alpha \geq 0.01$.

c. n = 200

   If samples are of same size, then

$$p_p' = (p_w + p_m)/2 = (0.88+0.75)/2 = 0.815$$

$$q_p' = 1 - p_p' = 1.000 - 0.815 = 0.185$$

$$z = [(p_w' - p_m') - (p_w - p_m)]/\sqrt{(p_p')(q_p')[(1/n_w)+(1/n_m)]}$$

$$z^* = (0.88 - 0.75)/\sqrt{(0.815)(0.185)[(1/200)+(1/200)]}$$

$$= 0.13/0.0388 = 3.35$$

$\mathbf{P} = 2P(z > 3.35);$
   Using Table 3, Appendix B, ES10-p810:
   $\mathbf{P} = 2(0.5000 - 0.4996) = 2(0.0004) = 0.0008$
   Using Table 5, Appendix B, ES10-p812:
   $2(\mathbf{P} = 0.0004);$ $P = 0.0008$
   Using a computer: p-value = 0.0008

There is a significant difference for $\alpha \geq 0.001$.

d. As the sample size increases, the standard error for the difference of two proportions became smaller; this in turn meant that the 88% and 75% become more standard errors apart as reflected in the increase in z^*. As n increased, the test became more sensitive meaning that the p-value kept decreasing.

10.157 Step 1: a. The difference in the proportion of accountants and lawyers who believe that the new burden-of-proof tax rules will cause an increase in taxpayer wins in court, $p_a - p_l$

   b. $H_o: p_a - p_l = 0$
      $H_a: p_a - p_l \neq 0$

Step 2:   a. n's > 20,   np's and nq's all > 5
          b. z   c. $\alpha = 0.01$

Step 3: a. $n_a = 175$, $x_a = 101$, $p_a' = 101/175 = 0.5771$

      $n_l = 165$, $x_l = 84$, $p_l' = 84/165 = 0.5091$

      $p_p' = (x_a + x_l)/(n_a + n_l) = (101+84)/(175+165)$
         $= 0.5441$

      $q_p' = 1 - p_p' = 1.000 - 0.5441 = 0.4559$

   b. $z = [(p_a' - p_l') - (p_a - p_l)]/\sqrt{(p_p')(q_p')[(1/n_a)+(1/n_l)]}$

$$z^* = (0.5771 - 0.5091)/\sqrt{(0.5441)(0.4559)[(1/175)+(1/165)]}$$

$$= 0.068/0.054 = 1.26$$

Step 4: -- using p-value approach --------------------
    a. $P = 2P(z > 1.26)$;
        Using Table 3, Appendix B, ES10-p810:
        $P = 2(0.5000 - 0.3962) = 2(0.1038) = 0.2076$
        Using Table 5, Appendix B, ES10-p812:
        $0.0968 < \frac{1}{2}P < 0.1056$; $0.1936 < P < 0.2112$
        Using a computer: p-value = 0.2077
    b. $P > \alpha$
  -- using classical approach ------------------
  a. critical region: $z \le -2.58$ and $z \ge 2.58$
  b. z* falls in the noncritical region

Step 5:a. Fail to reject H_O
    a. There is not sufficient evidence to show that
       accountants and lawyers differ in their beliefs
       about the burden of proof, at the 0.01 level.

10.159 Step 1: a. The ratio of the variance of time needed by
         men to that needed by women to assemble a product
      b. H_O: $\sigma_m^2 = \sigma_f^2$
      H_a: $\sigma_m^2 > \sigma_f^2$
Step 2: a. normality indicated, independence exists
    b. F   c. $\alpha = 0.05$
Step 3: a. $n_m = 15$, $s_m = 4.5$, $n_f = 15$, $s_f = 2.8$
    b. $F^* = s_m^2 / s_f^2 = (4.5)^2/(2.8)^2 = 2.58$

Step 4: -- using p-value approach --------------------
    a. $P = P(F > 2.58|df = 14,14)$;
        Using Tables 9, Appendix B, ES10-pp-816-821:
        $0.025 < P < 0.05$
        Using computer: p = 0.0435
    b. $P < \alpha$
  -- using classical approach ------------------
  a. critical region: $F \ge 2.53$
  b. F* falls in the critical region

Step 5: a. Reject H_O
    b. At the 0.05 level of significance, there is
       sufficient evidence to conclude that male
       assembly times are more variable.

10.161 Step 1: a. The difference between the variances in the
         threads of lug nuts and studs
      b. H_O: $\sigma_n^2 = \sigma_s^2$
      H_a: $\sigma_n^2 \ne \sigma_s^2$

Step 2: a. normality indicated, independence exists
        b. F   c. $\alpha = 0.05$
Step 3: a. $n_n = 60$, $s_n^2 = 0.00213$, $n_s = 40$, $s_s^2 = 0.00166$

        b. $F* = s_n^2 / s_s^2 = 0.00213/0.00166 = 1.28$

Step 4: -- using p-value approach --------------------
        a. **P** = $P(F > 1.28 | df = 59,39)$;
           Using Tables 9, Appendix B, ES10-pp-816-821:
           P > 0.10
           Using computer: **P** = 0.4160
        b. **P** $> \alpha$
        -- using classical approach ------------------
        a. critcal region: $F \geq 1.80$
        b. F* falls in the noncritical region

Step 5: a. Fail to reject H_O
        b. At the 0.05 level of significance, there is no
           difference in the variances of lug nuts and
           studs.

10.163 a.

	N	Mean	StDev
Cont	50	0.005459	0.000763
Test	50	0.003507	0.000683

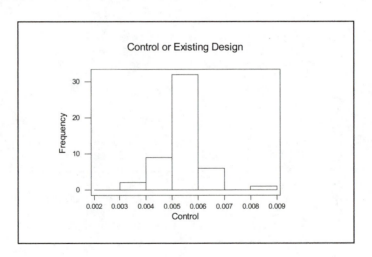

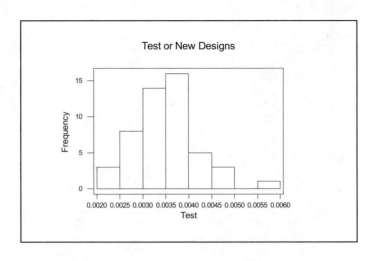

b. Both sets of data have an approximately normal distribution, therefore the assumptions are satisfied.

c. One-tailed – looking for a reduction

d.

Step 1: a. The difference between the variances in two different designs

b. H_O: $\sigma_c^2 = \sigma_t^2$

H_a: $\sigma_c^2 > \sigma_t^2$ (variance for control is greater)

Step 2: a. normality indicated, independence exists

b. F   c. $\alpha = 0.05$

Step 3: a. $n_c = 50$, $s_c = 0.000763$, $n_t = 50$, $s_t = 0.000683$

b. $F^* = \mathbf{s}_c^2/\mathbf{s}_t^2 = (0.000763^2/0.000683^2) = 1.248$

Step 4: -- using p-value approach --------------------
a. $\mathbf{P} = P(F > 1.25|df = 49,49)$;
   Using Tables 9, Appendix B, ES10-pp-816-821:
   $\mathbf{P} > 0.05$
   Using a computer: $\mathbf{P} = 0.44$

b. $\mathbf{P} > \alpha$
-- using classical approach ------------------
a. critical region: $F \geq 1.69$
b. F^* falls in the noncritical region

Step 5: a. Fail to reject H$_o$
   b. At the 0.05 level of significance, there is
      no difference in the variances for the two
      designs. The new design has not
      significantly reduced the variability.

e. Step 1: a. The difference between mean amount of force
      for two designs, $\mu_c - \mu_t$
   b. H$_o$: $\mu_c - \mu_t = 0$
      H$_a$: $\mu_c - \mu_t > 0$
   Step 2: a. normality indicated
      b. t   c. $\alpha = 0.05$
   Step 3: a. sample data given above
      b. t* =

$$[(0.005459-0.003507)-0]/\sqrt{((0.000763^2 / 50) + (0.000683^2 / 50))}$$
$$= 13.48$$

   Step 4: a. **P** = P(t > 13.48|df = 49);
         Using Table 6, Appendix B, ES10-p813:
         P = +0.000
         Using Table 7, Appendix B, ES10-p814:
         P = +0.000
         Using a computer: **P** = 0.000
      b. **P** < α
      -- using classical approach ------------------
      a. critical region: t ≤ -1.68, t ≥ 1.68
      b. t* falls in the critical region

   Step 5: a. Reject H$_o$
      b. At the 0.05 level of significance, there is
         sufficient evidence to show that the new
         design has reduced the mean amount of force

a. The mean force has been reduced, but not the
   variability.

CHAPTER 11 ∇ APPLICATIONS OF CHI-SQUARE

Chapter Preview

Chapter 11 demonstrates hypothesis tests, as did Chapters 8, 9, and 10. The difference lies in the type of data that is to be analyzed. Enumerative type data, that is, data which can be counted and placed into categories, will be discussed and investigated in three types of tests. Each test will compare actual (observed) results with expected (theoretical) results. One will use the comparisons to determine whether a "claimed" relationship exists, one will determine whether two factors or variables are independent, and the third will determine whether the proportions per variable are the same. Also, the chi square statistic, χ^2, will be reintroduced and utilized in performing these tests.

An article appearing in <u>USA Today</u> showing the ways that people cool their mouth after eating spicy hot food is used in the Chapter 11 opening section ``Cooling a Great Hot Taste''.

SECTION 11.1 EXERCISES

11.1 a. The name of their preferred way to "cool" their mouth after eating a delicious spicy favorite.

b. Population: US adults professing to love eating hot spicy food. Variable: method of cooling the heat.

c. Water: 73/200 = 0.365 = 36.5%;
Milk: 35/200 = 0.175 = 17.5%;
Soda: 20/200 = 0.10 = 10%;
Beer: 19/200 = 0.095 = 9.5%
Bread: 29/200 = 0.145 = 14.5%;
Other: 11/200 = 0.055 = 5.5%;
Nothing: 13/200 = 0.065 = 6.5%

d. The percentages seem to be fairly similar: the responses appear to break up into four levels: water has the greatest response, then milk and bread are similar and next most popular, while soda, beer and nothing all have drawn similar shares between 6 and 10 percent.

SECTION 11.2 EXERCISES

11.3 a. 23.2   b. 23.3   c. 3.94   d. 8.64

11.5 a. $\chi^2(14, 0.01) = 29.1$
       b. $\chi^2(25, 0.05) = 37.7$

Section 11.3

Characteristics of a Multinomial Experiment

1. There are **n** identical independent trials.

2. The outcome of each trial fits into exactly one of the **k** possible categories or cells.

3. The number of times a trial outcome falls into a particular cell is given by O_i (O - for observed, i for $1 \rightarrow k$).

4. $O_1 + O_2 + O_3 + \ldots + O_k = n$

5. There is a constant probability associated with each of the k cells in such a way that $p_1 + p_2 + \ldots + p_k = 1$.

NOTE: Variable - the characteristic about each item that is of interest.

Various levels of the variable - the k possible outcomes or responses.

11.7 a. One trial: asking one person
       b. Variable: birth day of the week
       c. Levels: 7 for the 7 days of the week

```
┌─────────────────────────────────────────────────────────────────────────┐
│           Writing Hypotheses for Multinomial Experiments                  │
│                                                                           │
│ The null hypothesis is written in a form to show that there is no         │
│ difference between the experimental (observed) frequencies and the        │
│ theoretical (expected) frequencies.                                       │
│ The alternative hypothesis is the "opposite" of the null                  │
│ hypothesis.  It is written in a form to show that a difference does       │
│ exist.                                                                     │
│                                                                           │
│ ex.: The marital status distribution for New York state is 21%,           │
│ 64%, 8%, and 7% for the possible categories of single, married,           │
│ widowed, and divorced.                                                     │
│         H_O: P(S) = .21, P(M) = .64, P(W) = .08, P(D) = .07               │
│         H_a: The percentages are different than specified in H_O.         │
│                                                                           │
│ ex.: A gambler thinks that a die may be loaded.                           │
│         H_O: P(1)=P(2)=P(3)=P(4)=P(5)=P(6) = 1/6   (not loaded)           │
│         H_a: The probabilities are different.  (is loaded)                │
└─────────────────────────────────────────────────────────────────────────┘
```

11.9   a. H_O: $P(1) = P(2) = P(3) = P(4) = P(5) = 0.2$
            H_a: The numbers are not equally likely.

        b. H_O: $P(1) = 2/8$, $P(2) = 3/8$, $P(3) = 2/8$, $P(4) = 1/8$
             H_a: At least one probability is different from H_O.

        c. H_O: $P(E) = 0.16$, $P(G) = 0.38$, $P(F) = 0.41$, $P(P) = 0.05$
             H_a: The percentages are different than specified in H_O.

```
┌─────────────────────────────────────────────────────────────────────────┐
│       The p-value is estimated using Table 8 (Appendix B, ES10-          │
│       p815):                                                              │
│       a) Locate df row.                                                   │
│       b) Locate χ²* between two critical values in the df row; the        │
│                  p-value is in the interval between the two               │
│                  corresponding probabilities at the top of the           │
│                  columns labeled area to the right.                       │
│       OR                                                                  │
│                                                                           │
│       The p-value can be calculated using computer and/or                │
│       calculator commands found in ES10-pp518-519.                        │
└─────────────────────────────────────────────────────────────────────────┘
```

```
┌─────────────────────────────────────────────────────────────────────────┐
│                    Determining the Test Criteria                        │
│                                                                         │
│  1. Draw a picture of the χ² distribution (skewed right, starting       │
│     at 0).                                                               │
│                                                                         │
│  2. Locate the critical region (based on α and Hₐ).                    │
│         Since we are testing H₀: "no difference" versus                │
│         Hₐ: "difference", all of the α is placed in the right tail     │
│         to represent a significant or large difference between the      │
│         observed and expected values.                                   │
│                                                                         │
│  3. Shade in the critical region (the area where you will reject H₀,    │
│         the right-hand tail)                                            │
│                                                                         │
│  4. Find the appropriate critical value from Table 8 (Appendix B,       │
│     ES10-p815), using χ²(df,α), where df = k - 1.  k is equal to the    │
│     number of cells or categories the data are classified into.         │
│                                                                         │
│  Remember this critical or boundary value divides the area under the    │
│  χ² distribution curve into critical and noncritical regions and is     │
│  part of the critical region.                                           │
└─────────────────────────────────────────────────────────────────────────┘
```

11.11 a.   b.

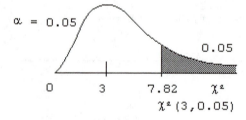

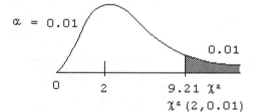

11.13 $E = n \cdot p = 556(9/16) = 312.75$

$O - E = 315 - 312.75 = 2.25$

$(O-E)^2/E = (2.25)^2/312.75 = 0.0162$

Steps for a Hypothesis/Significance Test
for Multinomial Experiments

Follow the steps outlined for p-value and classical hypothesis tests in: ES10-pp427&444, SSM-pp265&275.
The variations are noted below.

1. Distribute the sample information (observed frequencies) into the appropriate cells (data may be already categorized).

2. Calculate the expected frequencies using probabilities determined by H_O and the formula:
    $E_i = np_i$, where E_i is the expected frequency for cell i, n is the sample size and p_i is the probability for the ith cell

3. Use the observed and expected frequencies from each cell to calculate the test statistic, χ^2.

    Use the formula $$\chi^2 = \sum_{allcells} \frac{(O - E)^2}{E}$$

4.a) p-value approach:
    Since all of the α is placed in the right tail,
    the p-value = $P(\chi^2 > \chi^{2*})$.

The p-value is <u>estimated</u> using Table 8 (Appendix B, ES10-p815):
    a) Locate df row.
    b) Locate χ^{2*} between two critical values in the df row; the p-value is in the interval between the two corresponding probabilities at the top of the columns labeled *area to the right*.
    OR
    The p-value can be calculated using computer and/or calculator commands found in ES10-pp518-519.

  b) Classical approach:
    Follow the steps in determining the test criteria found in: ES10-pp623-624, SSM-p420. Then locate χ^{2*} on the χ^2 curve with respect to the critical value.

5. Make a decision and interpret it.
    a) If $P < \alpha$ or the calculated test statistic falls into the critical region, then reject H_O. There is sufficient evidence to indicate that there is a *difference* between the observed and expected frequencies.
    b) If $P > \alpha$ or the calculated test statistic falls into the noncritical region, fail to reject H_O. There is <u>not</u> sufficient evidence to indicate a *difference* between the observed and expected frequencies.

11.15   a. $E(\text{magenta}) = n \cdot p = 100[6/(6+3+1)] = \underline{60}$

b. $\underline{2}$

c. (1) & (2)

Step 1:   a. The proportions: P(magenta), P(chartreuse),
P(ochre)

b. H_o: 6:3:1 ratio vs. H_a: ratio other than 6:3:1

Step 2:   a. Assume that the 100 seeds represent a random
sample.

b. χ^2 with df = 2   c. $\alpha = 0.10$

Step 3:   a. sample information given in exercise

b. $\chi^2 = \Sigma[(O-E)^2/E]$ (as found on accompanying
table)

$$E(\text{chartreuse}) = n \cdot P(c) = 100(0.3) = 30$$
$$E(\text{orche}) = n \cdot P(o) = 100(0.1) = 10$$

Color	magenta	chartreuse	orche	Total
Observed	52	36	12	100
Expected	60	30	10	100
$(O-E)^2/E$	64/60	36/30	4/10	160/60

$$\chi^{2*} = 160/60 = 2.67$$

Step 4:   -- using p-value approach -------------------

a. **P** $= P(\chi^2 > 2.67 | df=2)$;
Using Table 8: $0.25 <$ **P** < 0.50
Using computer/calculator: **P** $= 0.263$

b. **P** $> \alpha$

-- using classical approach -----------------

a. critical region: $\chi^2 \geq 4.61$

b. χ^{2*} falls in the noncritical region

Step 5:   a. Fail to reject H_o

b. At the 0.10 level of significance, there is not
sufficient evidence to show the ratio is other
than 6:3:1.

11.17   Step 1: a. The proportions: P(Quality I), P(Quality II),
                   P(Quality III), P(Quality IV).
                b. H_o: The qualities of meats purchased are in the
                   proportions of 0.10, 0.30, 0.35, 0.25
                   H_a:The proportions are different than listed
        Step 2: a. Assume that the 500 purchases represent a random
                   sample.
                b. χ^2 with df = 3   c. α = 0.05
        Step 3: a. sample information given in exercise
                b. $\chi^2 = \Sigma[(O-E)^2/E]$ (as found on accompanying table)
            E(I) = n·p = 500(0.10) = 50;   E(II) = 500(0.30) = 150;
                E(III) = 500(0.35) = 175;   E(IV) = 500(0.25) = 125

Quality	I	II	III	IV	Total
Observed	46	162	191	101	500
Expected	50	150	175	125	500
$(O-E)^2/E$	16/50	144/150	256/175	576/125	7.35

                    χ^{2*} = 7.35
        Step 4: -- using p-value approach -------------------
                a. $P = P(\chi^2 > 7.35 | df=3)$;
                        Using Table 8:   0.05 < P < 0.10
                        Using computer/calculator: P = 0.062
                b. $P > \alpha$
                -- using classical approach -----------------
                a. critical region: $\chi^2 \geq 7.82$
                b. χ^{2*} falls in the noncritical region

        Step 5: a. Fail to reject H_o
                b. At the 0.05 level of significance, there is not
                   sufficient evidence to show the proportions of
                   meat qualities bought are different than listed.

11.19   Step 1: a. The proportions of staffing situations: P(1),
                   P(2), P(3), P(4), P(5)
                b. H_o: Opinions are distributed 0.12, 0.32, 0.38,
                   0.12,0.06

                   H_a: Opinions are distributed differently.

        Step 2: a. Assume that the 500 nurses surveyed represent a
                   random sample.
                b. χ^2 with df = 4   c. α = 0.05

Step 3: a. sample information given in exercise
b. $\chi^2 = \Sigma[(O-E)^2/E]$ (as found on accompanying table)
$E(1) = n\cdot p = 500(0.12) = 60;$ $E(2) = 500(0.32) = 160;$
$E(3) = 500(0.38) = 190;$ $E(4) = 500(0.12) = 60;$
$E(5) = 500(0.06) = 30$

Opinion	1	2	3	4	5	Total
Observed	165	140	125	50	20	500
Expected	60	160	190	60	30	500
$(O-E)^2/E$	183.75	2.50	22.24	1.67	3.33	213.49

$$\chi^{2*} = 213.49$$

Step 4: -- using p-value approach ------------------
a. **P** $= P(\chi^2 > 213.49 | df=4);$
Using Table 8:   **P** < 0.005
Using computer/calculator: **P** $= 0+$
b. **P** $< \alpha$
-- using classical approach -----------------
a. critical region: $\chi^2 \geq 9.49$
b. χ^{2*} falls in the critical region

Step 5: a. Reject H_o
b. The opinions of the 500 nurses are significantly different than the opinions of the original 1800 nurses, at the 0.05 level.

11.21 Step 1: a. The proportions of most valuable steps for climbing out of debt: P(1), P(2), P(3), P(4), P(5), P(6), P(7)
b. H_o: Opinions are distributed 0.30, 0.21, 0.17, 0.12, 0.07, 0.05, 0.08
H_a: Opinions are distributed differently.
Step 2: a. Assume that the 60 financial planners surveyed represent a random sample.
b. χ^2 with df = 6   c. $\alpha = 0.05$

Step 3: a. sample information given in exercise
   b. $\chi^2 = \Sigma[(O-E)^2/E]$ (as found on accompanying table)

   $E(1) = n \cdot p = 60(0.30) = 18$; $E(2) = 60(0.21) = 12.6$;
   $E(3) = 60(0.17) = 10.2$; $E(4) = 60(0.12) = 7.2$;
   $E(5) = 60(0.07) = 4.2$; $E(6) = 60(0.05) = 3$;
   $E(7) = 60(0.08) = 4.8$

Opinion	1	2	3	4	5	6	7	Total
Observed	10	13	13	8	9	3	4	60
Expected	18	12.6	10.2	7.2	4.2	3	4.8	60
$(O-E)^2/E$	3.56	0.013	0.768	0.089	5.49	0	0.133	10.05

   $\chi^{2*} = 10.05$

Step 4: -- using p-value approach -------------------
   a. $P = P(\chi^2 > 10.05 | df=6)$;
         Using Table 8: $0.10 < P < 0.25$
         Using computer/calculator: $P = 0.123$
   b. $P > \alpha$
   -- using classical approach -----------------
   a. critical region: $\chi^2 \geq 12.6$
   b. χ^{2*} falls in the noncritical region

Step 5: a. Fail to reject H_o
   b. At the 0.05 level of significance, there is not
      sufficient evidence to show the opinions are
      distributed differently than listed.

11.23 Step 1: a. The proportions of colors for Skittles: P(R),
      P(O), P(Y), P(G), P(P)
   b. H_o: Skittles colors are distributed all the
      same: P(each color) = P(R) = P(O) P(Y) = P(G) =
      P(P) = 0.20
      H_a: Skittles' colors are distributed
         differently.
Step 2: a. Assume that the 106 selected Skittles represent
      a random sample.
   b. χ^2 with df = 4   c. $\alpha = 0.05$

Step 3: a. sample information given in exercise
         b. $\chi^2 = \Sigma[(O-E)^2/E]$ (as found on accompanying table)

         E(each color) = n·p = 106(0.20)= 21.2

Birth	R	O	Y	G	P	Total
Observed	18	21	23	17	27	106
Expected	21.2	21.2	21.2	21.2	21.2	106
$(O-E)^2/E$	0.483	0.002	0.153	0.832	1.587	3.057

$$\chi^{2*} = 3.057$$

Step 4: -- using p-value approach -------------------
         a. $P = P(\chi^2 > 3.057 | df=4)$;
            Using Table 8: $0.50 < P < 0.75$
            Using computer/calculator: $P = 0.548$
         b. $P > \alpha$
         -- using classical approach -----------------
         a. critical region: $\chi^2 \geq 9.49$
         b. χ^{2*} falls in the noncritical region

Step 5: a. Fail to reject H_O
         b. At the 0.05 level of significance, there is not
            sufficient evidence to show the colors are
            distributed differently than with equal
            likeliness.

11.25 a.
Step 1: a. The proportions of types and number of guns in a
            household: P(rifle, shotgun, pistol), P(2 of the
            3 types), P(1 of the 3 types), P(decline).
         b. H_O: The proportions for all gun owners: 41%,
            27%, 29%, 3%
            H_a: The proportions are different than listed
Step 2: a. Assume that the 2000 individuals surveyed
            represent a random sample.
         b. χ^2 with df = 3   c. $\alpha = 0.05$
Step 3: a. sample information given in exercise
         b. $\chi^2 = \Sigma[(O-E)^2/E]$ (as found on accompanying table)

    E(3 types) = n·p = 2000(0.41) = 820,
    E(2 types) = 2000(0.27) = 540,
    E(1 type) = 2000(0.29) = 580, E(decline) = 2000(0.03)
            = 60

Handgun	3 types	2types	1 type	decline	Total
Observed	780	550	560	110	2000
Expected	820	540	580	60	2000
$(O-E)^2/E$	1.9512	0.1852	0.6897	41.6667	44.4928

$\chi^2* = 44.4928$

Step 4: -- using p-value approach ------------------
   a. $P = P(\chi^2 > 44.4928 | df=3)$;
         Using Table 8: $P < 0.005$
         Using computer/calculator: $P = 0+$
   b. $P < \alpha$
   -- using classical approach -----------------
   a. critical region: $\chi^2 \geq 7.82$
   b. χ^2* falls in the critical region

Step 5: a. Reject H_O
   b. The distribution of the number of types of guns
      owned in Memphis is different from those
      nationally at the 0.05 level of significance.

b. The 4th cell caused the calculated value of chi-square
    to be very large.

    The ''decline'' category would need to be broken down
    into more specific answer categories.

SECTION 11.4   EXERCISES

Contingency Tables

A contingency table is a table consisting of rows and columns used
to summarize and cross-classify data according to two variables.
Each row represents the categories for one of the variables, and
each column represents the categories for the other variable. The
intersections of these rows and columns produce cells. The data
will be in a form where two varieties of hypothesis tests are
possible. These are:

   1. Tests of independence
      - to determine if one variable is independent of the
        other variable, and   . . .

```
    2. Tests of homogeneity
        - to determine if the proportion distribution for one of
        the variables is the same for each of the categories of
        the second variable.


Writing Hypotheses for Tests of Independence and/or Homogeneity

The null hypothesis is written in a form to show that there is no
difference between the experimental (observed) frequencies and the
theoretical (expected) frequencies.  The alternative hypothesis is
the opposite of the null hypothesis.  It is written in a form to
show that a difference does exist.

Therefore, in tests of independence:

    H_O: variable is independent One of the other variable  and
        in tests of homogeneity:

    H_O: The proportions for one variable are distributed the
        same for all categories of the second variable.
```

11.27 a. H_O: Voters preference and voters party affiliation are
independent.

   H_a: Voters preference and party affiliation are not
   independent.

b. H_O: The distribution is the same for all three

   H_a: The distribution is not the same for all three

c. H_O: The proportion of yeses is the same in all
   categories sampled.

   H_a: The proportion of yeses is not the same in all
   categories.

11.29 E = (40)(50)/200 = 2000/200 = <u>10</u>

11.31 a. sample information given in exercise

$\chi^2 = \Sigma[(O-E)^2/E]$ (as found on accompanying table)

Expected values:

	Boat-Rel	Non-boat
Lee	18.84	29.16
Collier	12.16	18.84

$\chi^{2*} = 0.921 + 0.595 + 1.426 + 0.921 = 3.862$

b. $\mathbf{P} = P(\chi^2 > 3.862|df=1)$;

Using Table 8: $0.025 < \mathbf{P} < 0.05$
Using computer/calculator: $\mathbf{P} = 0.049$

c. $\mathbf{P} < \alpha$; Reject H_o
The proportion of boat-related deaths is not independent of the county at the 0.05 level of significance.

The computer or calculator can perform a hypothesis test for independence or homogeneity. Since these tests are completed in the same fashion, the same command may be used. See ES10-p641 for more information.

11.33 Step 1: a. The proportions of community size that married men were reared in: P(under 10,000 for present size categories), P(10,000 to 49,999 for present size categories),P(50,000 or over for present size categories).

b. H_o: Size of community of present residence is independent of the size of community reared in.
H_a: Size of community of present residence is not independent of the size of community reared in.

Step 2: a. Given a random sample.
b. χ^2 with df = 4
c. $\alpha = 0.01$

Step 3: a. sample information given in exercise
b. $\chi^2 = \Sigma[(O-E)^2/E]$ (as found on accompanying table)

Expected values:

	less than 10,000	10,000 – 49,999	50,000 or over
less than 10,000	14.36	37.16	62.47
10,000 – 49,999	19.15	48.55	83.30
50,000 or over	29.48	76.28	128.23

$$\chi^2{}^* = 6.471 + 1.654 + 4.886 +$$
$$0.069 + 4.214 = 2.124 +$$
$$2.439 + 6.508 + 7.384 = 35.749$$

Step 4: -- using p-value approach -------------------
a. $\mathbf{P} = P(\chi^2 > 35.749 | df=4)$;
     Using Table 8: $\mathbf{P} < 0.005$
     Using computer/calculator: $\mathbf{P} = 0.0+$
b. $\mathbf{P} < \alpha$
-- using classical approach -----------------
a. $\chi^2(4,0.01) = 13.3$

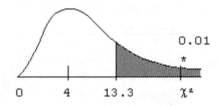

b. $\chi^2{}^*$ falls in the critical region

Step 5: a. Reject H_o
b. The sample evidence does present significant evidence to show the lack of independence.

11.35 Step 1: a. The proportions supervisor expectation responses: P(true for length of employment category), P(not true for length of employment category).
b. H_o: Response is independent of years of service.
H_a: Response is not independent of years of service.
Step 2: a. Assume a random sample.
b. χ^2 with df = 3
c. $\alpha = 0.10$

Step 3: a.   sample information given in exercise
         b. $\chi^2 = \Sigma[(O-E)^2/E]$ (as found on accompanying table)

Expected values:

	True	Not true
Less than 1 yr	21.94	9.06
1 to 3 years	19.82	8.18
3 to 10 years	26.18	10.82
10 or more years	24.06	9.94

$\chi^2* = 0.707 + 1.712 + 0.002 + 0.004 +$
$ 0.126 + 0.305 + 0.156 + 0.378 = 3.390$

Step 4: -- using p-value approach ------------------
         a. $\mathbf{P} = P(\chi^2 > 3.390 | df=3)$;
              Using Table 8: $0.25 < \mathbf{P} < 0.50$
              Using computer/calculator: $\mathbf{P} = 0.335$
         b. $\mathbf{P} > \alpha$
         -- using classical approach ----------------
         a. critical region: $\chi^2 \geq 6.25$
         b. χ^2* is not in the critical region

Step 5: a. Fail to reject H_o
         b. At the 0.10 level of significance, there is not
            sufficient evidence to conclude a lack of
            independence between the length of employment
            and response.

11.37 Step 1: a. The proportions: P(nondefective for 5 days of
               the week), P(defective for 5 days of the week)
           b. H_O: The number of defective items is independent
                of the day of the week
              H_a: The number of defectives is not independent
                of day
           Step 2: a. Assume a random sample.
                    b. χ^2 with df = 4   c. $\alpha = 0.05$

           Step 3: a. sample information given in exercise
                    b. $\chi^2 = \Sigma[(O-E)^2/E$ (as found on accompanying table)

Expected values:

	Mon	Tue	Wed	Thu	Fri	Total
Nondefective	91	91	91	91	91	455
Defective	9	9	9	9	9	45
						500

$$\chi^{2}* = \Sigma[(O-E)^2/E] = 0.396 + 0.011 + 0.176 + 0.176$$
$$+ 0.011 + 4.000 + 0.111 + 1.778 + 1.778 + 0.111$$

$$\chi^{2}* = 8.548$$

Step 4: -- using p-value approach ------------------
a. $\mathbf{P} = P(\chi^2 > 8.548 | df=4)$;
   Using Table 8: $0.05 < \mathbf{P} < 0.10$
   Using computer: $\mathbf{P} = 0.074$
b. $\mathbf{P} > \alpha$
-- using classical approach ----------------
a. critical region: $\chi^2 \geq 9.49$
b. $\chi^{2}*$ is in the noncritical region

Step 5: a. Fail to reject H_O
   b. At the 0.05 level of significance, there is not
      sufficient evidence to show a lack of
      independence between the number of defective
      articles and the day of the week.

11.39 Step 1: a. The proportions: P(Blog creators) and P(Blog
                    readers) are the same for all months.
                 b. H_O: Blog creators and Blog readers are
                       proportioned the same for each month.
                    H_a: Blog creators and Blog readers are not
                       proportioned the same for each month.
        Step 2: a. Assume a random sample.
                 b. χ^2 with df = 2   c. $\alpha = 0.05$
        Step 3: a. sample information given in exercise
                 b. $\chi^2 = \Sigma[(O-E)^2/E]$

```
           Expected counts are printed below observed
           counts
           Chi-Square contributions are printed below
           expected counts

                      Blog      Blog
                   creators   readers   Total
              1         74       205      279
                     62.78    216.23
                      2.007     0.583

              2         93       316      409
                     92.03    316.98
                      0.010     0.003

              3        130       502      632
                    142.20    489.80
                      1.047     0.304

           Total        297      1023     1320

           Chi-Sq = 3.954, DF = 2, P-Value = 0.138
```

$\chi^2* = 3.954$

Step 4: -- using p-value approach --------------------
a. $P = P(\chi^2 > 3.954 | df=2)$;
     Using Table 8: $0.10 < P < 0.25$
     Using computer: $P = 0.138$
b. $P > \alpha$
-- using classical approach -----------------
a. critical region: $\chi^2 \geq 5.99$
b. χ^2* is in the noncritical region

Step 5: a. Fail to reject Ho
     b. At the 0.05 level of significance, there is not
        sufficient evidence to show that the distribution
        of blog creators and blog readers are not
        proportioned the same for each month throughout
        the months listed.

11.41 Step 1: a. The proportions: P(who fear darkness) and P(who
                do not fear darkness) are the same for all age
                groups.
            b. Ho: Fear and Do Not Fear darkness are
                proportioned the same for each age group.
                Ha: Fear and Do Not Fear darkness are not
                    proportioned the same for each age group.
        Step 2: a. Assume a random sample.
            b. χ^2 with df = 4   c. $\alpha = 0.01$
        Step 3: a. sample information given in exercise
            b. $\chi^2 = \Sigma[(O-E)^2/E]$
        Expected values:

	Elem	J. H.	S. H.	Coll.	Adult
Fear	70.8	70.8	70.8	70.8	70.8
Do not	129.2	129.2	129.2	129.2	129.2

$\chi^2* =$ 2.102 + 0.020 + 6.712 + 17.105 + 26.359
          + 1.152 + 0.011 + 3.678 + 9.373 + 14.445 = 80.957

        Step 4: -- using p-value approach -------------------
            a. $P = P(\chi^2 > 80.957 | df=4)$;
                Using Table 8:   P < 0.005
                Using computer: P = 0+
            b. $P < \alpha$
            -- using classical approach ----------------
            a. critical region: $\chi^2 \geq 13.3$
            b. ••* is in the critical region

        Step 5: a. Reject Ho
            b. There is sufficient evidence to show that the
                age groups have different proportions which fear
                darkness, at the 0.01 level of significance.

11.43 Step 1: a. The proportions: P(females) and P(males) are the
                same for all drug dosage groups.
            b. Ho: Females and males are proportioned the same
                for each drug dosage.
                H_a: Females and males are not proportioned the
                same for each drug dosage.
        Step 2: a. Assume a random sample.
            b. χ^2 with df = 2   c. $\alpha = 0.01$

-- 434 --

Step 3: a. sample information given in exercise
  b. $\chi^2 = \Sigma[(O-E)^2/E]$

  Expected counts are printed below observed counts
  Chi-Square contributions are printed below expected counts

	10 mg drug	20 mg drug	placebo	Total
1	54	56	60	170
	57.33	55.33	57.33	
	0.194	0.008	0.124	
2	32	27	26	85
	28.67	27.67	28.67	
	0.388	0.016	0.248	
Total	86	83	86	255

  $\chi^{2*} = 0.978$

Step 4: -- using p-value approach --------------------
  a. $P = P(\chi^2 > 0.978 | df=2)$;
      Using Table 8: $0.50 < P < 0.75$
      Using computer: $P = 0.613$
  b. $P > \alpha$
  -- using classical approach ----------------
  a. critical region: $\chi^2 \geq 9.21$
  b. ••* is in the noncritical region

Step 5: a. Fail to reject Ho
  b. There is not sufficient evidence to show the distribution is not proportioned the same for each drug, at the 0.01 level of significance.

Step 1: a. The proportions: P(ages 40-49), P(ages 50-59) and P(ages 60-69) are the same for all drug dosage groups.
  b. Ho: Age groups are proportioned the same for each drug dosage.
     Ha: Age groups are not proportioned the same for each drug dosage.
Step 2: a. Assume a random sample.
  b. χ^2 with df = 4   c. $\alpha = 0.01$

Step 3: a. sample information given in exercise
b. $\chi^2 = \Sigma[(O-E)^2/E]$

Expected counts are printed below observed counts
Chi-Square contributions are printed below expected counts

	10 mg drug B	20 mg drug B	placebo B	Total
1	18	20	19	57
	19.22	18.55	19.22	
	0.078	0.113	0.003	
2	48	41	57	146
	49.24	47.52	49.24	
	0.031	0.895	1.223	
3	20	22	10	52
	17.54	16.93	17.54	
	0.346	1.521	3.239	
Total	86	83	86	255

$\chi^{2}* = 7.449$

Step 4: -- using p-value approach -------------------
a. $P = P(\chi^2 > 7.449 | df=4)$;
Using Table 8: $0.10 < P < 0.25$
Using computer: $P = 0.114$
b. $P > \alpha$
-- using classical approach -----------------
a. critical region: $\chi^2 \geq 13.3$
b. $\chi^{2}*$ is in the noncritical region

Step 5: a. Fail to reject H_O
b. There is not sufficient evidence to show the distribution is not proportioned the same for each drug, at the 0.01 level of significance.

11.45 a.

	West	N.Cent.	South	N.East	Total
1	55	46	47	41	190
	47.25	47.25	47.25	47.25	
2	45	54	53	59	210
	52.75	52.75	52.75	52.75	
Total	100	100	100	100	400

$$\chi^{2}* = 1.271 + 0.033 + 0.001 + 0.827 + 1.139 + 0.030 + 0.001 + 0.741 = \underline{4.043}$$

$$\mathbf{P} = P(\chi^{2}* > 4.043 | df = 3) = \underline{0.257}$$

b.

	West	N.Cent.	South	N.East	Total
1	110	92	94	82	380
	94.5	94.5	94.5	94.5	
2	90	108	106	118	420
	105.5	105.5	105.5	105.5	
Total	200	200	200	200	800

$$\chi^{2}* = 2.542 + 0.066 + 0.003 + 1.653 + 2.277 + 0.059 + 0.002 + 1.481 = \underline{8.083}$$

$$\mathbf{P} = P(\chi^{2}* > 8.083 | df = 3) = \underline{0.044}$$

	West	N.Cent.	South	N.East	Total
1	165	138	141	123	570
	141.75	141.75	141.75	141.75	
2	135	162	159	177	630
	158.25	158.25	158.25	158.25	
Total	300	300	300	300	1200

$$\chi^{2}* = 3.813 + 0.099 + 0.004 + 2.480 + 3.416 + 0.089 + 0.004 + 2.222 = \underline{12.127}$$

$$\mathbf{P} = P(\chi^{2}* > 12.127 | df = 3) = \underline{0.007}$$

c. Yes, as the sample sizes increase, the relative differences are the same but the actual counts are much larger and further apart.

CHAPTER EXERCISES

11.47 Step 1: a. The proportions: P(1st type), P(2nd type),
P(3rd type)

b. H_O: 1:3:4 proportions
H_a: proportions are other than 1:3:4

Step 2: a. Assume that the 80 hybrids represent a random sample.

b. χ^2 with df = 2   c. $\alpha = 0.05$

Step 3: a. sample information given in exercise

b. $\chi^2 = \Sigma[(O-E)^2/E]$ (as found on accompanying table)

$E(1st) = n \cdot p = 800[1/(1+3+4)] = 100$
$E(2nd) = 800[3/(1+3+4)] = 300$
$E(3rd) = 800[4/(1+3+4)] = 400$

	1st	2nd	3rd	Total
Observed	80	340	380	800
Expected	100	300	400	800
$(O-E)^2/E$	4.00	5.33	1.00	10.33

$\chi^{2}* = 10.33$

Step 4: -- using p-value approach --------------------

a. $P = P(\chi^2 > 10.33 | df=2)$;
Using Table 8: $0.005 < P < 0.01$
Using computer: $P = 0.006$

b. $P < \alpha$
-- using classical approach ----------------

a. $\chi^2(2,0.05) = 5.99$ $\chi^2 \geq 5.99$

b. $\chi^{2}*$ is in the critical region.

Step 5: a. Reject H_O

b. We have sufficient evidence to show that the ratio is not the hypothesized 1:3:4 ratio, at the 0.05 level of significance.

11.49 Step 1: a. The proportions: P(parent Home), P(campus), P(off-campus), P(own home), P(other)

       a. H_O: P(parent home) = 0.46, P(campus) = 0.26, P(off-campus) = 0.18, P(own home) = 0.09, P(other) = 0.02

       H_a: The percentages are different than listed

  Step 2: a. Assume that the 1000 individuals represent a random sample.

       b. χ^2 with df = 4   c. $\alpha = 0.05$

  Step 3: a. sample information given in exercise

       b. $\chi^2 = \Sigma[(O-E)^2/E]$ (as found on accompanying table)

$$E(parent) = n \cdot p = 1000(0.46) = 460,$$
$$E(campus) = 1000(0.26) = 260,$$
$$E(off\text{-}campus) = 1000(0.18) = 180,$$
$$E(own\ home) = 1000(0.09) = 90,$$
$$E(other) = 1000(0.02) = 20$$

	Par.	Cam.	Off.	Own.	Oth.	Total
Observed	484	230	168	96	22	1000
Expected	458	258	178	88	18	1000
$(O-E)^2/E$	1.476	3.039	0.562	0.727	0.889	6.693

$$\chi^2{}^* = 6.693$$

  Step 4: -- using p-value approach --------------------

      a. $\mathbf{P} = P(\chi^2 > 6.693 | df=4)$;

        Using Table 8: $0.10 < \mathbf{P} < 0.25$

        Using computer: $\mathbf{P} = 0.153$

      b. $\mathbf{P} > \alpha$

      -- using classical approach -----------------

      a. $\chi^2(4, 0.05) = 9.49$ $\chi^2 \geq 9.49$

      b. $\chi^2{}^*$ is in the noncritical region.

  Step 5: a. Fail to reject H_O

      b. We do not have sufficient evidence to show that the sample distribution is different than the newspaper distribution, at the 0.05 level of significance.

11.51 Step 1: a. The largest number of holes played by golfers in one day.

b. H_O: $P(18) = 0.05$, $P(19\text{-}27) = 0.12$,
$P(28\text{-}36) = 0.28$, $P(37\text{-}45) = 0.20$,
$P(46\text{-}54) = 0.18$, $P(55 \text{ or more}) = 0.17$

H_a: The percentages are different than listed

Step 2: a. Assume that the 200 individuals represent a random sample.

b. χ^2 with df = 5   c. $\alpha = 0.01$

Step 3: a. sample information given in exercise

b. $\chi^2 = \Sigma[(O\text{-}E)^2/E]$

$$\chi^{2}* = \frac{(12-10)^2}{10} + \frac{(35-24)^2}{24} + \frac{(60-56)^2}{56} + \frac{(44-40)^2}{40} + \frac{(35-36)^2}{36} + \frac{(14-34)^2}{34}$$

$$= 0.40 + 5.04 + 0.29 + 0.40 + 0.03 + 11.76 = 17.92$$
$$\chi^{2}* = 17.92$$

Step 4: -- using p-value approach --------------------

a. $\mathbf{P} = P(\chi^2 > 17.92 | df=5)$;

Using Table 8: $\mathbf{P} < 0.005$
Using computer: $\mathbf{P} = 0.003$

b. $\mathbf{P} < \alpha$

-- using classical approach -----------------

a. $\chi^2(5, 0.01) = 15.1$ $\chi^2 \geq 15.1$

b. χ^2* is in the critical region.

Step 5: a. Reject H_O

b. We do have sufficient evidence to show that the sample distribution is different than the Golf magazine distribution, at the 0.01 level of significance.

11.53 The probability associated with each weight interval can be found using the standard normal distribution on Table 3 (Appendix B, ES10-p810).

$P(x < 130) = P[z < (130\text{-}160)/15]$
$\qquad\qquad = P(z < -2.00) = 0.5000 - 0.4772 = \underline{0.0228}$

$P(130 < x < 145) = P[-2.00 < z < (145\text{-}160)/15]$
$\qquad\qquad = P(-2.00 < z < -1.00) = 0.4772 - 0.3413 = \underline{0.1359}$

$P(145 < x < 160) = P[-1.00 < z < (160\text{-}160)/15]$
$\qquad\qquad = P(-1.00 < z < 0.00) = \underline{0.3413}$

$P(160 < x < 175) = P[0.00 < z < (175-160)/15]$
$$= P(0.00 < z < 1.00) = \underline{0.3413}$$

$P(175 < x < 190) = P[1.00 < z < (190-160)/15]$
$$= P(1.00 < z < 2.00) = 0.4772 - 0.3413 = \underline{0.1359}$$

$P(x > 190) = P(z > 2.00) = 0.5000 - 0.4772 = \underline{0.0228}$

Step 1: a. The proportions of weight: $P(< 130)$, $P(130-144)$, $P(145-159)$, $P(160-174)$, $P(175-189)$, (190 and over).
       b. H_o: The weights are normally distributed about a mean of 160 with a standard deviation of 15 pounds
       H_a: The weights are not $N(160,15)$

Step 2: a. Assume that the 300 adult males represent a random sample.
       b. χ^2 with df = 5   c. $\alpha = 0.05$

Step 3: a. sample information given in exercise
       b. $\chi^2 = \Sigma[(O-E)^2/E]$ (as found on accompanying table)

Expected values = $n \cdot p$ = $300 \cdot p$

	<130	130-145	145-160	160-175	175-190	>190
Observed	7	38	100	102	40	13
Expected	6.84	40.77	102.39	102.39	40.77	6.84
$(O-E)^2/E$	0.004	0.188	0.056	0.001	0.015	5.548

$$\chi^2* = 5.812$$

Step 4: -- using p-value approach --------------------
       a. $\mathbf{P} = P(\chi^2 > 5.812 | df=5)$;
         Using Table 8: $0.25 < \mathbf{P} < 0.50$
         Using computer: $\mathbf{P} = 0.325$
       b. $\mathbf{P} > \alpha$
       -- using classical approach -----------------
       a. $\chi^2(5, 0.05) = 11.1$ $\chi^2 \geq 11.1$
       b. χ^2* is in the noncritical region.

Step 5: a. Fail to reject H_o
       b. We do not have sufficient evidence to show that this data is not normally distributed with $\mu = 160$ and $\sigma = 15$, at the 0.05 level of significance.

11.55 Step 1: a. The proportions for obtaining comfort food:
P(buy it), P(make it), P(ask someone to make it), P(don't know)

b. H_O: P(buy it) = 0.45, P(make it) = 0.36,
P(ask someone to make it) = 0.14,
P(don't know) = 0.05,

H_a: The percentages are different than listed

Step 2: a. Assume that the 120 individuals represent a random sample.

b. χ^2 with df = 3   c. α = 0.05

Step 3: a. sample information given in exercise

b. $\chi^2 = \Sigma[(O-E)^2/E]$ (as found on accompanying table)
E(buy it) = n·p = 120(0.45) = 54,
E(make it) = 120(0.36) = 43.2,
E(ask someone to make it) = 120(0.14) = 16.8,
E(don't know) = 120(0.05) = 6,

	Buy it	make it	ask someone	don't know	Total
Observed	54	48	17	1	120
Expected	54	43.2	16.8	6	120
$(O-E)^2/E$	0.000	0.533	0.002	4.167	4.70

$$\chi^{2*} = 4.70$$

Step 4: -- using p-value approach --------------------

a. $P = P(\chi^2 > 4.70 | df=3)$;
Using Table 8: $0.10 < P < 0.25$
Using computer: $P = 0.195$

b. $P > \alpha$

-- using classical approach -----------------

a. $\chi^2(3, 0.05) = 7.82$ $\chi^2 \geq 7.82$

b. χ^{2*} is in the noncritical region.

Step 5: a. Fail to reject H_O

b. We do not have sufficient evidence to show that the Midwest response distribution is different than the nation as a whole, at the 0.05 level of significance.

c. There is not sufficient evidence to say the East
   Coast and the Midwest obtain their comfort food
   in proportions that follow a distribution
   different from that of the nation as a whole.
   This does not necessarily say that their
   distributions are the same or the same as the
   nations - they're just not significantly
   different.

11.57 a. Step 1: a. The distribution of colors in a bag of M&M's.
                   b. H_o: P(brown) = 0.30, P(red & yellow) = 0.20,
                         P(blue, green & orange) = 0.10
                      H_a: Distributions are different
           Step 2: a. Assume a random sample.
                   b. χ^2 with df = 5   c. α = 0.05
           Step 3: a. sample information given in exercise
                   b. $\chi^2 = \Sigma[(O-E)^2/E]$ (as found on accompanying table)

Expected	red	green	blue	orange	yellow	brown
Probability	0.20	0.10	0.10	0.10	0.20	0.30
Expected Frequency	11.6	5.8	5.8	5.8	11.6	17.4
Observed Frequency	15	9	3	3	9	19

   $\chi^2* = 0.99655 + 1.76552 + 1.35172 + 1.35172 +$
          $0.58276 + 0.14713 = \underline{6.1954}$
   Step 4: -- using p-value approach -------------------
           a. **P** = P(χ^2 > 6.1954|df=5);
                        Using Table 8:   **P** > 0.25
                        Using computer/calculator: **P** = 0.2877
           b. **P** > α
           -- using classical approach -----------------
           a. critical region: $\chi^2 \geq 11.1$
           b. χ^2* falls in the critical region

   Step 5: a. Fail to reject H_o
           b. There is not significant evidence to show the
              distribution is different than the target
              distribution, at the 0.05 level of
              significance.

b. Step 1: a. The distribution of colors in a bag of M&M's.
      b. H_o: P(brown) = 0.30, P(red & yellow) = 0.20,
            P(blue, green & orange) = 0.10
         H_a: Distributions are different
Step 2: a. Assume a random sample.
      b. χ^2 with df = 5   c. α = 0.05
Step 3: a. sample information given in exercise
      b. $\chi^2 = \Sigma[(O-E)^2/E]$ (as found on accompanying table)

Expected	red	green	blue	orange	yellow	brown
Probability	0.20	0.10	0.10	0.10	0.20	0.30
Expected Frequency	23.4	11.7	11.7	11.7	23.4	35.1
Observed Frequency	24	26	22	6	12	27

$\chi^{2}* = 0.0154 + 17.4778 + 9.0675 + 2.7769 + 5.5538 + 1.8692 = \underline{36.761}$

Step 4: -- using p-value approach -------------------
      a. **P** = P(χ^2 > 36.761|df=5);
            Using Table 8:   **P** < 0.005
            Using computer/calculator: **P** = 0+
      b. **P** > α
      -- using classical approach ----------------
      a. critical region: $\chi^2 \geq 11.1$
      b. $\chi^{2}*$ falls in the critical region

Step 5: a. Reject H_o
      b. There is significant evidence to show the
         distribution is different than the target
         distribution, at the 0.05 level of
         significance.

c. Step 1: a. The distribution of colors in a bag of M&M's.
      b. H_o: P(brown) = 0.30, P(red & yellow) = 0.20,
            P(blue, green & orange) = 0.10
         H_a: Distributions are different
Step 2: a. Assume a random sample.
      b. χ^2 with df = 5   c. α = 0.05
Step 3: a. sample information given in exercise
      b. $\chi^2 = \Sigma[(O-E)^2/E]$ (as found on accompanying table)

Expected	red	green	blue	orange	yellow	brown
Probability	0.20	0.10	0.10	0.10	0.20	0.30
Expected Frequency	342.6	171.3	171.3	171.3	342.6	513.9
Observed Frequency	288	222	217	199	413	374

$$\chi^{2}\text{*} = 08.7016 + 15.0058 + 12.1920 + 4.4792 + 14.4663 + 38.0853 = \underline{92.93}$$

Step 4: -- using p-value approach ------------------

a. $\mathbf{P} = P(\chi^2 > 92.93|df=5)$;

Using Table 8: $\mathbf{P} < 0.005$

Using computer/calculator: $\mathbf{P} = 0+$

b. $\mathbf{P} > \alpha$

-- using classical approach -----------------

a. critical region: $\chi^2 \geq 11.1$

b. $\chi^{2}\text{*}$ falls in the critical region

Step 5: a. Reject $H_{\circ}$

b. There is significant evidence to show the distribution is different than the target distribution, at the 0.05 level of significance.

d. Chi-square is not very sensitive to the variability within small samples, and becomes more sensitive to those variations as the sample size gets larger.

11.59   Step 1: a. The distributions of housing starts: P(northeast for years category), P(south for years category), P(midwest for years category), P(west for years category).

b. H_O: The distribution of housing starts across the regions is the same for all years.

H_a: The distributions are different.

Step 2: a. Assume a random sample.

b. χ^2 with df = 6   c. $\alpha = 0.05$

Step 3: a. sample information given in exercise

b. $\chi^2 = \Sigma[(O-E)^2/E]$ (as found on accompanying table)

Expected counts are printed below observed counts

	1996-2000	2001-2005	2006-2010	Total
1	145	161	170	476
	160.20	158.05	157.75	
2	710	687	688	2085
	701.71	692.31	690.97	
3	331	314	313	958
	322.42	318.10	317.48	
4	382	385	373	1140
	383.67	378.53	377.80	
Total	1568	1547	1544	4659

Chi-Sq = 1.442 + 0.055 + 0.952 + 0.098 +
0.041 + 0.013 + 0.228 + 0.053 +
0.063 + 0.007 + 0.111 + 0.061 = 3.123

$\chi^2* = 3.123$

Step 4: -- using p-value approach ------------------
a. $P = P(\chi^2 > 3.123 | df=6)$;
Using Table 8: $0.75 < P < 0.90$
Using computer: $P = 0.793$
b. $P > \alpha$
-- using classical approach ----------------
a. $\chi^2(6,0.05) = 12.6$
b. χ^2* is in the noncritical region.

Step 5: a. Fail to reject H_O
b. There is not sufficient evidence to show the
distribution of housing starts across the
regions are different, at the 0.05 level of
significance.

11.61 Step 1: a. The proportion of each political preference for
each age group and the proportion of each age
group who answer with each political preference.
b. H_o: Political preference is independent of age.
H_a: Political preference is not independent of
age.

Step 2: a. Assume a random sample.
   b. χ^2 with df = 4   c. $\alpha = 0.01$
Step 3: a. sample information given in exercise
   b. $\chi^2 = \Sigma[(O-E)^2/E]$ (as found on accompanying table)

Expected values:

	20-35	36-50	Over 50
Conservative	28.00	30.00	22.00
Moderate	73.50	78.75	57.75
Liberal	38.50	41.25	30.25

$$\chi^{2}* = \begin{array}{l} 2.286 + 3.333 + 0.182 + \\ 0.575 + 0.496 + 2.815 + \\ 0.058 + 6.402 + 7.192 = 23.339 \end{array}$$

Step 4: -- using p-value approach -------------------
   a. $\mathbf{P} = P(\chi^2 > 23.339 | df=4)$;
      Using Table 8: $\mathbf{P} < 0.005$
      Using computer/calculator: $\mathbf{P} = 0.000+$
   b. $\mathbf{P} < \alpha$
   -- using classical approach -----------------
   a. $\chi^2(4, 0.01) = 13.3$ $\chi^2 \geq 13.3$
   b. $\chi^{2}*$ falls in the critical region

Step 5: a. Reject H_o
   b. There is sufficient evidence to reject the null
      hypothesis and conclude that political
      preference is not independent of age, at the
      0.01 level of significance.

11.63 Step 1: a. The proportions of popcorn that popped and did
                not pop:P(Brand A), P(Brand B), P(Brand C),
                P(Brand D)
             b. H_O: Proportion of popcorn that popped is the
                same for all brands.
                H_a: The proportions are not the same for all
                       brands.
          Step 2: a. Assume a random sample.
             b. χ^2 with df = 3   c. $\alpha = 0.05$
          Step 3: a. sample information given in exercise
             b. $\chi^2 = \Sigma[(O-E)^2/E]$

Expected values:

	A	B	C	D	Total
Popped	88	88	88	88	352
Not popped	12	12	12	12	48
Totals	100	100	100	100	400

$$\chi^{2*} = \Sigma[(O-E)^2/E] = 0.333 + 1.333 + 0.083 + 0.750$$
$$+ 0.045 + 0.182 + 0.011 + 0.102$$
$$\chi^{2*} = 2.839$$

Step 4: -- using p-value approach -------------------

    a. $\mathbf{P} = P(\chi^2 > 2.839 | df=3)$;

        Using Table 8: $0.25 < \mathbf{P} < 0.50$

        Using computer: $\mathbf{P} = 0.417$

    b. $\mathbf{P} > \alpha$

    -- using classical approach ----------------

    a. $\chi^2(3,0.05) = 7.82$ $\chi^2 \geq 7.82$

    a. χ^{2*} falls in the noncritical region

Step 5: a. Fail to reject H_o

    b. There is not sufficient evidence to show that
the proportions of popped corn are not the same
for all brands, at the 0.05 level of
significance.

11.65   a. 2003: $18650/(18650+6812) = 0.732 = 73.2\%$ deceased donors

        2004: $20018/(20018+6966) = 0.742 = 74.2\%$ deceased donors

    These percentages do not seem significantly different.

b.

Step 1: a. The proportions for organ donors from a deceased
person and living person, over 2003 and 2004.

    b. H_o: The ratio of deceased donor to living donor
is the same.

    H_a: The ratio of organ donors is not the same.

Step 2: a. Given a random sample.

    b. χ^2 with df = 1   c. $\alpha = 0.05$

Step 3: a. sample information given in exercise

    b. $\chi^2 = \Sigma[(O-E)^2/E]$ (as found on accompanying table)

Expected counts are printed below observed
counts

Chi-Square contributions are printed below
expected counts

	From a deceased donor	From a living donor	Total
1	18650	6812	25462
	18772.92	6689.08	
	0.805	2.259	
2	20018	6966	26984
	19895.08	7088.92	
	0.759	2.131	
Total	38668	13778	52446

$$\chi^{2*} = 5.955$$

Step 4: -- using p-value approach ------------------
    a. $P = P(\chi^2 > 5.955 | df=1)$;
        Using Table 8: $0.01 < P < 0.025$
        Using computer/calculator: $P = 0.015$
    b. $P < \alpha$
    -- using classical approach ----------------
    a. $\chi^2(1, 0.05) = 3.84$ $\chi^2 \geq 3.84$
    b. χ^2* falls in the critical region

Step 5: a. Reject H_o
        b. The ratio of deceased versus living donors did
           change significantly between 2003 and 2004, at
           the 0.05 level of significance.

    c. The decision reached in the hypothesis is the opposite of
       what was predicted in part a. With very large sample
       sizes, differences in proportions must be very small to
       be considered non-existent.

11.67 Conditions for rolling a balanced die 600 times:
The critical value is $\chi^2(5, 0.05) = 11.1$ (6 possible outcomes)

With 600 rolls, the expected value for each cell is 100.

Many combinations of observed frequencies are possible to
cause us to reject the equally likely hypothesis. The
combinations will have to have a calculated χ^2 value greater
than 11.1 or a p-value less than 0.05. Two possibilities
are presented.

1. If each observed frequency is different from the
   expected by the same amount, then $11.1/6 = 1.85$ is the
   amount of chi-square that would come from each cell.

   $(O-E)^2/E = (O - 100)^2/100 = 1.85$

   $(O - 100)^2 = 185$

   $O - 100 = \pm 13.6$

   $O = 86$ or 114

That is, if three of the observed frequency values are
86 and the other three are 114, the faces of the die
will be declared not to be equally likely.

Row	P	OBS	EXP	CHI-SQ
1	0.166667	86	100.000	1.96005
2	0.166667	86	100.000	1.96005
3	0.166667	86	100.000	1.96005
4	0.166667	114	100.000	1.95994
5	0.166667	114	100.000	1.95994
6	0.166667	114	100.000	1.95994

Row	SUM(P)	SUN(OBS)	SUM(EXP)	CHI-SQ*
1	1.00000	600	600.001	11.7600

DF   5

p-value   0.038

Note χ^2 value and p-value

2. Now suppose just one is different and the other five all occur with the same frequency:
Remember, the total observed most be 600. Therefore, for every five one outcome is different from the expected, the other five each must be different by one to balance. If the five are each different from the expected by x, then the one that is very different is off by 5x. The sum of 5 - x's squared and 5x squared is $30x^2$. Thus,

$30x^2 = 11.1$

$x^2 = 11.1/30 = 0.37$

$(O-E)^2/E = (O-100)^2/100 = 0.37$

$(O-100)^2 = 37$

$O - 100 = \pm 6.08$ (round-up)

O = either 93 or 107   for the five cells, and

O for the other cell must be off by 5(7) or 35; it is either 65 or 135.

Row	P	OBS	EXP	CHI-SQ
1	0.166667	93	100.000	0.4900
2	0.166667	93	100.000	0.4900
3	0.166667	93	100.000	0.4900
4	0.166667	93	100.000	0.4900
5	0.166667	93	100.000	0.4900
6	0.166667	135	100.000	12.2498

Row	SUM(P)	SUN(OBS)	SUM(EXP)	CHI-SQ*
1	1.00000	600	600.001	14.7000

DF   5

p-value   0.012

Note χ^2 and p-value.

CHAPTER 12 ▽ ANALYSIS OF VARIANCE

Chapter Preview

In Chapters 8 and 9, hypothesis tests were demonstrated for testing
a single mean. Hypothesis tests between two means was subsequently
demonstrated in Chapter 10. To continue in this fashion, Chapter
12 introduces the concept of the analysis of variance technique
(ANOVA) so that a hypothesis test for the equality of several means
can be completed. The F-distribution will be utilized in this
test, since we will be comparing the measures of variation
(variance) among the different sets of data and the measure of
variation (variance) within the sets the data.

An article, featured in USA Today on time spent reading the
newspaper, is used in this chapter's opening section ''Time Spent
Commuting to Work''.

CHAPTER 12 EXERCISES

12.1 a.

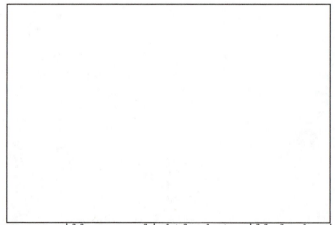

b. Answers will vary slightly but will look similar to
above.

c. Yes, it does appear that the city has an effect on the average amount of time spent by workers who commute to work, as not all of the means are the same value.

d. Yes, it also appears that the city has an effect on the variation. Boston and St. Louis have larger spreads to their datasets.

SECTION 12.2 EXERCISES

12.3 Units produced per hour at each temperature level

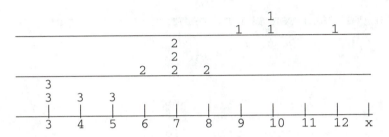

Yes, there appears to be a difference between the three sets.

12.5 a. df(error) = 20 - 3 = 17; SS(Factor) = 164.2 - 40.4 = 123.8
MS(Factor) = 123.8/3 = 41.2667
MS(Error) = 40.4/17 = 2.3765

b. F* = 41.2667/2.3765 = 17.36

SECTION 12.3 EXERCISES

12.7 The boxplot shows greater variability between the four levels. The means for each are all quite different.

12.9 a. The category has a slight effect on the average number of students per computer. Students in senior high have 1 less person per computer than elementary students.

b. The number of students per computer is separated into categories of levels of school just as ANOVA subdivides data into categories. The data could be the number of students per computer from each school in a sample of schools with schools at each different level.

SECTION 12.4 EXERCISES

12.11   a. 0   b. 2   c. 4   d. 31   e. 393

WRITING HYPOTHESES FOR THE DIFFERENCE AMONG SEVERAL MEANS

null hypothesis: H_O: $\mu_1 = \mu_2 = \mu_3 = \ldots = \mu_n$

   or

H_O: The mean values for all n levels of the experiment are the same.
[factor has no effect]

alternative hypothesis: H_a: The means are not all equal.
[factor has an effect]

   or

H_a: At least one mean value is different from the others.

Use subscripts on the population means that correspond to the different levels or sources of the experiment.

12.13   a. H_O: $\mu_1 = \mu_2 = \mu_3 = \mu_4 = \mu_5$ vs. H_a: Means not all equal
[mean scores are all same]

   b. H_O: $\mu_1 = \mu_2 = \mu_3 = \mu_4$ vs. H_a: Means not all equal
[mean scores are all same]

   c. H_O: $\mu_1 = \mu_2 = \mu_3 = \mu_4$ vs. H_a: Means not all equal
[factor has no effect]   [has an effect]

   d. H_O: $\mu_1 = \mu_2 = \mu_3$ vs. H_a: Means not all equal
[no effect]   [has an effect]

Use Tables 9A, B or C (Appendix B, ES10-pp816-821) depending on α. Locate the critical value using the degrees of freedom for the numerator (factor) and the degrees of freedom for the denominator(error). Reviewing how to determine the test criteria in: ES10-p594, SSM-p394&395, as it is applied to the F-distribution may be helpful.

12.15   a.   b.

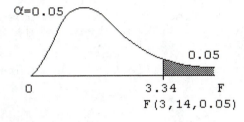

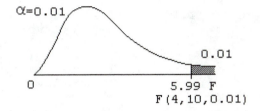

c.

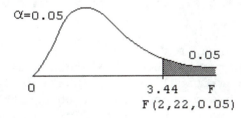

Review the rules for calculating the p-value in: ES10-p597, SSM-p394, if necessary. Remember to use the F-distribution, therefore either Tables 9A, B or C will be used to find probabilities.

12.17 a. 0.04 of the probability distribution associated with F
and a true null hypothesis is more extreme than F*.
That is, area under the curve and to the right of F*.

      b. Reject the null hypothesis; since the p-value is less
than the previously set value for alpha.

      c. Fail to reject the null hypothesis; since the p-value
is greater than the previously set value for alpha.

12.19 a. The test factor has no effect on the mean at the
tested levels or the mean value of the variable is the
same at all levels of the test factor.

      b. The test factor does have an effect on the mean at the
tested levels

      c. For the p-value approach, $\mathbf{P} = P(F > F^*)$ must be $\leq \alpha$.
For the classical approach, the calculated value of F
must fall in the critical region; that is, the variance
between levels of the factor must be significantly
larger than variance within the levels.

      d. The tested factor has a significant effect on the
variable.

      e. For the p-value approach, $\mathbf{P} = P(F > F^*)$ must be $> \alpha$.
For the classical approach, the calculated value of F
must fall in the non-critical region; that is, the
variance between levels of the factor must not be
significantly larger than variance within the levels.

      f. The tested factor does not have a significant effect on
the variable.

Hypothesis Test for the Difference Among Several Means,
Independent Samples

Review the parts to a hypothesis test as outlined in: ES10-pp427&444-445, SSM-pp265&275, if needed. Slight changes will occur in:

1. **the hypotheses**: (see box before exercise 12.13)
2. **the calculated test statistic**

$$F = \frac{MS(factor)}{MS(error)}$$

3. If H_O is rejected, a significant difference among the means is indicated, the various levels of the factor do have an effect.

   If H_O is not rejected, no significant difference among the means is indicated, the various levels of the factors do not have an effect.

FORMULAS FOR ANOVA TABLE

$$SS(total) = \sum(x^2) - \frac{(\sum x)^2}{n}$$

$$SS(factor) = \left(\frac{C_1^2}{k_1} + \frac{C_2^2}{k_2} + \frac{C_3^2}{k_3} + \ldots \right) - \frac{(\sum x)^2}{n} \quad , \text{ where } C_i = \text{column total}$$

k_i = number of data values in the ith column

(check: $n = \sum k_i$)

$$SS(error) = \sum x^2 - \left(\frac{C_1^2}{k_1} + \frac{C_2^2}{k_2} + \frac{C_3^2}{k_3} + \ldots \right) \quad \text{or} \quad = SS(total) - SS(factor)$$

$df_{total} = n-1, \qquad df_{factor} = c-1, \qquad df_{error} = n - c$

$df_{total} = df_{factor} + df_{error}$

. . .

```
ANOVA Table
Source     df              SS              MS          MS(factor) =
Factor                                                     SS(factor)/df(factor)
Error                                                  MS(error) =
Total                                                      SS(error)/df(error)
```

12.21   a. n = 17
            b. df(Group) = 2 = c - 1; c = 3 groups
            c. MS(Group) = 11.0/2 = 5.50
                MS(Error) = 36.53/14 = 2.609 = 2.61
            d. F = 5.50/2.61 = 2.107 = 2.11
            e. $P = P(F > 2.11 | df_n = 2, df_d = 14)$
                using Tables 9: P > 0.05
                using a computer/calculator: $P = 0.1581 \cdot 0.159$
            f. Most likely with a p-value of 0.159, the decision would
                be to Fail to Reject Ho. The conclusion would be that
                there is not sufficient evidence to show the group means
                are not all the same.

12.23   Verify -- answers given in exercise.


```
Computer and/or calculator commands to ANOVA hypothesis tests can
be found in ES10-p674. Explanation of the output is found in ES10-
p675.
```

12.25   Step 1:   a. The mean level of work for a new worker, the
                          mean level of work for worker A, the mean level
                          of work for worker B.
                      b. H$_o$: The mean values for workers are all equal.
                          H$_a$: The mean values for workers are not all
                          equal.
            Step 2:   a. Assume the data were randomly collected and are
                          independent, and the effects due to chance and
                          untested factors are normally distributed.
                      b. F   c. $\alpha = 0.05$

Step 3: a. $n = 15$, $C_1 = 46$, $C_2 = 58$, $C_3 = 57$, $T = 161$,
$\Sigma x^2 = 1771$

Source	df	SS	MS	F*
Work	2	17.73	8.87	4.22
Error	12	25.20	2.10	
Total	14	42.93		

$F* = 8.87/2.10 = 4.22$

Step 4: -- using p-value approach ---------------
a. $\mathbf{P} = P(F > 4.22 | df_n = 2, df_d = 12)$;
Using Table 9: $0.025 < \mathbf{P} < 0.05$
Using computer: $\mathbf{P} = 0.041$
b. $\mathbf{P} < \alpha$
-- using classical approach -------------
a. critical region: $F \geq 3.89$
b. F* falls in the critical region

Step 5: a. Reject H_o.
b. There is significant difference between the
workers   with regards to mean amount of work
produced, at the 0.05 level of significance.

12.27 Step 1: a. The mean ratings obtained by the restaurants in
each of the three categories.
b. H_o: $\mu_F = \mu_D = \mu_S$ (no difference in ratings)
Ha: The means of the ratings obtained by the
restaurants in the three categories are not all
equal.

Step 2: a. Assume the data were randomly collected and are
independent, and the effects due to chance and
untested factors are normally distributed.
b. F   c. $\alpha = 0.05$

Step 3: a. $n = 15$, $C_1 = 95$, $C_2 = 82$, $C_3 = 84$, $T = 261$,
$\Sigma x^2 = 4593$

b.

Source	DF	SS	MS	F	P
Factor	2	19.60	9.80	3.68	0.057
Error	12	32.00	2.67		
Total	14	51.60			

$F* = 9.80/2.67 = 3.68$

Step 4: -- using p-value approach ---------------
        a. $P = P(F > 3.68|df_n = 2, df_d = 12)$;
           Using Table 9: $P > 0.05$
           Using computer: $P = 0.057$
        b. $P > \alpha$
        -- using classical approach -------------
        a. critical region: $F \geq 3.89$
        b. F^* is in the noncritical region

Step 5: a. Fail to reject Ho.
        b. The data shows no significant evidence that the
           means of the three categories of ratings given
           to the restaurants are different, at the 0.05
           level of significance.

12.29   Step 1: a. The means of the won/loss percentages
                    obtained by the teams playing on the road
                    in each of the three divisions.
                 b. Ho: $\mu E = \mu C = \mu W$

                    Ha: The means of the won/loss percentages
                    obtained by the teams playing on the road
                    in the three divisions are not all equal.
        Step 2: a. Assume the data were randomly collected and are
                    independent, and the effects due to chance and
                    untested factors are normally distributed.
                 b. F   c. $\alpha = 0.05$
        Step 3: a. $n = 30$, $C_1 = 468.4$, $C_2 = 513.1$, $C_3 = 413.4$,

                    $T = 1394.9$, $\Sigma x^2 = 67455.2$

Source	DF	SS	MS	F	P
Factor	2	4.3	2.1	0.02	0.978
Error	27	2592.8	96.0		
Total	29	2597.0			

                    $F^* = 0.02$
        Step 4: -- using p-value approach ---------------
                 a. $P = P(F > 0.02|df_n = 2, df_d = 27)$;

                    Using Table 9: $P > 0.05$
                    Using computer: $P = 0.978$
                 b. $P > \alpha$

```
                -- using classical approach -------------
                a. F(2,27,0.05) = 3.39          F ≥ 3.39
                b. F* is not in the critical region
                -------------------------------------------
        Step 5:  a. Fail to reject H_O.
                b. At the 0.05 level of significance, the data
                   shows no evidence for concluding that the means
                   won/loss percentages obtained by the teams
                   playing on the road are not all equal.
```

12.31   Step 1: a. The mean age of three test groups: the mean age
                     for the TTS group, the mean age for the
                     Antivert group, the mean age for the placebo
                     group.
                  b. H_O: The mean age for groups are all equal.
                     H_a: The mean age for groups are not all equal.
          Step 2: a. Assume the data were randomly collected and are
                     independent, and the effects due to chance and
                     untested factors are normally distributed.
                  b. F   c. α = 0.05
          Step 3: a. n = 58, k_1 = 18, C_1 = 846, k_2 = 21, C_2 = 894,
                     k_3 = 19, C_3 = 805, T = 2545, Σx^2 = 120,549

Source	df	SS	MS	F*
Group	2	254.591	127.30	0.81
Error	55	8621.564	156.76	
Total	57	8876.155		

```
        Step 4:  -- using p-value approach ---------------
```
                a. $\mathbf{P}$ = P(F > 0.81|df_n = 2, df_d = 55);

                   Using Table 9: $\mathbf{P}$ > 0.05
                   Using computer: $\mathbf{P}$ = 0.449
                b. $\mathbf{P}$ > α
```
                -- using classical approach -------------
                a. F(2,55,0.05) ≈ 3.15          F ≥ 3.15
                b. F* is not in the critical region
                -------------------------------------------
```

Step 5: a. Fail to reject H_O.

          b. The data does not show a significant difference
between the mean ages of the groups, at the
0.05 level of significance.

12.33 a.

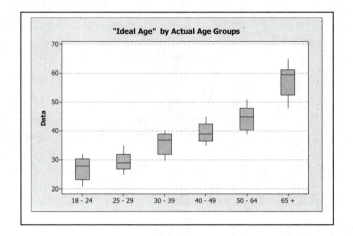

The graph suggests that each age group has its own
''ideal age'' and it increases with age, but each age
group seems to have a range of ideal ages within a 10
year range.

b. Step 1: a. The mean " ideal age" for different age groups.

          b. Ho: The mean " ideal age" is the same for all
age groups.

          Ha: The mean ''ideal age'' is not the same for
all age groups.

Step 2: a. Assume the data were randomly collected and are
independent, and the effects due to chance and
untested factors are normally distributed.

          b. F   c. $\alpha = 0.05$

Step 3: a. n = 36, C_{18} = 163, C_{25} = 207, C_{30} = 250, C_{40} =
197, C_{50} = 222, C_{65} = 346,

          T = 1385, Σx^2 = 57583

      b.

Source	DF	SS	MS	F	P
Factor	5	3765.3	753.1	42.33	0.000
Error	30	533.7	17.8		
Total	35	4299.0			

      F^* = 753.1/17.8 = 42.33

Step 4:   -- using p-value approach ---------------
          a. $P = P(F > 42.33 | df_n = 5, df_d = 30)$;

             Using Table 9: $P < 0.01$
             Using computer: $P = 0.000$
          b. $P < \alpha$
          -- using classical approach -------------
          a. critical region: $F \geq 2.53$
          b. F^* is in the critical region

Step 5:   a. Reject Ho.
             b. The data shows significant evidence that the
             mean ''ideal age'' is not the same for each of
             the age groups, at the 0.05 level of
             significance.

       c. Answers will vary but it appears that a person's ''ideal
          age'' depends on their current age.

       d. The boxplot showed that the mean ''ideal ages'' were
          different since the means do not line up horizontally.
          The idea that each age group has its own ideal age and
          it increases with current age is suggested by the fact
          that the means do line up on a nearly straight diagonal
          line.

12.35 a. Step 1: a. The mean hourly wages paid per month to
                     production workers.
                  b. H_o: The mean hourly wage is the same for each
                       of the twelve months.
                     H_a: The mean hourly wage is not the same for
                       each of the twelve months.
          Step 2: a. Assume the data were randomly collected and are
                     independent, and the effects due to chance
                     and untested factors are normally
                     distributed.
                  b. F   c. $\alpha = 0.05$
          Step 3: a.

Source	DF	SS	MS	F	P
Factor	11	0.78	0.07	0.03	1.000
Error	112	304.27	2.72		
Total	123	305.05			

Step 4: -- using p-value approach ---------------
a. $P = P(F > 0.03 | df_n = 11, df_d = 112)$;

Using Table 9: $P > 0.05$
Using computer: $P = 1.000$
b. $P > \alpha$
-- using classical approach -------------
a. $F(11, 112, 0.05) = 1.99$
b. $F*$ is not in the critical region

Step 5: a. Fail to reject H_o.
b. There is not sufficient evidence to show that at least one monthly mean is significantly different from the others at the 0.05 level of significance.

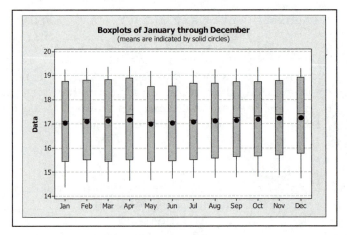

b. Step 1: a. The mean hourly wage paid each year from 1995 - 2005 to production workers.
b. H_o: The mean hourly wage is the same for each of the years.
H_a: The mean hourly wage is not the same for each of the years.
Step 2: a. Assume the data were randomly collected and are independent, and the effects due to chance and untested factors are normally distributed.
b. F   c. $\alpha = 0.05$

Step 3: a.

Source	DF	SS	MS	F	P
Factor	10	302.3136	30.2314	1250.59	.000
Error	113	2.7316	0.0242		
Total	123	305.0452			

Step 4: -- using p-value approach ---------------

a. $\mathbf{P} = P(F > 1250.59 | df_n = 10, df_d = 113)$;

Using Table 9: $\mathbf{P} < 0.01$
Using computer: $\mathbf{P} = 0.000$

b. $\mathbf{P} < \alpha$

-- using classical approach -------------

a. $F(10,113,0.05) = 1.99$

b. F* is in the critical region

Step 5: a. Reject H_o.

b. There is sufficient evidence to show that at least one mean is significantly different from the others at the 0.05 level of significance.

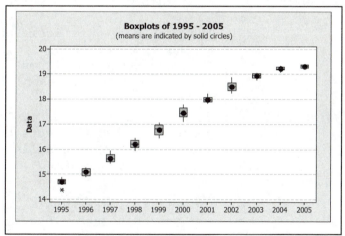

12.37 a. Answers will vary but it would make sense that the number of items purchased would increase with an increase in number of customers. This being the case, the months of November and December should have the higest volume.

b.

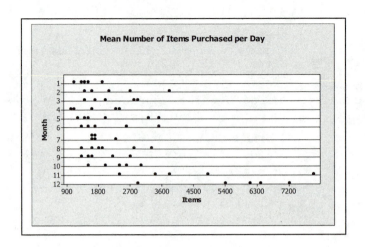

c. Answers may vary but yes, the graph supports the
   conjecture that the months with the larger number of
   customers would be the same months for the large
   number of items purchased per day.

d. Step 1: a. The mean number of items purchased per day by
                month.
           b. H_O: The mean number of items purchased per day
                is the same for each of the twelve months.
              H_a: The mean number of items purchased per day
                is not the same for each of the twelve
                months.
   Step 2: a. Assume the data were randomly collected and are
                independent, and the effects due to chance and
                untested factors are normally distributed.
           b. F   c. $\alpha = 0.05$
   Step 3: a.

Source	DF	SS	MS	F	P
Month	11	84869019	7715365	7.56	0.000
Error	50	51003447	1020069		
Total	61	135872465			

   Step 4: -- using p-value approach ---------------
           a. $P = P(F > 7.56 | df_n = 11, df_d = 50)$;

              Using Table 9: $P < 0.01$
              Using computer: $P = 0.000$
           b. $P < \alpha$

   a. $F(11,50,0.05) = 2.08$
   b. $F*$ is in the critical region

   Step 5: a. Reject H_o.
            b. There is sufficient evidence to show that at
               least one monthly mean is significantly
               different from the others, that is that the
               month does effect the mean number of items
               purchased per day, at the 0.05 level of
               significance.

   e. Answers may vary but based on answers given above, the
      reasoning in part a is supported by the F-test results.

12.39 a. H_o: The mean amount of salt is the same in all tested
             brands of peanut butter.
          Ha: The mean amount of salt is not the same in all tested
             brands of peanut butter.
       b. Assumptions: samples were randomly selected and are
          independent, and the effects due to chance and untested
          factors are normally distributed.
          $\alpha = 0.05$; test statistic: F

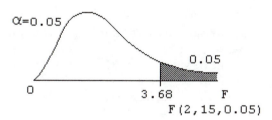

       c. Fail to reject H_o. There is no evidence to show a
          significant difference in the mean amounts of salt in
          the tested brands.

       d. Since the p-value is quite large (much larger than 0.05),
          it tells us the sample data is quite likely to have
          occurred under the assumed conditions and a true null
          hypothesis. Therefore, we 'fail to reject H_o.'

12.41 a. H_o: The mean amount spent is the same for all four supermarkets.

   H_a: The mean amount spent is not the same for all four supermarkets.

b. Fail to reject H_o. There is not sufficient evidence to reject the null hypothesis - that is, there is not sufficient evidence to show the mean amount spent is not the same for all four supermarkets. In other words, with regards to the mean amount spent, it does not matter financially where you grocery shop.

c. No, there does not seem to be a difference among the mean grocery cost for these four supermarkets - the means range form $2.47 to $2.72, that is only 25 cent diference.

d. No, the 4 standard deviations listed are all between 2.17 and 2.93 which seem very close in value, therefore showing little difference.

e. The means are nearly identical and the standard deviations are also very close in value. Almost all of the variation in this data is within the stores (factor levels) thus totally supporting the results of the hypothesis test.

12.43 Step 1: a. The mean stopping distance on wet pavement for each brand of tire.

   b. H_o: The mean stopping distance is not affected by the brand of tire.

   H_a: The mean stopping distance is affected by the brand of tire.

Step 2: a. Assume the data were randomly collected and are independent, and the effects due to chance and untested factors are normally distributed.

   b. F   c. $\alpha = 0.05$

Step 3: a. n = 23, C_A = 217, C_B = 194, C_C = 216, C_D = 245

   T = 872, Σx^2 = 33,282

Source	df	SS	MS	F*
Brand	3	95.36	31.79	4.78
Error	19	126.47	6.66	
Total	22	221.83		

Step 4: -- using p-value approach ---------------
a. $P = P(F > 4.78 | df_n = 3, df_d = 19)$;

Using Table 9: $0.01 < P < 0.025$
Using computer: $P = 0.012$
b. $P < \alpha$
-- using classical approach -------------
a. critical value: $F(3, 19, 0.05) = 3.13$
critical region: $F \geq 3.13$
b. F^* is in the critical region

Step 5: a. Reject H_o.
b. There is a significant difference between the mean stopping distance for the different tire brands, at the 0.05 level of significance.

12.45 Step 1: a. The mean amounts of soft drink dispensed: the mean amount for machine A, the mean amount for machine B, the mean amount for machine C, the mean amount for machine D, the mean amount for machine E.
b. Ho: The mean amounts dispensed by the machines are all equal.
Ha: The mean amounts dispensed by the machines are not all equal.
Step 2: a. Assume the data were randomly collected and are independent, and the effects due to chance and untested factors are normally distributed.
b. F   c. $\alpha = 0.01$
Step 3: a. $n = 18$, CA = 16.5, CB = 20.6, $C_C = 16.9$, $C_D = 19.1$,

$C_E = 21.8$, $T = 94.9$, $\Sigma x^2 = 523.49$

Source	df	SS	MS	F*
Machine	4	20.998	5.2495	31.6
Error	13	2.158	0.166	
Total	17	23.156		

Step 4:   -- using p-value approach ---------------
           a. $\mathbf{P} = P(F > 31.6 | df_n = 4, df_d = 13)$;

              Using Table 9: $\mathbf{P} < 0.01$
              Using computer: $\mathbf{P} = 0.000$
           b. $\mathbf{P} < \alpha$
           -- using classical approach -------------
           a. critical value: $F(4,13,0.01) = 5.21$
              critical region: $F \geq 5.21$
           b. F^* is in the critical region

Step 5:   a. Reject H_o.
           b. There is a significant difference between the
              machines with regards to mean amount of soft
              drink dispensed, at the 0.01 level of
              significance.

12.47 Step 1:   a. The mean points scored by teams representing
                         each division.
                      b. Ho: $\mu E = \mu N = \mu S = \mu W$
                         Ha: The mean points scored by teams
                         representing each division are not all equal.
           Step 2:   a. Assume the data were randomly collected and are
                         independent, and the effects due to chance and
                         untested factors are normally distributed.
                      b. F   c. $\alpha = 0.05$
           Step 3:   a. n = 32,

Source	DF	SS	MS	F	P
Factor	3	3055	1018	0.19	0.900
Error	28	147817	5279		
Total	31	150872			

           Step 4:   -- using p-value approach ---------------
                      a. $\mathbf{P} = P(F > 0.19 | df_n = 3, df_d = 28)$;

                         Using Table 9: $\mathbf{P} > 0.05$
                         Using computer: $\mathbf{P} = 0.900$
                      b. $\mathbf{P} > \alpha$
                      -- using classical approach -------------
                      a. $F(3,28,0.05) \approx 2.99$
                      b. F^* is not in the critical region

Step 5:   a. Fail to reject H_o.
          b. At the 0.05 level of significance, the
             data does not show sufficient evidence to
             conclude that the points scored by the
             teams in each division are not all equal.

Step 1:   a. The mean points scored by the opponents of the
             teams representing each division.
          b. Ho: $\mu E = \mu N = \mu S = \mu W$
             Ha: The mean points scored by the opponents of
             the teams representing each division are not
             all equal.
Step 2:   a. Assume the data were randomly collected and are
             independent, and the effects due to chance and
             untested factors are normally distributed.
          b. F   c. $\alpha = 0.05$
Step 3:   a. n = 32,

Source	DF	SS	MS	F	P
Factor	3	22599	7533	2.40	0.089
Error	28	87827	3137		
Total	31	110426			

Step 4:   -- using p-value approach ---------------
          a. $\mathbf{P} = P(F > 2.40 | df_n = 3, df_d = 28)$;

             Using Table 9: $\mathbf{P} > 0.05$
             Using computer: $\mathbf{P} = 0.089$
          b. $\mathbf{P} > \alpha$
          -- using classical approach -------------
          a. $F(2,28,0.05) \approx 2.99$
          b. F* is not in the critical region

Step 5:   a. Fail to reject H_o.
          b. The data does not show sufficient evidence
             to conclude the mean points scored by the
             opponents of each division are not all
             equal, at the 0.05 level of significance.

12.49 a.
Step 1:   a. The mean nominal comparison for five
             competitors: the mean nominal comparison for A,
             the mean nominal comparison for B, the mean
             nominal comparison for C, the mean nominal
             comparison for D, the mean nominal comparison
             for E.

    b. H_O: The mean nominal comparison is the same for
       all five competitors.
       H_a: The mean nominal comparison is not the same
       for all five competitors.

Step 2:   a. Assume the data were randomly collected and are
       independent, and the effects due to chance and
       untested factors are normally distributed.
       b. F   c. $\alpha = 0.01$

Step 3:   a.

Source	DF	SS	MS	F	P
Factor	4	0.001830	0.000458	1.05	0.385
Error	105	0.045732	0.000436		
Total	109	0.047563			

Step 4:   -- using p-value approach ---------------
       a. $P = P(F > 1.05 | df_n = 4, df_d = 105)$;

       Using Table 9: $P > 0.05$
       Using computer: $P = 0.385$
       b. $P > \alpha$
       -- using classical approach -------------
       a. critical value: $F(4, 105, 0.01) = 3.65$
       critical region: $F \geq 3.65$
       b. F* is not in the critical region

Step 5:   a. Fail to reject H_O.
       b. There is no significant difference between the
       mean
       nominal comparisons for the five competitors,
       at the 0.01 level of significance.

b.

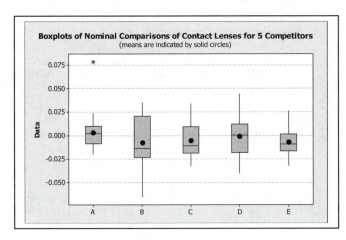

c. The graph portrays the results found in the F-test, the fact the all of the means are the same. The solid circles shown on the graph are the means for each of the competitors. It shows that the means are all at about the same level of nominal comparison (they all lay close to a horizontal straight line across).

12.51   Step 1:   a. The mean durability for different brands of golf balls: the mean durability for Brand A, the mean durability for Brand B, the mean durability for Brand C, the mean durability for Brand D, the mean durability for Brand E, the mean durability for Brand F.

b. H_o: The six different brands of golf balls withstood the durability test equally well, as measured by the mean number of hits before failure.

H_a: The six different brands of golf balls do not withstand the durability test equally well.

Step 2:   a. Assume the data were randomly collected and are independent, and the effects due to chance and untested factors are normally distributed.

b. F   c. $\alpha = 0.05$

Step 3:   a.

Source	df	SS	MS	F*
Brand	5	75047	15009.4	5.30
Error	36	101899	2830.5	
Total	41	176946		

Step 4:   -- using p-value approach ----------------
a. **P** = P(F > 5.30|df_n = 5,df_d = 36);

Using Table 9: **P** < 0.01
Using computer: **P** = 0.001
b. **P** < α
-- using classical approach -------------
a. critical value: F(5,36,0.05) = 2.48
critical region: F ≥ 2.48
b. F* is in the critical region

Step 5: a. Reject H_o.
        b. There is a significant difference between the mean number of hits before failure for the six brands of golf balls tested, at the 0.05 level of significance.

12.53 a.

Step 1: a. The mean petal width for three species of irises.
        b. H_o: The mean petal width is the same for each specie of iris
          H_a: The mean petal width is not the same for each specie of iris.

Step 2: a. Assume the data were randomly collected and are independent, and the effects due to chance and untested factors are normally distributed.
        b. F   c. $\alpha = 0.05$

Step 3: a.

Source	df	SS	MS	F
Specie	2	1671.56	835.78	118.06
Error	27	191.14	7.08	
Total	29	1862.70		

Step 4: -- using p-value approach ---------------
        a. $P = P(F > 118.06 | df_n = 2, df_d = 27)$;

          Using Table 9: $P < 0.01$
          Using computer: $P = 0.000$
        b. $P < \alpha$
        -- using classical approach -------------
        a. critical value: F(2,27,0.05) • 3.37
        b. F* is in the critical region

Step 5: a. Reject H_o.
        b. There is sufficient evidence to show that at least one specie's petal width is significantly different from the others, at the 0.05 level of significance.

b.
  Step 1: a. The mean sepal width for three species of
              irises.
           b. H_o: The mean sepal width is the same for each
              specie of iris
              H_a: The mean sepal width is not the same for
              each specie of iris.
  Step 2: a. Assume the data were randomly collected and are
              independent, and the effects due to chance and
              untested factors are normally distributed.
           b. F   c. $\alpha = 0.05$
  Step 3: a.

Source	df	SS	MS	F
Specie	2	197.1	98.6	7.78
Error	27	342.2	12.7	
Total	29	539.4		

  Step 4: -- using p-value approach ---------------
           a. $P = P(F > 7.78 | df_n = 2, df_d = 27)$;

              Using Table 9: $P < 0.01$
              Using computer: $P = 0.002$
           b. $P < \alpha$
           -- using classical approach -------------
           a. critical value: $F(2,27,0.05) \cdot 3.37$
           b. F* is in the critical region

  Step 5: a. Reject H_o.
           b. There is sufficient evidence to show that at
              least one specie's sepal width is significantly
              different from the others, at the 0.05 level of
              significance.

c. Type 0 has the shortest PW and the longest SW. Type 1 has
   the longest PW and the middle SW. Type 2 has the middle
   PW and the shortest SW.

12.55 a. The graphical evidence that not all days of the week
          are the same is illustrated by the various ranges of
          data values per day of the week. Mondays appears to
          have the smallest range and number of customers,
          whereas Tuesday and Wednesdays have some of the
          highest values and largest spreads of data values.

b. The 5 points located to the right and separate from the rest of the data all occurred during the months of November and December. The Tuesday and Wednesday both had November months, the days right before Thanksgiving.

c. Based on the F statistic shown on the ANOVA table, not all of the days of the week are the same when it comes to the number of customers if a level of significance of 0.10 is used.

d. No, one can not tell which days of the week are different based on the ANOVA table output.

e. Verify - answers given in exercise.

12.57 a. Answers will vary but it would make sense that the total cost of items purchased would increase with an increase in number of customers and number of items purchased. This being the case, the months of November and December should have the highest amounts. Whether the day of the week will have a significant effect needs to be seen and will probably be borderline based on the other results.

b.

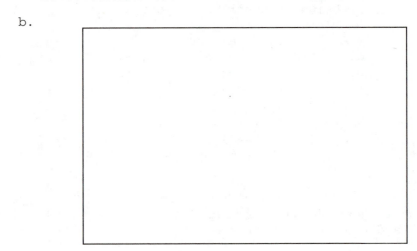

c. Answers may vary but yes, the graph supports the conjecture that the days of the week with the larger total cost of items would be the same days that were large for number of customers and number of items purchased per day. Again each of these are in the holiday months of November and December.

d. Step 1: a. The mean total cost of items purchased per day
by day of the week.
b. H_O: The mean total cost of items purchased per
day is the same for each day of the week.
H_a: The mean total cost of items purchased per
day is not the same for each day of the
week. That is; day of the week does have
effect on the mean number of items
purchased.
Step 2: a. Assume the data were randomly collected and are
independent, and the effects due to chance and
untested factors are normally distributed.
b. F   c. $\alpha = 0.05$
Step 3: a.

Source	DF	SS	MS	F	P
Day	5	2657284622	531456924	2.24	.063
Error	56	13311874185	237712039		
Total	61	15969158806			

Step 4: -- using p-value approach ----------------
a. $P = P(F > 2.24 | df_n = 5, df_d = 56)$;

Using Table 9: $P > 0.05$
Using computer: $P = 0.063$
b. $P > \alpha$
-- using classical approach -------------
a. $F(5,56,0.05) \cdot 2.45$
b. F* is in the noncritical region

Step 5: a. Fail to reject H_O.
b. There is not sufficient evidence to show that
the day of the week has an effect on the mean
total cost of items purchased per day, at the
0.05 level of significance.

e. Answers may vary but based on answers given above, the
reasoning in part a is not supported by the F-test
results. The p-value is very close to the level of
significance. With a higher level of significance, the F-
test results would correspond closer to the answers given
based on the dotplot.

f. The answers in exercises 12.55, 12.56 and 12.57 all support each other and make sense. The more customers, the larger the number of items purchased and the larger the total cost of items purchased per day. All of these variables are related to each other, but the day of the week does not have a significant effect unless a level of significance higher than 0.05 is used.

12.59   Sample information:
$k_1 = 3$, $k_2 = 3$, $k_3 = 3$, $n = 9$,
$C_1 = 24$, $C_2 = 39$, $C_3 = 27$, $T = 90$, $\Sigma x^2 = 960$

Using formula (12.3):
$$SS(factor) = [(24^2/3)+(39^2/3)+(27^2/3)]-(90^2/9)$$
$$= 942 - 900 = \underline{42}$$

Using formula given in exercise:
$x_1 = 24/3 = 8$, $x_2 = 39/3 = 13$, $x_3 = 27/3 = 9$,
$\overline{X} = 90/9 = 10$

$$SS(factor) = 3(8-10)^2 + 3(13-10)^2 + 3(9-10)^2$$
$$= 12 + 27 + 3 = \underline{42}$$

12.61   Using the formula in exercise 12.58:

$$SS(error) = 6(9.48^2) + 6(8.91^2) + 5(9.06^2) = \underline{1425.969}$$

Using formula in exercise 12.59:

$$\Sigma x = 7(21.43) + 7(20.00) + 6(20.83) = 414.99$$

$$\overline{X} = 414.99/20 = 20.75$$

$$SS(factor) = 7(21.43-20.75)^2+7(20.00-20.75)^2+6(20.83-20.75)^2$$
$$= 7.2127$$

ANOVA Table			
Factor	7.213	2	3.61
Error	1425.969	17	83.88
Total	1433.182	19	

CHAPTER 13 ▽ LINEAR CORRELATION AND REGRESSION ANALYSIS

Chapter Preview

Chapter 3 introduced the concepts of correlation and regression analysis for bivariate data. In Chapter 13, we will look at these concepts in a more detailed manner using confidence intervals and hypothesis tests. These inference tests will be utilized on the correlation coefficient, the slope of the regression line and the regression line.

Wheat planted and harvasted in the United States during 2002 is the focus of this chapter's opening section "Wheat! Beautiful Golden Wheat!".

SECTION 13.1 EXERCISES

13.1 a.

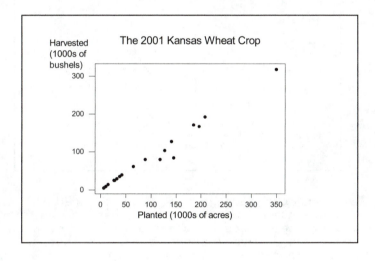

b.

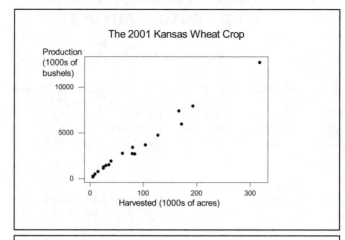

The 2001 Kansas Wheat Crop

c.

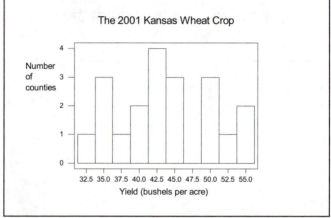

The 2001 Kansas Wheat Crop

d. As the number of acres of wheat planted increases so does the number of bushels harvested. As the number of harvested acres of wheat increases so does the production in thousands of bushels increase. The yield rate has an approximately normal distribution.

13.3   a. $\bar{x}$ and $\bar{y}$ are the mean values of the two variables. In Chapter Two we learned that the summation of the deviations about the mean was zero.

Algebraically: $\Sigma(x - \bar{x}) = \Sigma x - \Sigma\bar{x} = \Sigma x - n \cdot \bar{x} = 0$

and $\Sigma(y - \bar{y}) = \Sigma y - \Sigma\bar{y} = \Sigma y - n \cdot \bar{y} = 0$

b.   divides data into 4 quadrants

c. 1.) The set of data will be predominantly ordered pairs which have coordinates such that both the x and y values are larger than $\bar{x}$ and $\bar{y}$, and both smaller than $\bar{x}$ and $\bar{y}$; this will result in the product (x-$\bar{x}$)(y-$\bar{y}$) being positive. Graphically, the points will be mostly located in the upper right and the lower left of the four quarters of the graph formed by the vertical line x = $\bar{x}$ and the horizontal line y = $\bar{y}$.

2.) The set of data will be predominantly ordered pairs which have coordinates such that either the x value is larger than $\bar{x}$ and y is smaller than $\bar{y}$, or x is smaller than $\bar{x}$ and y is larger than $\bar{y}$; this will result in the product (x-$\bar{x}$)(y-$\bar{y}$) being negative. Graphically, the points will be mostly located in the upper left and the lower right of the four quarters of the graph formed by the vertical line x = $\bar{x}$ and the horizontal line y = $\bar{y}$.

3.) The set of data will be ordered pairs which have coordinates such that the product (x-$\bar{x}$)(y-$\bar{y}$) being distributed between positive, negative and zero so that the sum is near zero. Graphically, the points will be approximately evenly distributed between the four quarters of the graph formed by the vertical line x = $\bar{x}$ and the horizontal line y = $\bar{y}$.

13.5   a.

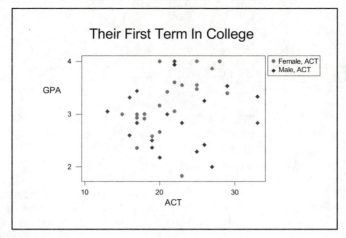

b. Patterns are somewhat similar in that the points
   pretty much cover the full area of the diagram. The
   females, with one exception, are all located in the
   top part of the diagram.

c.

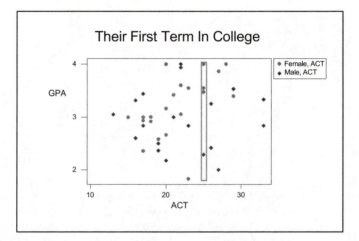

Prediction: any value from 1.8 to 4.0, not much of a
prediction.

d. No, knowing the ACT score does not help at all.

13.7 a.

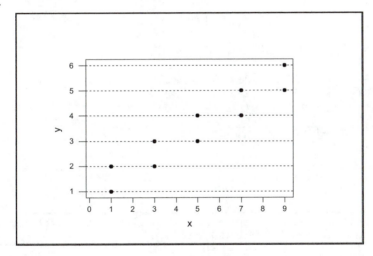

b.

x	y	x-x̄	y-ȳ	(x-x̄)(y-ȳ)
1	1	-4	-2.5	+10
1	2	-4	-1.5	+ 6
3	2	-2	-1.5	+ 3
3	3	-2	-0.5	+ 1
5	3	0	-0.5	0
5	4	0	+0.5	0
7	4	+2	+0.5	+ 1
7	5	+2	+1.5	+ 3
9	5	+4	+1.5	+ 6
9	6	+4	+2.5	+10
50	35	0	0.0	40

$\bar{x} = 50/10 = 5.0$ and $\bar{y} = 35/10 = 3.5$

$covar(x,y) = [\Sigma(x-\bar{x})(y-\bar{y})]/(n-1) = 40/9 = \underline{4.44}$

Summary of data: n = 10, $\Sigma x = 50$, $\Sigma y = 35$,

$\Sigma x^2 = 330$, $\Sigma xy = 215$, $\Sigma y^2 = 145$

c. $s_x = \sqrt{[330 - (50^2/10)]/9} = \sqrt{8.889} = \underline{2.981}$

$s_y = \sqrt{[145 - (35^2/10)]/9} = \sqrt{2.50} = \underline{1.581}$

d. $r = 4.444/[(2.981)(1.581)] = \underline{0.943}$

e. $SS(x) = 330 - (50^2/10) = 80$
$SS(y) = 145 - (35^2/10) = 22.5$
$SS(xy) = 215 - [(50)(35)/10] = 40$

$r = 40/\sqrt{(80)(22.5)} = \underline{0.943}$

13.9   Verify -- answers given in exercise.

Computer and/or calculator commands to calculate r, the correlation coefficient can be found in ES10-p166.

13.11   n = 32, $\Sigma x = 10996$, $\Sigma y = 11000$, $\Sigma x^2 = 3929712$,

$\Sigma xy = 3780951$,

$\Sigma y^2 = 3891114$

$SS(x) = 151211.5$, $SS(y) = 109864$, $SS(xy) = 1076$

a. $r = SS(xy)/\sqrt{SS(x) \cdot SS(y)} = 1076/\sqrt{(151211.5)(109864)} = \underline{0.008}$

b. no linear relationship

c.

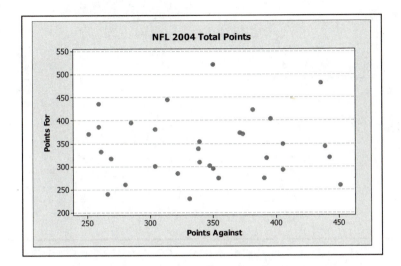

NFL 2004 Total Points

Points are scattered with no upward or downward trend, causing close to a zero correlation coefficient, or no correlation.

13.13 a.

x	y	x-$\overline{x}$	y-$\overline{y}$	(x-$\overline{x}$)(y-$\overline{y}$)
20	10	-50	-20	1000
30	50	-40	20	-800
60	30	-10	0	0
80	20	10	-10	-100
110	60	40	30	1200
120	10	50	-20	-1000
420	180	0	0	300

$\overline{x} = 420/6 = 70$, $\overline{y} = 180/6 = 30$

covar(x,y) = 300/5 = $\underline{60}$

b. $s_x = 40.99$, $s_y = 20.98$
c. $r = 60/(40.99)(20.98) = 0.0698 = \underline{0.07}$
d. The value for r is the same.

13.15

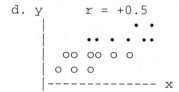

a. y   r = -1   b. y   r = 0   c.   y   r = +1

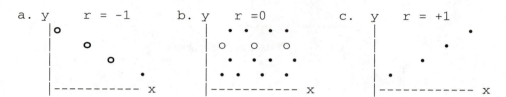

d. y   r = +0.5   e. y   r = -0.6

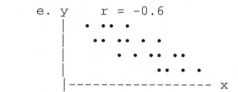

        Estimating ρ - the population correlation coefficient
1. point estimate: r
2. confidence interval: use Table 10 in Appendix B, ES10-p822, in
   the text to determine a 95% confidence interval.
   Locate the r value on the horizontal axis. Follow a vertical
   line up until the corresponding sample size band intersects.
   This value is the lower limit. To find the upper limit, locate
   the r value on the vertical axis. Follow a horizontal line
   until it intersects the corresponding sample size.

13.17   a. 0.17 to 0.52
         b. The interval becomes more narrow.

13.19   a. 0.40 to 0.74   b. -0.78 to +0.15
         c. 0.05 to 0.93   d. -0.65 to -0.45

Reminder: $r = SS(xy)/\sqrt{SS(x)\cdot SS(y)}$ where: $SS(x) = \Sigma x^2 - [(\Sigma x)^2/n]$
$SS(y) = \Sigma y^2 - [(\Sigma y)^2/n]$
$SS(xy) = \Sigma xy - [(\Sigma x)(\Sigma y)/n]$

13.21 Summary of data: $n = 10$, $\Sigma x = 746$, $\Sigma y = 736$,
$\Sigma x^2 = 57,496$, $\Sigma xy = 56,574$, $\Sigma y^2 = 55,826$

$SS(x) = 57496 - (746^2/10) = 1844.4$
$SS(y) = 55826 - (736^2/10) = 1656.4$
$SS(xy) = 56574 - [(746)(736)/10] = 1668.4$

$r = SS(xy)/\sqrt{SS(x)\cdot SS(y)} = 1668.4/\sqrt{(1844.4)(1656.4)} = \underline{0.955}$

From Table 10: $\underline{0.78 \text{ to } 0.98}$, the 0.95 interval for ρ

13.23 a. Pearson corre lation of Score and Price \$: $r = 0.985$

b. Using Table 10: Confidence interval: $0.55 < \rho < 1.00$

c. With 95% confidence, the population correlation coefficient for Wine Spectator scores and the prices of Portuguese red wine is between 0.55 and 1.00.

d. The interval is very wide due to the small sample size.

Hypotheses for the correlation coefficient are written with the same rules before. Now in place of μ or σ, the population correlation coefficient, ρ, will be used. The standard form is using 0 as the test value, unless some other information is given in the exercise.
0 indicates that there is no linear relationship.
(ex. H_o: $\rho = 0$ vs. H_a: $\rho \neq 0$)

13.25 a. H_o: $\rho = 0$ vs. H_a: $\rho > 0$

b. H_o: $\rho = 0$ vs. H_a: $\rho \neq 0$

c. H_o: $\rho = 0$ vs. H_a: $\rho < 0$

d. H_o: $\rho = 0$ vs. H_a: $\rho > 0$

13.27 a. $0.05 < P < 0.10$ b. $0.025 < P < 0.05$

Test criteria

1. Draw a bell-shaped distribution locating 0 at the center, -1 at the far left and +1 at the far right.
2. Shade in the critical region(s) based on the alternative hypothesis (H_a).
3. Find the critical value(s) from Table 11, Appendix B, ES10-p823:
    a. degrees of freedom (n - 2) is the row id #
    b. α, is the column id #;
       1) use the given α for a two-tailed test
       2) use 2α for a one-tailed test
    c. all values given in the table are positive. Negate the value if the critical region or part of the critical region is to the left of 0.

13.29 a. ± 0.444
    b. -0.378, if left tail critical region; 0.378, if right tail

13.31 a. The linear correlation coefficient, r = 0.58 is significant for all levels of $\alpha > 0.008$.
    b. n = 25, df = 23: $P < 0.01$ for a two-tailed test. 0.008 is less than 0.01, therefore significant.
    c. ± 0.537, using next smaller table value;
      ± 0.507, using interpolation
    d. r is significant at the $\alpha = 0.01$ level.

Hypothesis tests will be completed using the same format as before. You may want to review: ES10-pp427 & 444-445, SSM-pp265&275. The only differences are:
    1. **writing hypotheses**: (see box before ex. 13.25)
    2. **using Table 11**: (see box before ex. 13.29 for the classical approach and ES10-p706-707 for the p-value approach)
    3. **the calculated test statistic**: r*, the sample correlation coefficient

13.33 Step 1: a. The linear correlation coefficient for the
                       population, ρ.
                   b. H_o: $\rho = 0.0$
                       H_a: $\rho \neq 0.0$
         Step 2: a. random sample, assume normality for y at each x
                   b. r, df = n − 2 = 20 − 2 = 18
                   c. $\alpha = 0.10$
         Step 3: a. n = 20, r = 0.43
                   b. r* = 0.43
         Step 4: −− using p-value approach −−−−−−−−−−−−−−−−−−−−
                   a. **P** = P(r < 0.43) + P(r > 0.43)
                          = 2P(r > 0.43|df = 18);
                             Using Table 11, ES10-p823:
                                    0.05 < **P** < 0.10
                   b. **P** < α
                 −− using classical approach −−−−−−−−−−−−−−−−−−
                   a. ±r(18, 0.10) = ±0.378

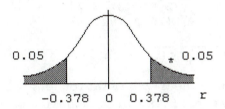

                       0.05   * 0.05

                   −0.378   0   0.378   r

                   b. r* falls in the critical region, see Step 4a.
                 −−
         Step 5: a. Reject H_o.
                   b. There is sufficient reason to reject the null
                       hypothesis, at the 0.10 level of significance.

13.35 Step 1: a. The linear correlation coefficient for the
                       population, ρ.
                   b. H_o: $\rho = 0.0$
                       H_a: $\rho \neq 0.0$
         Step 2: a. random sample, assume normality for y at each x
                   b. r, df = n − 2 = 5 − 2 = 3
                   c. $\alpha = 0.05$
         Step 3: a. n = 5, r = 0.532
                   b. r* = 0.532

Step 4: -- using p-value approach ---------------------
        a. $\mathbf{P}$ = P(r < -0.532) + P(r > 0.532)
           = 2P(r > 0.532|df = 3);
                Using Table 11, ES10-p823: $\mathbf{P}$ > 0.10
        b. $\mathbf{P} > \alpha$
        -- using classical approach -----------------
        a. $\pm$r(3, 0.05) = $\pm$0.878
        b. r* falls in the noncritical region.

Step 5: a. Fail to Reject H_o.
        b. There is insufficient reason to conclude that
           the relationship is significant, at the 0.05
           level of significance.

13.37 Summary of data: n = 10, Σx = 26.2, Σy = 82.5,
              Σx^2 = 174.88, Σxy = 256.41, Σy^2 = 704.61

SS(x) = 174.88 - (26.2^2/10) = 106.236
SS(y) = 704.61 - (82.5^2/10) = 23.985
SS(xy) = 256.41 - [(26.2)(82.5)/10] = 40.26

r = SS(xy)/$\sqrt{\text{SS}(x) \cdot \text{SS}(y)}$ = 40.26/$\sqrt{(106.236)(23.985)}$ = <u>0.798</u>

Step 1: a. The linear correlation coefficient for size of a
           metropolitan area and its crime rate, ρ.
        b. H_o: ρ = 0.0
           H_a: $\rho \neq$ 0.0
Step 2: a. random sample, assume normality for y at each x
        b. r, df = n - 2 = 10 - 2 = 8
        c. α = 0.05
Step 3: a. n = 10, r = 0.798
        b. r* = 0.798
Step 4: -- using p-value approach ---------------------
        a. $\mathbf{P}$ = P(r < 0.798) + P(r > 0.798)
           = 2P(r > 0.798);
           Using Table 11, ES10-p823: $\mathbf{P}$ < 0.01
           Using computer: $\mathbf{P}$ = 0.006
        b. $\mathbf{P} < \alpha$
        -- using classical approach -----------------
        a. critical region: r $\leq$ -0.632 and r $\geq$ 0.632
        b. r* falls in the critical region

Step 5: a Reject H_o
   b. There is sufficient reason to conclude that the correlation coefficient is different than zero, at the 0.05 level of significance.

13.39 **a.** Summary of data: $n = 7$, $r = \underline{0.855}$

Step 1: a. The linear correlation coefficient for a state's total personal income and value of housing units, ρ.

   b. H_o: $\rho = 0.0$

   H_a: $\rho \neq 0.0$

Step 2: a. random sample, assume normality for y at each x
   b. r, df = n - 2 = 7 - 2 = 5
   c. $\alpha = 0.05$

Step 3: a. $n = 7$, $r = 0.855$
   b. $r* = 0.855$

Step 4: -- using p-value approach --------------------
   a. **P** = P(r < -0.855) + P(r > 0.855)
      = 2P(r > 0.855);
      Using Table 11, ES10-p823: 0.01 < P < 0.02
      Using a computer: **P** = 0.014

   b. **P** < α
   -- using classical approach -----------------
   a. critical region: r ≤ -0.754 and r ≥ 0.754
   b. r* falls in the critical region

Step 5: a Reject H_o
   b. There is sufficient reason to conclude that there is significant correlation between a state's personal income and the value of newly ouwned housing units, at the 0.05 level of significance.

Sample Regression Line: $\hat{y} = b_o + b_1 x$, where

$$b_1 = \frac{SS(xy)}{SS(x)} = \frac{\sum(xy) - [(\sum x)(\sum y) / n]}{\sum x^2 - (\sum x)^2 / n} \quad \text{and} \quad b_o = \frac{1}{n}\left(\sum y - b_1 \cdot \sum x\right)$$

Population Regression Line: $y = \beta_0 + \beta_1 x + \varepsilon$

Sample estimate of $\varepsilon = e = (y - \hat{y})$

Variance of y about the regression line = variance of the error e

$$s_e^2 = \frac{(\sum y^2) - [(b_0)(\sum y)] - [(b_1)(\sum xy)]}{n - 2} = \frac{SSE}{n - 2}$$

Standard Deviation of the error $= s_e = \sqrt{s_e^2}$

Explore the relationship between residuals and the line of best fit with the Chapter 13 Skillbuilder Applet ''Residuals & Line of Best Fit'' on your CD..

Slight variations in sums of squares and further calculations can result from round-off errors.

13.41 Verify -- answers given in exercise.

13.43

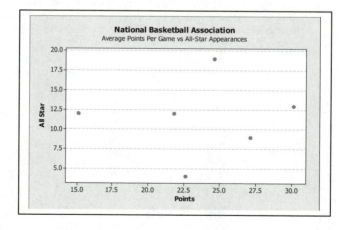

Summary of data: n = 6, Σx = 141.3, Σy = 69

Σx^2 = 3459.59, Σxy = 1635.8, Σy^2 = 915

SS(x) = 3459.59 - (141.3²/6) = 131.975
SS(y) = 915 - (69²/6) = 121.5
SS(xy) = 1635.8 - [(141.3)(69)/6] = 10.85

$r = SS(xy)/\sqrt{SS(x)\cdot SS(y)}$
   $= 10.85/\sqrt{(131.975)(121.5)}$ = <u>0.086</u>

Using formula 3.6:
  b_1 = 10.85/131.975 = 0.082

Using formula 3.7:
  b_0 = [69 - (0.082)(141.3)]/6 = 9.57

Best fit line: $\hat{y}$ = 9.6 + 0.082x

There is not a strong linear relationship shown by
this data. The scatter diagram shows no definite
trend or pattern to indicate a relationship. The
correlation coefficient is close to zero and with a
p-value of 0.872 would be considered insignificant
with just about all levels of significance.

Computer and/or calculator commands to construct a scatter diagram
can be found in ES10-p155. Commands to calulate the equation of
the line of best fit can be found in ES10-pp180&181.

13.45 a.

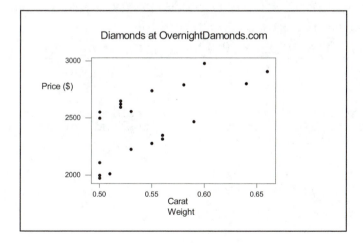

b. There is a linear pattern to the data, however the data falls into two groups forming two parallel linear patterns, one forming the top and the other forming the bottom of the total pattern.

c. Only have data in this weight range, can not predict with confidence outside this range. Smaller values and larger value decrease and increase, respectively exponentially.

d. Summary of data: $n = 20$, $\Sigma x = 10.92$, $\Sigma y = 49428$
$\Sigma x^2 = 6.005$, $\Sigma xy = 27166.9$, $\Sigma y^2 = 123927308$

SS(x) = 6.005 - (10.92²/20) = 0.04268
SS(y) = 123927308 - (49428²/20) = 1770949
SS(xy) = 27166.9 - [(10.92)(49428)/20] = 179.212

Using formula 3.6:
b_1 = 179.212/0.04268 = 4198.969

Using formula 3.7:
b_0 = [49428 - (4198.969)(10.92)]/20 = 178.763

Best fit line: $\hat{y}$ = 179 + 4199x

e. $\hat{y}$ = 179 + 4199x = 179 + 4199(0.50) = $2278.50

f. $41.99,   x = 0.50 carats to 0.66

g. s = 237.9, s^2 = (237.9)² = 56596.41 (by computer)
Using formula by hand:
$$s_e^2 = \frac{\sum y^2 - (b_0)\sum y - (b_1)\sum xy}{n-2} = \frac{123927308 - 179.28(49428) - 4198.03(27166}{18}$$
$= 56577.49739$

The scatter diagram shows a sizeable amount of vertical distance between the top and bottom points along the line of best fit.

13.47 a. Summary of data: $n = 10$, $\Sigma x = 50$, $\Sigma y = 35$,

$$\Sigma x^2 = 330, \ \Sigma xy = 215, \ \Sigma y^2 = 145$$

SS(x) = 330 - (50²/10) = 80
SS(xy) = 215 - [(50)(35)/10] = 40

Using formula 3.6:
b_1 = 40/80 = 0.50

Using formula 3.7:
b_0 = [35 - (0.50)(50)]/10 = 1.0

$\hat{y}$ = <u>1.0 + 0.5x</u>

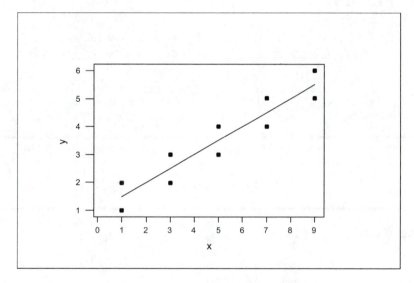

b.   If x = 1, then $\hat{y}$ = 1.0 + 0.5(1) = <u>1.5</u>
     If x = 3, then $\hat{y}$ = 1.0 + 0.5(3) = <u>2.5</u>
     If x = 5, then $\hat{y}$ = 1.0 + 0.5(5) = <u>3.5</u>
     If x = 7, then $\hat{y}$ = 1.0 + 0.5(7) = <u>4.5</u>
     If x = 9, then $\hat{y}$ = 1.0 + 0.5(9) = <u>5.5</u>

c.

point	A	B	C	D	E	F	G	H	I	J
x	1	1	3	3	5	5	7	7	9	9
y	1	2	2	3	3	4	4	5	5	6
$\hat{y}$	1.5	1.5	2.5	2.5	3.5	3.5	4.5	4.5	5.5	5.5
e=y-$\hat{y}$	-.5	.5	-.5	.5	-.5	.5	-.5	.5	-.5	.5

d. $s_e^2 = 2.50/8 = \underline{0.3125}$

e. $s_e^2 = [145 - (1.0)(35) - (0.5)(215)]/8$
$= 2.50/8 = \underline{0.3125}$

SECTION 13.5 EXERCISES

13.49 From the scatter diagram, n = 72

a. $S_1 = -3953.85 + 3.13(12,600) = 35484.15 \approx 35,500$

b. From the scatter diagram, n = 72
$b_1 \pm t(df, \alpha/2) \cdot s_{b1}$
$3.130 \pm (1.99)(0.065) = 3.130 \pm 0.129$
$\underline{3.001 \text{ to } 3.259}$ (difference due to round-off)

13.51 $s_{b1} = \sqrt{s_e^2/SS(x)} = \sqrt{1.213/49.6} = \underline{0.1564}$

The Confidence Interval Estimate of β_1
$b_1 \pm t(df, \alpha/2) \cdot s_{b1}$ with df = n - 2

13.53 a. $\hat{y} = -348 + 2.04x$

b. The 95% confidence interval for β_1 is 1.60 to 2.48.
$2.04 \pm (2.31)(0.1894)$
2.04 ± 0.44
$1.60 \text{ to } 2.48$

c. With 95% confidence, it is believed that the slope of the line of best fit for the population is between 1.60 and 2.48.

Hypotheses for the slope of the regression line are written with
the same rules as before. Now in place of μ or σ, the population
slope, β_1, will be used. The standard form is using 0 as the test
value, unless some other information is given in the exercise. 0
indicates that the line has no value in predicting y for given x.
H_a represents what the experimenter wants to show, i.e. that the
slope is meaningful and valuable in predicting y for a given x.
(ex. H_0: $\beta_1 = 0$ vs. H_a: $\beta_1 \neq 0$)

To determine the p-value for the test of the slope of the
regression line, the t-distribution is used. Review its use in:
ES10-p483-484, SSM-p305-306, if necessary. The only difference
required is **df = n - 2**, since the data is bivariate. The test
statistic is $t^* = (b_1 - \beta_1)/s_{b1}$.

13.55 a. **P** = $P(t>2.40 | df=16)$ = <u>0.0145</u>

b. **P** = $2 \cdot P(t>2.00 | df=13)$ = $2(0.0334)$ = <u>0.0668</u>

c. **P** = $P(t<-1.57 | df=22)$ = <u>0.0653</u>

Draw a picture of a t-distribution curve. Shade in the critical
regions based on the alternative hypothesis (H_a). Using α and df =
n - 2, find the critical value(s) using Table 6 (Appendix B, ES10-
p813).

13.57 a. $\hat{y}$ = <u>5936.79 + 30.732x</u>

b. t = $(30.732 - 0)/17.158$ = 1.79

c. **P** = $P(t > 1.79 | df = 8)$;
      Using Table 6, ES10-p813:
            $0.10 <$ **P** < 0.20
      Using Table 7, ES10-p814:
            $0.11 <$ **P** < 0.128
      Using computer: P = 0.111
      P > most levels of significance
      Based on the p-value, horsepower does not appear to be
      an effective predictor of base price

d. $b_1 \pm t(df, \alpha/2) \cdot s_{b1}$

    $30.732 \pm (2.31)(17.158) = 30.732 \pm 39.635$

    <u>-8.903</u> <u>to</u> <u>70.367</u>, the 0.95 interval for β_1
    (discrepancies due to rounding)

Hypothesis tests will be completed using the same format as before. You may want to review: ES10-pp427&444-445, SSM-pp265&275. The only differences are:

    1. **writing hypotheses**: (see box before ex. 13.55)
    2. **using Table 6 or 7**: df = n - 2

    3. **the calculated test statistic**: $t* = \dfrac{b_1 - \beta_1}{s_{b1}}$

Computer and/or calculator commands to determine the line of best fit and also perform a hypothesis test concerning the slope can be found in ES10-pp722&723. Excel also constructs the confidence interval for the slope.

Slight variations in sums of squares and further calculations can result from round-off errors.

13.59 a.

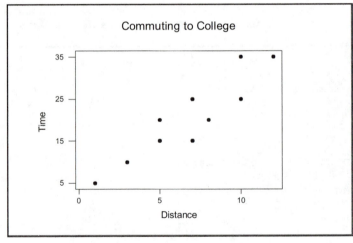

b. Summary of data: $n = 10$, $\Sigma x = 68$, $\Sigma y = 205$,

$\quad\quad\quad\Sigma x^2 = 566$, $\Sigma xy = 1670$, $\Sigma y^2 = 5075$

SS(x) = 566 - (68²/10) = 103.6
SS(xy) = 1670 - [(68)(205)/10] = 276.0
b_1 = 276.0/103.6 = 2.664
b_0 = [205 - (2.664)(68)]/10 = 2.38

$\hat{y}$ = __2.38 + 2.664x__

c.

Step 1: a. The slope β_1 of the line of best fit for the population of distances and their corresponding times required for students to commute to college.

      b. H_0: $\beta_1 = 0$ (no value)
        H_a: $\beta_1 > 0$

Step 2: a. random sample, assume normality for y at each x
      b. t, df = 8   c. $\alpha = 0.05$

Step 3: a. n = 10, b_1 = 2.664,
        s_e^2 = [5075 - (2.38)(205) - (2.664)(1670)]/8
           = 137.192/8 = 17.149
        s_{b1} = $\sqrt{s_e^2/SS(x)}$ = $\sqrt{17.149/103.6}$ = 0.407
      b. t = $(b_1 - \beta_1)/s_{b1}$
        t* = (2.664 - 0)/0.407 = 6.55

Step 4: -- using p-value approach --------------------
      a. **P** = P(t > 6.55|df = 8);
         Using Table 6, ES10-p813:   **P** < 0.005
         Using Table 7, ES10-p814:   **P** < 0.002
         Using computer: **P** = 0.000

      b. **P** < α
     -- using classical approach ------------------
      a. critical region: t ≥ 1.86
      b. t* falls in the critical region

Step 5: a. Reject H_0.
      b. The slope is significantly greater than zero, at the 0.05 level of significance.

d. $b_1 \pm t(df, \alpha/2) \cdot s_{b1}$
   2.66 ± (2.90)(0.407) = 2.66 ± 1.18
   __1.48 to 3.84__, the 0.98 interval for β_1

13.61 a.

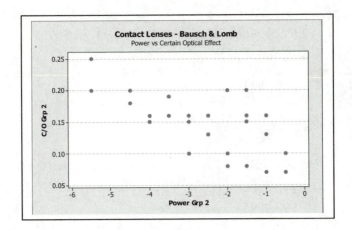

b. Pearson correlation of Power Grp 2 and C/O Grp 2
   = -0.674   P-Value = 0.000

c. Step 1: a. The linear correlation coefficient between
            lens power and a certain optical effect, ρ.

          b. H_{o}: ρ = 0.0

             H_{a}: $\rho \neq 0.0$

    Step 2: a. random sample, assume normality for y at
            each x
          b. r, df = n - 2 = 30 - 2 = 28
          c. α = 0.05

    Step 3: a. n = 30, r = -0.674
          b. r* = -0.674

    Step 4: -- using p-value approach --------------------
          a. **P** = P(r < -0.674) + P(r > 0.674)
             = 2P(r > 0.674);
               Using Table 11, ES10-p823: **P** < 0.01
             Using a computer: **P** = 0.000

          b. **P** < α
          -- using classical approach ------------------
          a. critical region: r $\leq$ -0.381 and r $\geq$ 0.381
          b. r* falls in the critical region

    Step 5: a. Reject H_{o}
          b. At the 0.05 level of significance, the
             linear correlation coefficient is
             significantly different from zero.

d. The regression equation is: $\hat{y} = 0.0881 - 0.0221x$

e. Step 1: a. The slope β_1 of the line of best fit for lens
        power and a certain optical effect.
      b. H_o: $\beta_1 = 0$ (no value)
        H_a: $\beta_1 < 0$
 Step 2: a. random sample, assume normality for y at each x
      b. t, df = 28   c. $\alpha = 0.05$
 Step 3: a. n = 30, $b_1 = -0.022120$, $s_{b1} = 0.004587$
      b. $t = (b_1 - \beta_1)/s_{b1}$
        $t* = (-0.022120 - 0)/0.004587 = -4.82$
 Step 4: -- using p-value approach --------------------
      a. **P** = P(t < -4.82|df = 28) = P(t > 4.82|df = 28);
        Using Table 6, ES10-p813:   **P** < 0.005
        Using Table 7, ES10-p814:   **P** = 0+
        Using a computer/calculator: **P** = 0.000

      b. **P** < α
     -- using classical approach ------------------
      a. critical region: t • -1.70
      b. t* falls in the critical region

 Step 5: a. Reject H_o.
      b. The slope is significantly less than zero, at
        the   0.05 level of significance.

SECTION 13.6 EXERCISES

Estimating $\mu_{y|x_o}$ and y_{x_o}

$\mu_{y|x_o}$ - the mean of the population y-values at a given x

y_{x_o} - the individual y-value selected at random for a given x

1. point estimate for $\mu_{y|x_o}$ and y_{x_o}: $\hat{y}$ ($\hat{y} = b_o + b_1 x$)

2. confidence interval for $\mu_{y|x_o}$:

$$\hat{y} \pm t(n-2, \alpha/2) \cdot s_e \sqrt{(1/n) + [(x_o - \overline{x})^2 / SS(x)]}$$

3. prediction interval for y_{x_o}:

$$\hat{y} \pm t(n-2, \alpha/2) \cdot s_e \sqrt{1 + (1/n) + [(x_o - \overline{x})^2 / SS(x)]}$$

Review the Five-Step Procedure in: ES10-p404, SSM-p256, if
necessary.

13.63 Step 1: $\mu_{y|x=70}$, the mean crutch length for individuals who say they are 70 in. tall.

Step 2: a. random sample, normality assumed for y at each x

        b. t   c. $1 - \alpha = 0.95$

Step 3: $n = 107$, $x_o = 70$, $\bar{x} = 68.84$, $s_e = \sqrt{0.50} = 0.707$

      $SS(x) = (n-1)s^2 = 106(7.35^2) = 5726.385$

      $\hat{y} = 4.8 + 0.68(70) = 52.4$

Step 4: a. $\alpha/2 = 0.05/2 = 0.025$; df = 105; t(105,0.025) = 1.96

        b. $E = t(n-2, \alpha/2) \cdot s_e \sqrt{(1/n) + [(x_o - \bar{x})^2/SS(x)]}$

          $E = (1.96)(0.707) \cdot$

        $\sqrt{(1/107) + [(70-68.84)^2/5726.385]}$

          $E = (1.96)(0.707)\sqrt{0.0095808} = 0.14$

        c. $\hat{y} \pm E = 52.4 \pm 0.14$

Step 5: <u>52.3 to 52.5</u>, the 0.95 interval for $\mu_{y|x=70}$

Computer and/or calculator commands for calculating the regressions line can be found in ES10-pp180&181.
MINITAB also provides confidence interval belts and prediction interval belts. See commands in ES10-p732-733.

13.65 From exercise 13.59:

$n = 10$, $\Sigma x = 68$, $SS(x) = 103.6$

$\hat{y} = 2.38 + 2.664x$, $s_e^2 = 17.149$

a. When x = 4, then $\hat{y} = 2.38 + 2.664(4) = 13.04$

   Point estimate for $\mu_{y|x=4}$ = <u>13.04</u>

b. Step 1: $\mu_{y|x=4}$, the mean travel time required to commute four miles.

     Step 2: a. random sample, normality assumed for y at each x

           b. t   c. $1 - \alpha = 0.90$

     Step 3: $n = 10$, $x_o = 4$, $\bar{x} = \Sigma x/n = 68/10 = 6.8$,

        $s_e^2 = 17.149$, $s_e = \sqrt{17.149} = 4.141$,

        $\hat{y} = 13.04$

     Step 4: a. $\alpha/2 = 0.10/2 = 0.05$; df = 8; t(8,0.05) = 1.86

         b. $E = t(n-2, \alpha/2) \cdot s_e \sqrt{(1/n) + [(x_o - \bar{x})^2/SS(x)]}$

           $E = (1.86)(4.141)(\sqrt{(1/10) + [(4-6.8)^2/103.6]})$

           $E = (1.86)(4.141)\sqrt{0.175676} = 3.23$

         c. $\hat{y} \pm E = 13.04 \pm 3.23$

     Step 5: <u>9.81 to 16.27</u>, the 0.90 interval for $\mu_{y|x=4}$

c. Step 1: $y_{x=4}$, the travel time required for one person to commute four miles.

   Step 2: a. random sample, normality assumed for y at each x

   b. t   c. $1 - \alpha = 0.90$

   Step 3: $n = 10$, $x_O = 4$, $\bar{x} = \Sigma x/n = 68/10 = 6.8$,

   $s_e^2 = 17.149$, $s_e = \sqrt{17.149} = 4.141$

   $\hat{y} = 13.04$

   Step 4: a. $\alpha/2 = 0.10/2 = 0.05$; $df = 8$; $t(8,0.05) = 1.86$

   b. $E = t(n-2,\alpha/2) \cdot s_e \sqrt{1+(1/n)+[(x_O-\bar{x})^2/SS(x)]}$

   $E = (1.86)(4.141)\sqrt{1 + 0.175676}$

   c. $\hat{y} \pm E = 13.04 \pm 8.35$

   Step 5: <u>4.69 to 21.39</u>, the 0.90 interval for $y_{x=4}$

d. When $x = 9$, then $\hat{y} = 2.38 + 2.664(9) = 26.36$

   Point estimate for $\mu_{y|x=9} = \underline{26.36}$

   $26.36 \pm (1.86)(\sqrt{17.149})(\sqrt{(1/10)+[(9-6.8)^2/103.6]})$

   $26.36 \pm (1.86)(4.141)\sqrt{0.146718} = 26.36 \pm 2.95$

   <u>23.41 to 29.31</u>, the 0.90 interval for $\mu_{y|x=9}$

   $26.36 \pm (1.86)(4.141)\sqrt{1 + 0.146718} = 26.36 \pm 8.25$

   <u>18.11 to 34.61</u>, the 0.90 interval for $y_{x=9}$

13.67 Summary of data: $n = 10$, $\Sigma x = 16.25$, $\Sigma y = 152$,

$\Sigma x^2 = 31.5625$, $\Sigma xy = 275$, $\Sigma y^2 = 2504$

$SS(x) = 31.5625 - (16.25^2/10) = 5.15625$
$SS(xy) = 275 - [(16.25)(152)/10] = 28.0$
$b_1 = 28.0/5.15625 = 5.4303$
$b_0 = [152 - (5.4303)(16.25)]/10 = 6.3758$

$\hat{y} = 6.3758 + 5.4303x$

$s_e^2 = [2504 - (6.3758)(152) - (5.4303)(275)]/8$
       $= 5.19324$

$s_e = \sqrt{5.19324} = 2.279$

When x = 2.0, then $\hat{y}$ = 6.3758 + 5.4303(2.0) = 17.24
t(8,0.025) = 2.31   and $\bar{x}$ = 16.25/10 = 1.625

a. Step 1: $\mu_y|_{x=2.00}$, the mean heart-rate reduction for a dose of 2.00 mg.

    Step 2: a. random sample, normality assumed for y at each x

    b. t   c. 1 - α = 0.95
    Step 3: n = 10, x_o = 2.00, $\bar{x}$ = 1.625, s_e = 2.279
    $\hat{y}$ = 17.24

    Step 4: a. α/2 = 0.05/2 = 0.025; df = 8; t(8,0.025) = 2.31

    b. $E = t(n-2,\alpha/2)\cdot s_e\sqrt{(1/n)+[(x_o-\bar{x})^2/SS(x)]}$
    E = (2.31)(2.279)·
    $\sqrt{(1/10)+[(2.0-1.625)^2/5.15625]}$
    E = (2.31)(2.279)$\sqrt{0.127273}$ = 1.88

    c. $\hat{y}$ ± E = 17.24 ± 1.88
    Step 5: 15.4 to 19.1, the 0.95 interval for $\mu_y|_{x=2}$

b. Step 1: $Y_{x=2.00}$, the heart-rate reduction expected for an individual receiving a dose of 2.00 mg.

    Step 2: a. random sample, normality assumed for y at each x

    b. t   c. 1 - α = 0.95
    Step 3: n = 10, x_o = 2.00, $\bar{x}$ = 1.625, s_e = 2.279
    $\hat{y}$ = 17.24

    Step 4: a. α/2 = 0.05/2 = 0.025; df = 8; t(8,0.025) = 2.31

    b. $E = t(n-2,\alpha/2)\cdot s_e\sqrt{1+(1/n)+[(x_o-\bar{x})^2/SS(x)]}$
    E = (2.31)(2.279)$\sqrt{1 + 0.127273}$ = 5.59

    c. $\hat{y}$ ± E = 17.24 ± 5.59
    Step 5: 11.6 to 22.8, the 0.95 interval for $Y_{x=2}$

13.69 a. The overall pattern is elongated in that as the number of customers increases, so does the number of items. The linearity stops with the last upper right 6 points.

b. In the regression analysis printout, the t-test for a significant slope is verified with a t = 27.71 and a corresponding p-value of 0.000. This would indicate that the linear model does fit the data.

c. Upon reviewing the entire dataset, the 6 points are all days in November and December - months in which sales are different due to the holiday season.

13.71 a.

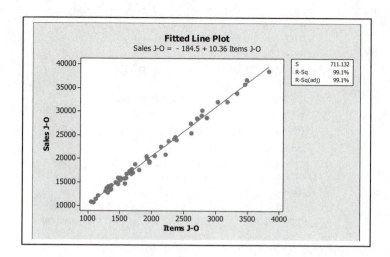

b. The data for January through October follow very closely
   to a straight line, linear, relationship. There do not
   appear to be any ordered pairs different from the others.

c. The relationship between the number of items purchased per
   day and the total daily sales is a strong positive linear
   relationship. As the number of items purchased per day
   increased so did the total daily sales. The corresponding
   correlation coefficient of 0.995 is significant as noted
   below with its p-value:

   Pearson correlation of Items J-O and Sales J-O
      = 0.995   P-Value = 0.000

d.

  Step 1: a. The slope β_1 of the line of best fit for the
             population of number of items purchased per day
             and the total daily sales.
          b. H_o: $\beta_1 = 0$ (no value)
             H_a: $\beta_1 \bullet 0$
  Step 2: a. random sample, normality assumed for y at each x
          b. t, df = 50   c. $\alpha = 0.05$
  Step 3: a. n = 52, b_1 = 10.3555, s_{b1} = 0.1405
          b. t = $(b_1 - \beta_1)/s_{b1}$
             t* = (10.3555 - 0)/0.1405 = 73.68

Step 4: -- using p-value approach --------------------
       a. **P** = 2P(t > 73.68|df = 50);
         Using Table 6, ES10-p813:
         P < 0.01
         Using Table 7, ES10-p814:
         P =0+
         Using a computer: **P** = 0.000
       b. **P** < α
    -- using classical approach ------------------
       a. critical region: t ≤ -2.01, t ≥ 2.01
       b. t* falls in the critical region

Step 5: a. Reject H_o.
       b. The slope is significantly different than zero,
         at the 0.05 level of significance.

d. Using the equation:
   When x = 3000, then $\hat{y}$ = -185 + 10.3555(3000) = 30881.5

   Step 1: $y_{x=3000}$, the daily total sales if the number of
          items purchased per day is equal to 3000.
   Step 2: a. random sample, normality assumed for y at
          each x
         b. t   c. 1 - α = 0.95
   Step 3: n = 52, x_o = 3000, $\overline{x}$ = 1991.02,
          s_e = 711.132 , SS(x) = 25600216.98
   Step 4: a. α/2 = 0.05/2 = 0.025; df = 50;
          t(50,0.025) = 2.01

      b. $E = t(n-2,\alpha/2) \cdot s_e \sqrt{1+(1/n)+[(x_o-\overline{x})^2/SS(x)]}$
      E = (2.01)(711.132)·
$$\sqrt{1 + (1 / 52) + (3000 - 1991.02)^2 / 25600216.98)}$$
      c. $\hat{y}$ ± E = 30881.5 ± 1470.94
   Step 5: <u>29410.6 to 32352.4</u>, the 0.95 interval for
      $y|_{x=3000}$

13.73 The standard error for $\overline{x}$'s is much smaller than the
standard deviation for individual x's (CLT). Thus the
confidence interval will be narrower in accordance.

13.75   a. Always.

b. Never.   r = 0.99 only indicates a strong linear
correlation.   It never indicates cause-effect.

c. Sometimes.   An r value greater than zero indicates
that as x increases, y tends to increase. However,
there may be a few high x-values with low y-values.

d. Sometimes.   The two coefficients measure two completely
different concepts.   Their signs are unrelated.

e. Always.

13.77   Step 1:   a. The linear correlation coefficient for the
population, ρ.

b. $H_{o:}$ ρ = 0.0

$H_{a:}$ ρ > 0.0
Step 2:   a. random sample, assume normality for y at each x
b. df = n - 2 = 45 - 2 = 43

c. α = 0.05
Step 3:   a. n = 45
b. r* = 0.69
Step 4:   -- using p-value approach --------------------
a.**P** = P(r > 0.69)
Using Table 11, ES10-p823: **P** < 0.005

b. **P** < α
-- using classical approach -----------------
a. critical region: r ≥ 0.29
b. r* falls in the critical region

Step 5:   a. Reject H_o
b. There is sufficient reason to conclude that the
correlation coefficient is positive, at the
0.05 level of significance.

13.79   a. Step 1:   a. The linear correlation coefficient for the
population, ρ.

b. $H_{o:}$ ρ = 0.0

$H_{a:}$ ρ ≠ 0.0

Step 2:   a. random sample, assume normality for y at
        each x
      b. df = n - 2 = 17 - 2 = 15
      c. α = 0.05

Step 3:   a. n = 17
      b. r* = 0.61

Step 4:   -- using p-value approach --------------------
      a. **P** = P(r > 0.61)
         Using Table 11, ES10-p823: **P** < 0.01
      b. **P** < α
      -- using classical approach ------------------
      a. critical region: r ≤ -0.482, r ≥ 0.482
      b. r* falls in the critical region

Step 5:   a. Reject H_O.
      b. Yes, the correlation coefficient is
         significantly different from zero, at the
         0.05 level of significance.

b. $\hat{y}$ = 1.8(50) + 28.7 = <u>118.7</u>

13.81 a.

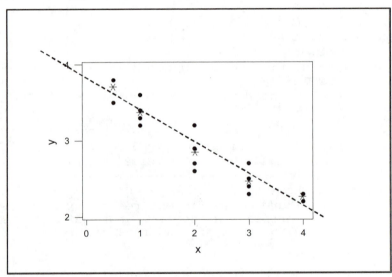

b. See dashed line on graph in (a).

c. See *'s on graph. The *'s seem to follow a curved path, not a straight line. The *'s are above the line at the ends and below the line in the middle.

d. Summary of data: n = 18 Σx = 37.5, Σy = 52.7,
   Σx^2 = 104.75, Σxy = 98.75, Σy^2 = 159.49

   SS(x) = 104.75 - (37.5²/18) = 26.625
   SS(xy) = 98.75 - [(37.5)(52.7)/18] = -11.0417

   b_1 = -11.0417/26.625 = -0.4147
   b_0 = [52.7 - (-0.4147)(37.5)]/18 = 3.792

   $\hat{y}$ = 3.79 - 0.415x

e. s_e^2 = [159.49 - (3.79)(52.7) - (-0.4147)(98.75)]/16
        = 0.04429

   s_e = $\sqrt{0.04429}$ = 0.21045

f. s_{b1} = $\sqrt{s_e^2/SS(x)}$ = $\sqrt{0.04429/26.625}$ = 0.041

   -0.415 ± 2.12(0.041) = -0.415 ± 0.087

   -0.502 to -0.328, the 0.95 interval for β_1

g. At x = 3.0:
   Step 1: $\mu_{y|x=3.0}$, the mean value of y at x = 3.0.
   Step 2: a. random sample, normality assumed for y at
              each x
            b. t   c. 1 - α = 0.95
   Step 3: a. n = 18, x_o = 3.0, $\bar{x}$ = 37.5/18 = 2.08,
              s_e = 0.21045
            b. $\hat{y}$ = 3.79 - 0.415(3.0) = 2.55
   Step 4: a. $\alpha/2$ = 0.05/2 = 0.025; df = 16;
              t(16,0.025) = 2.12
            b. E = t(n-2,$\alpha/2$)·$s_e\sqrt{(1/n)+[(x_o-\bar{x})^2/SS(x)]}$·
              E =
              (2.12)(0.21045)$\sqrt{(1/18) + [(3-2.08)^2/26.625]}$
              E = 0.13
            c. $\hat{y}$ ± E = 2.55 ± 0.13
   Step 5: 2.42 to 2.68, the 0.95 interval for $\mu_{y|x=3}$

At x = 3.5:
Step 1: $\mu_y|_{x=3.5}$, the mean value of y at x = 3.5.

Step 2: a. random sample, normality assumed for y at
each x

b. t   c. 1 - α = 0.95

Step 3: a. n = 18, x_o = 3.5, $\overline{x}$ = 2.08,
s_e = 0.21045

b. $\hat{y}$ = 3.79 - 0.415(3.5) = 2.34

Step 4: a. α/2 = 0.05/2 = 0.025; df = 6;
t(16,0.025) = 2.12

b. E = t(n-2,α/2)·$s_e\sqrt{(1/n)+[(x_o-\overline{x})^2/SS(x)]}$
E =
(2.12)(0.21045)$\sqrt{(1/18) + [(3.5-2.08)^2/26.625]}$
E = 0.16

c. $\hat{y}$ ± E = 2.34 ± 0.16

Step 5: <u>2.18 to 2.50</u> the 0.95 interval for $\mu_y|_{x=3.5}$

h. At x = 3.0:
Step 1: $y|_{x=3.0}$, the value of y at x = 3.0.
Step 2: a. random sample, normality assumed for y at
each x

b. t   c. 1 - α = 0.95

Step 3: a. n = 18, x_o = 3.0, $\overline{x}$ = 37.5/18 = 2.08,
s_e = 0.21045

b. $\hat{y}$ = 3.79 - 0.415(3.0) = 2.55

Step 4: a. α/2 = 0.05/2 = 0.025; df = 16;
t(16,0.025) = 2.12

b. E = t(n-2,α/2)·$s_e\sqrt{1+(1/n)+[(x_o-\overline{x})^2/SS(x)]}$
E = (2.12)(0.
21045)$\sqrt{1 + (1/18) + [(3-2.08)^2/26.625]}$
E = 0.47

c. $\hat{y}$ ± E = 2.55 ± 0.47

Step 5: <u>2.08 to 3.02</u>, the 0.95 interval for $y_{x=3}$

At x = 3.5:
Step 1: $y|_{x=3.5}$, the value of y at x = 3.5.
Step 2: a. random sample, normality assumed for y at
each x

b. t   c. 1 - α = 0.95

Step 3: a. n = 18, x_o = 3.5, $\overline{x}$ = 2.08,
s_e = 0. 21045

b. $\hat{y}$ = 3.79 - 0.415(3.5) = 2.34

Step 4:   a. $\alpha/2 = 0.05/2 = 0.025$; df = 16;
          $t(16,0.025) = 2.12$
      b. $E = t(n-2,\alpha/2)\cdot s_e\sqrt{1+(1/n)+[(x_0-\overline{x})^2/SS(x)]}$
        $E =$
      $(2.12)(0.21045)\sqrt{(1+1/18)+[(3.5-2.08)^2/26.625]}$
        $E = 0.47$
      c. $\hat{y} \pm E = 2.34 \pm 0.47$
Step 5:   <u>1.87 to 2.81</u> the 0.95 interval for $y_{x=3.5}$

13.83   Summary of data: n = 21 $\Sigma x = 1177$, $\Sigma y = 567$,
            $\Sigma x^2 = 70033$, $\Sigma xy = 32548$, $\Sigma y^2 = 15861$
  $SS(x) = 70033 - (1177^2/21) = 4064.95$
  $SS(xy) = 32548 - [(1177)(567)/21] = 769.0$
  $SS(y) = 15861 - (567^2/21) = 552.0$

  $r = 769.0/\sqrt{(4064.95)(552.0)} = 0.5133$

a. Step 1: a.   The linear correlation coefficient for the number of stamens and the number of carpels in a particular species of flowers, ρ.

      b. H_o: $\rho = 0.0$

      c. H_a: $\rho \neq 0.0$

Step 2:   a. random sample, assume normality for y at each x

      b. df = n - 2 = 21 - 2 = 19

      c. $\alpha = 0.05$

Step 3:   a. n = 21
      b. $r* = 0.513$

Step 4:   -- using p-value approach --------------------
      a. **P** = 2P(r > 0.513)
        Using Table 11, ES10-p823: 0.01 < **P** < 0.02
        Using computer: **P** = 0.017

      b. **P** < α
      -- using classical approach ------------------
      a. critical region: r ≤ -0.433, r ≥ 0.433
      b. r* is in the critical region

Step 5:   a. Reject H_o.
      b. There is sufficient reason to conclude that there is linear correlation.

b. $b_1 = 769.0/4064.95 = 0.1892$
   $b_0 = [567 - (0.1892)(1177)]/21 = 16.3958$

$\hat{y} = \underline{16.40 + 0.189x}$

c. Step 1:   a. The slope β_1 of the line of best fit for the population of number of stamens per flower and the corresponding number of carpels.

   b. H_o: $\beta_1 = 0$ (no value)
      H_a: $\beta_1 > 0$

   Step 2:   a. random sample, assume normality for y at each x

   b. t, df = 19   c. $\alpha = 0.05$

   Step 3:   a. n = 21, $b_1 = 0.1892$,
             $s_e^2 = [15861 - (16.3958)(567) - (0.1892)(32548)]/19$
             $s_e^2 = 21.3947$
             $s_{b1} = \sqrt{s_e^2/SS(x)}$
                    $= \sqrt{21.3947/4064.95} = 0.07255$

   b. $t = (b_1 - \beta_1)/s_{b1}$
      $t* = 0.189/0.07255 = 2.61$

   Step 4:   -- using p-value approach --------------------
   a. $P = P(t > 2.61 | df = 19)$;
      Using Table 6, ES10-p813: $0.005 < P < 0.01$
      Using Table 7, ES10-p814: $0.007 < P < 0.009$
      Using computer: $P = 0.0086$

   b. $P < \alpha$
   -- using classical approach ------------------
   a. critical region: $t \geq 1.73$
   b. t* is in the critical region

   Step 5:   a. Reject H_o.
   b. The slope is significantly greater than zero.

d. Step 1: $y|_{x=64}$, the number of carpels found in a mature flower if the number of stamens is 64.
   Step 2:   a. t   b. $1 - \alpha = 0.95$
   Step 3:   a. n = 21, $x_o = 64$, $\overline{x} = 1177/21 = 56.05$
             $s_e = \sqrt{21.3947}$
   b. $\hat{y} = 16.40 + 0.189(64) = 28.50$

Step 4:   a. $\alpha/2 = 0.05/2 = 0.025$; df = 19;
            $t(19,0.025) = 2.09$

          b. $E = t(n-2,\alpha/2) \cdot s_e\sqrt{1+(1/n)+[(x_o-\overline{x})^2/SS(x)]}$
    $E = (2.09)(\sqrt{21.3947}$
         $)\sqrt{1+(1/21)+[(64-56.05)^2/4064.95]}$
          $E = 9.97$

          c. $\hat{y} \pm E = 28.50 \pm 9.97$
Step 5:   <u>18.53 to 38.47</u>, the 0.95 interval for one $y_{x=64}$

13.85   Summary of data: n = 24 $\Sigma x = 1560$, $\Sigma y = 554$,
              $\Sigma x^2 = 114000$, $\Sigma xy = 40930$, $\Sigma y^2 = 15186$

     $SS(x) = 114000 - (1560^2/24) = 12600$
     $SS(xy) = 40930 - [(1560)(554)/24] = 4920$
     $SS(y) = 15186 - (554^2/24) = 2397.83$

     $r = 4920/\sqrt{(12600)(2397.83)} = 0.895$

   a. Step 1:   a. The linear correlation coefficient for the
                  pounds of fertilizer and the pounds of wheat
                  harvested, ρ.

                b. H_o: $\rho = 0.0$

                c. H_a: $\rho > 0.0$
      Step 2:   a. random sample, assume normality for y at
                  each x

                b. df = n - 2 = 24 - 2 = 22

                c. $\alpha = 0.05$
      Step 3:   a. n = 24
                b. r* = 0.895
      Step 4:   -- using p-value approach --------------------
                a. **P** = P(r > 0.895)
                   Using Table 11, ES10-p823: **P** < 0.005
                   Using computer: **P** = 0.000

                b. **P** < α
                -- using classical approach ------------------
                a. critical region: r ≥ 0.34
                b. r* is in the critical region

      Step 5:   a. Reject H_o.
                b. There is sufficient reason to conclude that
                   there is a positive linear correlation, at
                   the 0.05 level of significance.

b. b_1 = 4920/12600 = 0.39048 = 0.39
   b_0 = [554 - (0.39)(1560)]/24 = -2.298 = -2.30

   $\hat{y}$ = -2.30 + 0.39x

Step 1: $\mu_y|_{x=50}$, the mean yield that could be expected
         if 50 pounds of fertilizer were used per plot.
Step 2: a. t   b. 1 - α = 0.98
Step 3: a. n = 24, x_o = 50, $\bar{x}$ = 1560/24 = 65
            $s_e{}^2$ = [15186-(-2.298)(554)-
            (0.39048)(40930)]/22
               = 21.667
            s_e = $\sqrt{21.667}$ = 4.655
         b. $\hat{y}$ = -2.298 + 0.3905(50) = 17.23
Step 4: a. α/2 = 0.02/2 = 0.01; df = 22;
            t(22,0.01) = 2.51
         b. E = t(n-2,α/2)$\cdot s_e \sqrt{(1/n)+[(x_o-\bar{x})^2/SS(x)]}$

            E = (2.51)(4.655) $\sqrt{(1/24)+[(50-65)^2/12600]}$

            E = 2.85
         c. $\hat{y}$ ± E = 17.23 ± 2.85
Step 5: <u>14.38 to 20.08</u>, the 0.98 interval for $\mu_y|_{x=50}$

c. Step 1: $\mu_y|_{x=75}$, the mean yield that could be expected
           if 75 pounds of fertilizer were used per plot.
   Step 2: a. t   b. 1 - α = 0.98
   Step 3: a. n = 24, x_o = 75, $\bar{x}$ = 1560/24 = 65
               s_e = 4.655
            b. $\hat{y}$ = -2.298 + 0.3905(75) = 26.99
   Step 4: a. α/2 = 0.02/2 = 0.01; df = 22;
               t(22,0.01) = 2.51
            b. E = t(n-2,α/2)$\cdot s_e \sqrt{(1/n)+[(x_o-\bar{x})^2/SS(x)]}$

               E = (2.51)(4.655) $\sqrt{(1/24)+[(75-65)^2/12600]}$

               E = 2.60
            c. $\hat{y}$ ± E = 26.99 ± 2.60
   Step 5: <u>24.39 to 29.59</u>, the 0.98 interval for $\mu_y|_{x=75}$

13.87 Exercises 13.70 (customers vs items) and 13.86 (customers vs sales) both show a linear relationship but have several ordered pairs that appear to be different from the others. Therefore the fit is not as strong as in Exercise 13.71 (items vs sales). There is a stronger relationship with number of items purchased and the total daily sales. If more items are purchased, the sales have to also increase. When customers are involved the relationship is not as strong due to the fact that you can have a customer that does not make a purchase. This would effect the number of items purchased as well as the total daily sales.

13.89 Summary of data: $n = 5$, $\Sigma x = 16$, $\Sigma y = 38$,

$$\Sigma x^2 = 66, \ \Sigma xy = 145, \ \Sigma y^2 = 326$$

$SS(x) = 66 - (16^2/5) = 14.8$
$SS(xy) = 145 - [(16)(38)/5] = 23.4$
$SS(y) = 326 - (38^2/5) = 37.2$

$b_1 = 23.4/14.8 = 1.5811$

$r = 23.4/\sqrt{(14.8)(37.2)} = \underline{0.9973}$ [Formula 13.3]

$r = 1.5811\sqrt{14.8/37.2} = \underline{0.9973}$ [Formula in exercise]

CHAPTER 14 ∇ ELEMENTS OF NONPARAMETRIC STATISTICS

Chapter Preview

Chapter 14 introduces the concept of nonparametric statistics. Up to this point, especially in chapters 8, 9 and 10, the methods used were parametric methods. Parametric methods rely on the normality assumption through knowledge of the parent population or the central limit theorem. In nonparametric (distribution-free) methods, few assumptions about the parent population are required yet the methods are only slightly less efficient than their parametric counterparts. Chapter 14 will demonstrate nonparametric methods for hypothesis tests concerning one mean, two independent means, two dependent means, correlation and randomness.

A survey conducted by NFO Research, Inc. on the attitudes of teenagers toward social and moral values is used in this chapter's opening section "Teenagers' Attitudes".

SECTION 14.1 EXERCISES

14.1 a. There appears to be a "general" agreement; boys and girls agreed on the two most important, they agreed on the four least important, and they scrambled the order for the middle six.

      b.

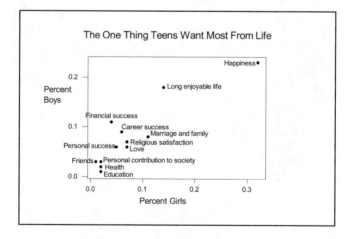

c. The importance placed on these twelve items appears to correlate as described in (a).

SECTION 14.4 EXERCISES

14.3 a. The Sign Test only involves the counts of plus and minus signs.
  b. The population median can be tested using the sign test. By nature of a median, half of the data is above the median and half of the data is below the median.

Sign test $(1 - \alpha)$ Confidence Interval for M - the population median

1. Arrange data in ascending order (smallest to largest)

2. Assign the notation: x_1 (smallest), x_2, x_3 ... x_n (largest) to the data.

3. Critical value = k (value from Table 12 using n and α)

4. $(1 - \alpha)$ confidence interval extends from the data values:

$$x_{k+1} \quad \text{to} \quad x_{n-k}$$

14.5   Ranked data:

33	34	35	36	38	39	40	40	42	45
46	46	46	46	47	47	48	54	59	65

For n = 20 and $1 - \alpha = 0.95$, the critical value from Table 12 is k = 5.

$x_{k+1} = x_6 = 39$ and $x_{n-k} = x_{15} = 47$

39 to 47, the 0.95 interval for median M

14.7

-30	-16	-14	-14	-13	-12	-12	-11	-10	-10
-9	-9	-8	-8	-8	-7	-6	-6	-5	-4
-4	-4	-3	-3	-2	-1	-1	0	1	2
2	5	6	6	6	6	6	9	12	12
13	16	18	19						

For $n = 44$ and $1 - \alpha = 0.95$, the critical value from Table 12 is $k = 15$

$x_{k+1} = x_{16} = -7$ and $x_{n-k} = x_{29} = 1$

$\underline{-7 \text{ to } 1}$ points of change in scores is the 95% confidence interval for the median.

The Sign Test

The Sign Test is used to test one mean (median) or the median difference between paired data (dependent samples). Refer to the Five-Step Hypothesis Test Procedure in: ES10-pp427&444-445, SSM-pp265&275, if necessary. The only changes are in:

1. **the hypotheses**
   a. null hypothesis
      H_o: M = # or H_o: p(preference) = 0.5 or H_o: p(sign) = 0.5
   b. possible alternative hypotheses:
      H_a: M ≠ # or H_o: p(preference) ≠ 0.5 or H_o: p(sign) ≠ 0.5
      H_a: M < # or H_o: p(preference) < 0.5 or H_o: p(sign) < 0.5
      H_a: M > # or H_o: p(preference) > 0.5 or H_o: p(sign) > 0.5

2. **the critical value of the test statistic,**
   a. for sample sizes ≤ 100, x; Table 12 (Appendix B, ES10-p824)
   b. for sample sizes > 100, z; Table 4 (Appendix B, ES10-p811)

3. **the calculated test statistic,**
   a. for n ≤ 100; x = the number of the less frequent sign
   b. for n > 100; x = n(sign of preference)

$$z^* = (x' - (n/2))/(\tfrac{1}{2}\sqrt{n}) \qquad \text{where}$$

$$x' = x - \tfrac{1}{2}, \text{ if } x > (n/2) \quad \text{or} \quad x' = x + \tfrac{1}{2}, \text{ if } x < (n/2)$$

14.9 a. H_o: Median = 18   vs. H_a: Median < 18

b. H_o: Median = 32   vs. H_a: Median < 32

c. H_o: Median = 4.5   vs. H_a: Median $\neq$ 4.5

14.11 a. **P** = P(x $\leq$ 3 for n = 18) = 0.01

b. P = P(x $\leq$ 30 for n = 78) = ½(0.05 < P < 0.10):
    0.025 < P < 0.05

c. **P** = P(x $\leq$ 10 for n = 38) = ½(0.01) = 0.005

d. **P** = 2P(z < -2.56) = 2[0.5000 - 0.4948] = 2[0.0052]
     = 0.0104

Table 12, Critical Values of the Sign Test, gives the maximum
allowable number of the less frequent sign, k, that will cause
rejection of H_o. k is based on n (the total number of signs,
excluding zeros) and α.

Therefore, if x $\leq$ k, reject H_o and if x > k, fail to reject H_o.

14.13 Step 1: a. Median age of the population of all leukemia
                  patients who receive stem cell transplants.
               b. H_o: Median = 42 years
                  H_a: Median $\neq$ 42 years
        Step 2: a. Assume sample is random. Age is continuous.
               b. x = n(least frequent sign)
               c. α = 0.05
        Step 3: a. + = over 42 years of age; n = 100
                  n(+) = 40, n(0) = 0, n(-) = 60
               b. x = n(+) = 40
        Step 4: -- using p-value approach ---------------
               a. **P** = 2P(x $\leq$ 40|n = 100);
                  Using Table 12, Appendix B, ES10-p824:
                  0.05 < **P** < 0.10
               b. **P** > α
               -- using classical approach ------------
               a. critical region: n(least freq sign) $\leq$ 39
               b. The test statistic is not in the critical
                      region.

Step 5: a. Fail to reject H_O.
　　　　b. The evidence is not sufficient to show the
　　　　　 median age is not equal to 42 years, at the 0.05
　　　　　 level of significance.

14.15　Step 1: a. Proportion of boys who wear protective clothing
　　　　　 b. H_O: $P(+) = 0.5$
　　　　　　 H_a: $P(+) \neq 0.5$
　　　 Step 2: a. Assume random sample. x is binomial and
　　　　　　 approximately normal.
　　　　　 b. x = n(least freq sign)
　　　　　 c. α is unspecified
　　　 Step 3: a. + = correct solution;　 n = 75
==

a.　　　　　　　　 b. x = n(+) = 20
　　　 Step 4:　 -- using p-value approach ---------------
　　　　　　 a. **P** = 2P(x ≤ 20|n = 75);
　　　　　　　 Using Table 12, Appendix B, ES10-p824:
　　　　　　　 P < 0.01
----

b.　　　　　　　　 b. x = n(+) = 27
　　　 Step 4:　 -- using p-value approach ---------------
　　　　　　 a. **P** = 2P(x ≤ 27|n = 75);
　　　　　　　 Using Table 12, Appendix B, ES10-p824:
　　　　　　　 0.01 < **P** < 0.05
----

c.　　　　　　　　 b. x = n(+) = 30
　　　 Step 4:　 -- using p-value approach ---------------
　　　　　　 a. **P** = 2P(x ≤ 30|n = 75);
　　　　　　　 Using Table 12, Appendix B, ES10-p824:
　　　　　　　 0.10 < **P** < 0.25
----

d.　　　　　　　　 b. x = n(+) = 33
　　　 Step 4:　 -- using p-value approach ---------------
　　　　　　 a. **P** = 2P(x ≤ 33|n = 75);
　　　　　　　 Using Table 12, Appendix B, ES10-p824:
　　　　　　　 P > 0.25
==
　　　　　　 b. If **P** < α

Step 5: a. Reject H_O.
   b. The evidence does show that the proportion is significantly different than one-half.
----
   b. If **P** > α
Step 5: a. Fail to reject H_O.
   b. The evidence does not show that the proportion is significantly different than one-half.

For testing the median difference between paired data, subtract corresponding pairs of data and use the signs of the differences. Hypotheses can be written in three forms.

null hypothesis:

H_O: No difference between the pairs   <u>or</u> H_O: M = 0

Or H_O: p(+) = 0.5

possible alternative hypotheses:

H_a: There is a difference between the pairs
   H_a: M ≠ #   <u>or</u> H_a: p(+) ≠ 0.5
H_a: One of the pairs is greater than the other
   (subtract greater - smaller)   <u>or</u>
   H_a: M > #   <u>or</u> H_a: p(+) > 0.5
H_a: One of the pairs is less than the other
   (subtract smaller - greater)   <u>or</u>
   H_a: M < #   <u>or</u> H_a: p(+) < 0.5

14.17 a.

Nation	1995	1999	2003	d=99-95	Sign	d2=03-99	Sign2
Belgium-Flemish	533	535	516	2	+	-19	-
Bulgaria	545	518	479	-27	-	-39	-
Cyprus	452	460	441	8	+	-19	-
Hong Kong	510	530	556	20	+	26	+
Hungary	537	552	543	15	+	-9	-
Iran, Islamic Republic	463	448	453	-15	-	5	+
Japan	554	550	552	-4	-	2	+
Korea, Republic of	546	549	558	3	+	9	+
Latvian	476	503	513	27	+	10	+
Lithuania	464	488	519	24	+	31	+
Netherlands	541	545	536	4	+	-9	-
New Zealand	511	510	520	-1	-	10	+
Romania	471	472	470	1	+	-2	-
Russian Federation	523	529	514	6	+	-15	-
Singapore	580	568	578	-12	-	10	+
Slovak Republic	532	535	517	3	+	-18	-
United States	513	515	527	2	+	12	+

d = 1999 - 1995
b. 12 countries improved and 5 had lower scores:
   x* = n(-) = 5

Step 1: a. Median change in science scores.
         b. H_o: M = 0   There is no difference between
            the 1995 and 1999 science scores.
            H_a: M > 0   The science scores improved.
Step 2: a. Assume sample is random. Test scores are
            numerical.
         b. x = n(least frequent sign)
         c. α = 0.05
Step 3: a. + = positive, - = negative; n = 17
            n(+) = 12, n(0) = 0, n(-) = 5
         b. x = n(-) = 5

Step 4: -- using p-value approach ----------------
        a. P = P(x ≤ 5|n = 17);
           Using Table 12, Appendix B, ES10-p824:
           P ≈ 0.125
           Using a computer: P = 0.0717
        b. P > α
        -- using classical approach -------------
        a. critical region: n(least freq sign) ≤ 4
        b. The test statistic is not in the critical
              region.

Step 5: a. Fail to reject H_o.
         b. Insufficient evidence to support the claim
            that the science scores show a significant
            overall increase thoughout the world, at the
            0.05 level of significance.

      d = 2003 - 1999
    b. **9 countries improved and 8 had lower scores:**
        x* = n(-) = 8

Step 1: a. Median change in science scores.
         b. H_o: M = 0   There is no difference between
                the 1999 and 2003 science scores.
            H_a: M > 0   The science scores improved.
Step 2: a. Assume sample is random. Test scores are
            numerical.
         b. x = n(least frequent sign)
         c. α = 0.05
Step 3: a. + = positive, - = negative;   n = 17
            n(+) = 9,   n(0) = 0, n(-) = 8
         b. x = n(-) = 8
Step 4: -- using p-value approach ----------------
        a. P = P(x ≤ 8|n = 17);
           Using Table 12, Appendix B, ES10-p824:
           P > 0.125
           Using a computer: P = 0.5000
        b. P > α
        -- using classical approach -------------
        a. critical region: n(least freq sign) ≤ 4
        b. The test statistic is not in the critical
              region.

-- 523 --

Step 5: a. Fail to reject H_o.
   b. Insufficient evidence to support the claim that the science scores show a significant overall increase thoughout the world, at the 0.05 level of significance.

14.19   Step 1: a. Preference for the taste of a new cola.
   b. H_o: There is no preference; $p = P(\text{prefer}) = 0.5$
   H_a: There is a preference for the new; $p > 0.5$
Step 2: a. x is binomial and approximately normal
   b. z   c. $\alpha = 0.01$
Step 3: a. + = prefer new;   n = 1228;
   $n(+) = 645$, $n(0) = 272$, $n(-) = 583$
   b. $x = n(+) = 645$; $x' = 644.5$
   $z = (x' - (n/2))/(\frac{1}{2}\sqrt{n})$
   $z^* = (644.5 - (1228/2))/(\frac{1}{2}\sqrt{1228})$
   $= (644.5 - 614)/17.5214 = 1.74$
Step 4: -- using p-value approach ---------------
   a. **P** $= P(z > 1.74) = 0.5000 - 0.4591 = 0.0409$
   b. **P** $> \alpha$
   -- using classical approach -------------
   a. critical region: $z \geq 2.33$
   b. The test statistic is not in the critical region.

Step 5: a. Fail to reject H_o.
   b. The evidence does not allow us to conclude that there is a significant preference for the new cola, at the 0.01 level of signficance.

14.21   Step 1: a. Proportion of high school seniors that can solve problems involving decimals, fractions, percentages and simple equations.
   b. H_o: $P(+) = 0.5$
   H_a: $P(+) > 0.5$
Step 2: a. Assume random sample. x is binomial and approximately normal.
   b. x = n(least frequent sign)   c. $\alpha = 0.05$
Step 3: a. n = 1500
Step 4: a. $-z(0.05) = -1.65$

b. $z = (x' - (n/2))/(\frac{1}{2}\sqrt{n})$
         $-1.65 = (x' - (1500/2))/(\frac{1}{2}\sqrt{1500})$
         $-1.65 = (x' - 750)/19.3649$
         $x' = 718.048;$ critical value is <u>718</u>

SECTION 14.5 EXERCISES

14.23 a. Difference between two independent means
          b. The actual size of the data is not used, only its rank.

The Mann-Whitney U Test

The Mann-Whitney U Test is used to test the difference between two independent means. Refer to the Five-Step Hypothesis Test Procedure in: ES10-pp427&444-445, SSM-pp265&275, if necessary. The only changes are in:

1. **the hypotheses**
   a. null hypothesis
      H_O: The average value is the same for both groups.
   b. possible alternative hypotheses:
      H_a: The average value is not the same for both groups.
      H_a: The average value of one group is greater than that of the other group.
      H_a: The average value of one group is less than that of the other group.

2. **the critical value of the test statistic,**
   a. for sample sizes ≤ 20, U; Table 13 (Appendix B, ES10-p825)

   b. for sample sizes > 20, z; Table 4 (Appendix B, ES10-p811)

3. **the calculated test statistic,**
   a. for n $\leq$ 20; U* = smaller of U_a and U_b, where

$$U_a = n_a \cdot n_b + \frac{(n_b)(n_b + 1)}{2} - R_b \quad \text{and}$$

$$U_b = n_a \cdot n_b + \frac{(n_a)(n_a + 1)}{2} - R_a \quad \ldots$$

$$R_a = \text{sum of ranks for sample A,} \quad R_b = \text{sum of ranks for sample B}$$

b. for $n > 20$; $U^* = $ smaller of U_a and U_b

$$z^* = (U - \mu_u) / (\sigma_u) \quad \text{where}$$

$$\mu_u = \frac{n_a \cdot n_b}{2} \qquad \sigma_u = \sqrt{\frac{n_a n_b (n_a + n_b + 1)}{12}}$$

14.25   a. H_o: The distributions are the same for both groups
       H_a: The distributions are different for the groups

     b. H_o: The average value is the same for both groups
       H_a: The average value is not the same for the groups

     c. H_o: The distribution of blood pressure is the same for
           both groups
       H_a: The distribution of blood pressure for group A is
           higher than for group B

14.27   a.   **P** > 0.05
     b.   **P** < 0.05
     c.   **P** $= 0.0089$

Table 13, Critical Values of U in the Mann-Whitney Test, gives only critical values for the left-hand tail. $U(n_1, n_2, \alpha)$ is based on the two sample sizes and the amount of α for a one or two-tailed test.

If $U^* \leq U(n_1, n_2, \alpha)$, reject H_o and

      if $U^* > U(n_1, n_2, \alpha)$, fail to reject H_o.

14.29   a. Critical region: $U \leq 88$

     b. Critical region: $z \leq -1.65$

14.31   MINITAB verify -- answers given in exercise.

14.33 Step 1: a. Number of preoperative glaucoma medications for patients receiving combined cataract surgery and those receiving just the surgery.

b. H_o: Number of medications is the same for both groups.

H_a: Number of medications is not the same for both groups.

Step 2: a. Independent samples and number values are numerical.

b. U   c. $\alpha = 0.05$

Step 3: a. $n_{CS} = 6$; $n_S = 5$

b. $U^* = 12.5$

Step 4: -- using p-value approach ---------------

a. $\mathbf{P} = 2P(U \leq 12.5| n_{CS}=6, n_S=5)$;

Using Table 13, Appendix B, ES10-p825:

$\mathbf{P} > 0.10$

Using a computer: P = 0.7066

b. $\mathbf{P} > \alpha$

-- using classical approach -------------

a. U(6,5,0.05); Crit. reg. ≤ 3

b. The test statistic is not in the critical region.

Step 5: a. Fail to reject H_o.

b. The evidence does allow us to reject the null hypothesis that the two groups are the same with respect to number of medications, at the 0.05 level of significance.

14.35 Step 1: a. Rainfall amounts based on cloud unseeding and cloud seeding.

b. H_o: Rainfall amount is the same for the two methods.

H_a: Rainfall amount is higher with cloud seeding.

Step 2: a. Independent samples and rainfall amounts are numerical.

b. U   c. $\alpha = 0.05$

Step 3: a. $n_U = n_S = 25$; $R_U = 503$, $R_S = 772$,
$\quad\quad\quad$ $U_U = 178$, $U_S = 447$
$\quad\quad\quad$ $\mu_u = (25 \cdot 25)/2 = 312.5$,
$\quad\quad\quad$ $\sigma_u = \sqrt{[(25)(25)(25+25+1)]/12} = 51.54$
$\quad\quad$ b. $U^* = 178$
$\quad\quad$ c. $z^* = (U - \mu_u)/\sigma_u$
$\quad\quad\quad$ $z^* = (178 - 312.5)/51.54 = -2.61$

Step 4: $\quad$ -- using p-value approach ---------------
$\quad\quad$ a. $\mathbf{P} = P(z > 2.61) = 0.5000 - 0.4955 = 0.0045$
$\quad\quad$ b. $\mathbf{P} < \alpha$
$\quad\quad$ -- using classical approach -------------
$\quad\quad$ a. critical values: $z(0.05) = 1.65$
$\quad\quad$ b. The test statistic is in the critical region.
$\quad\quad$ ---

Step 5: a. Reject H_O.
$\quad\quad$ b. The evidence does allow us to conclude that
$\quad\quad\quad$ there is a significant increase in the average
$\quad\quad\quad$ amount of rainfall with cloud seeding, at the
$\quad\quad\quad$ 0.05 level of significance.

SECTION 14.6 EXERCISES

The Runs Test

The Runs Test is used to test the randomness of a set of data.
Refer to the Five-Step Hypothesis Test Procedure in: ES10-
pp427&444-445,SSM-pp265&275, if necessary. The only changes are
in:

1. **the hypotheses**
   $\quad$ H_O: The data occurred in a random order
   $\quad$ H_a: The data is not in random order

2. **the critical value of the test statistic,**
   $\quad$ a. for sample sizes ≤ 20, V; Table 14 (Appendix B, ES10-p826)

   $\quad$ b. for sample sizes > 20, z; Table 4 (Appendix B, ES10-p811)
   $\quad\quad\quad\quad\quad\quad\quad\quad\quad\quad\quad\quad\quad\quad\quad\quad\quad\quad\quad\quad\quad$ . . .

14.37 a. H_O: The data did occur in a random order
           H_a: The data did not occur in a random order

   b. H_O: Sequence of odd/even is in random order
      H_a: Not in random order.

   c. H_O: The order of entry by gender was random
      H_a: The order of entry was not random

Table 14, Critical Values for Total Number of Runs, gives two
critical values, V, for each sample size category.

If $V* \leq$ the smaller V given or if $V* >$ the larger V given, reject
H_O.

14.39 a. Critical regions: $V \leq 9$ or $V \geq 22$

   b. Critical regions: $z \leq -1.96$ or $z \geq +1.96$

14.41 Step 1: a. Randomness; P(women) and P(men).
                   b. H_O: The hiring sequence is random.
                       H_a: The hiring sequence is not of random order
           Step 2: a. Each data fits one of two categories.
                   b. V   c. $\alpha = 0.05$
           Step 3: a. $n(M) = 15$, $n(F) = 5$
                   b. $V* = 9$
           Step 4: -- using p-value approach ---------------
                   a. Using Table 14, Appendix B, ES10-p826
                       P > 0.05
                   b. **P** $> \alpha$

```
              -- using classical approach -------------
           a. Critical regions:  V ≤ 4  or  V ≥ 12
           b. The test statistic is not in the critical
              region.
              ------------------------------------------
  Step 5:  a. Fail to reject H_O.
           b. The evidence is not significant, we can not
              conclude that this sequence is not random, at
              the 0.05 level of significance.
```

14.43 Step 1: a. Randomness; P(late bus).
           b. H_O: Random order of increase and decrease in
                 value from previous value.
              H_a: Lack of randomness (a trend, an increase in
                 wait time)
   Step 2: a. Each data fits one of two categories.
            b. V   c. α = 0.05
   Step 3: a. n(decreases) = 4, n(increases) = 13
            b. V* = 8
   Step 4:   -- using p-value approach ----------------
            a. Using Table 14, Appendix B, ES10-p826
               P > 0.05
            b. **P** > α
               -- using classical approach -------------
            a. Critical regions: V ≤ 3 or V ≥ 10
            b. The test statistic is not in the critical
               region.

   Step 5: a. Fail to reject H_O.
            b. The evidence is not significant, we can not
               conclude that there is an increase in wait
               time, at the 0.05 level of significance.

14.45 a. Median = 4.4; V = 5
           b. Step 1: a. Randomness; P(occurrence).
                       b. H_O: Random reported ages above and below
                          median.
                          H_a: The data did not occur randomly.
                  Step 2: a. Each data fits one of two categories.
                           b. V   c. α = 0.05

Step 3: a. n(a) = 6,   n(b) = 7
        [**Note**: When the median is one of the data, the two
        categories are 'above the median' and 'below or equal to
        the median']
     b. V* = 5
Step 4: -- using p-value approach ----------------
     a. Using Table 14, Appendix B, ES10-p826
        P > 0.05
     b. **P** > α
     -- using classical approach -------------
     a. Critical regions: V ≤ 3 or V ≥ 12
     b. The test statistic is not in the critical
        region.

Step 5: a. Fail to reject H_O.
     b. The evidence is not significant, we are
        unable to conclude that this sequence lacks
        randomness, at the 0.05 level of
        significance.

14.47   a. MINITAB verify -- answers given in exercise.
       b. z* = <u>-3.76</u>
          P = 2P(z < -3.76) = 2(0.0001) = <u>0.0002</u>
       c. Yes, reject the hypothesis of random runs above and
          below the median.

       d.

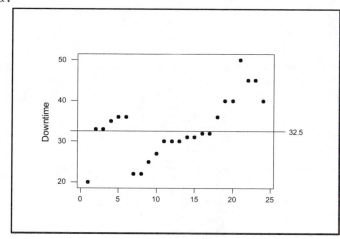

14.49　a. Step 1:　a. Randomness of data above and below the median.

　　　　　　　b. H_O: Randomness in number of absences.
　　　　　　　 H_a: The data did not occur randomly.

　　　Step 2:　a. Each data fits one of two categories.
　　　　　　　b. V　　　　　　　c. $\alpha = 0.05$

　　　Step 3:　a. $\tilde{x} = 10.5$;　n(above) = 13,　n(below) = 13
　　　　　　　b. V* = 9

　　　Step 4:　-- using p-value approach ---------------
　　　　　　　a. Using Table 14, Appendix B, ES10-p826
　　　　　　　　　P > 0.05
　　　　　　　b. **P** > α
　　　　　　　-- using classical approach -------------
　　　　　　　a. Critical regions:　V $\leq$ 8　or　V $\geq$ 20
　　　　　　　b. The test statistic is not in the critical region.

　　　Step 5:　a. Fail to reject H_O.
　　　　　　　b. The evidence is not significant, we can not conclude that this sequence is not random, at the 0.05 level of significance.

　b. Step 1:　a. Randomness of data above and below the median.

　　　　　　　b. H_O: Randomness in number of absences.
　　　　　　　 H_a: The data did not occur randomly.

　　　Step 2:　a. Each data fits one of two categories.
　　　　　　　b. V　　　　　　　c. $\alpha = 0.05$

　　　Step 3:　a. $\tilde{x} = 10.5$;　n(above) = 13,　n(below) = 13
　　　　　　　b. V* = 9
　　　　　　　c. n(1) = 125,　n(2) = 125

$$\mu = [(2n_1 n_2)/(n_1 + n_2)] + 1$$
$$\mu_V = [(2)(13)(13)/(13+13)] + 1 = 14$$

$$\sigma_V = \sqrt{\frac{(2n_1 n_2)(2n_1 n_2 - n_1 - n_2)}{(n_1 + n_2)^2(n_1 + n_2 - 1)}}$$

$$= \sqrt{[(2)(13)(13)][2(13)(13)-13-13]/(13+13)^2(13+13-1)}$$
$$= 2.498$$

$$z = (V - \mu_V)/\sigma_V$$
$$z* = (9 - 14)/2.498 = -2.00$$

　　　　　　　a. **P** = 2P(z > 2.00)
　　　　　　　　 Using Table 3, Appendix B, ES10-p810
　　　　　　　　 P = 2(0.5000 - 0.4772) = 0.0456
　　　　　　　b. **P** < α
　　　　　　　-- using classical approach -------------
　　　　　　　a. Critical values: z(0.025) = ±1.96
　　　　　　　b. The test statistic is in the critical region.
　　　　　　　--
Step 5: a. Reject H$_O$.
　　　　　　　b. The evidence is significant, we can conclude
　　　　　　　　 that this sequence is not random, at the 0.05
　　　　　　　　 level of significance.

14.51 a. By comparing the number of actual occurrences to the
expected number of occurrences using a multinomial or
contingency table test the relative frequency of
occurrences can be tested.

　b. The runs test will test the order, or sequence, of
occurrence for the numbers generated.

　c. The correlation will test the independence of side-
by-side outcomes to be sure there is no influence of
one part of a game with another part of the same
game.

　d. When testing for randomness, it is the null
hypothesis that states random, thereby making the
"fail to reject" decision the desired outcome. The
probability associated with that result is 1 - α, not
the level of significance, and 1 - α is known as the
level of confidence.

SECTION 14.7 EXERCISES

The Spearman Rank Correlation Test

The Rank Correlation Test is used to test for the correlation or
relationship between two variables. Refer to the Five-Step
Hypothesis Test Procedure in: ES10-pp427&444-445, SSM-pp265&275, if
necessary. The only changes are in:　　　　　　. . .

---

1. **the hypotheses**
    a. null hypothesis
       H_O: There is no correlation or relationship between the
           two variables or $\rho_s = 0$
    b. possible alternative hypotheses:
       H_a: There is a correlation or relationship between the
           two variables or $\rho_s \neq 0$.
       H_a: There is a positive correlation or $\rho_s > 0$.
       H_a: There is a negative correlation or $\rho_s < 0$.

2. **the critical value of the test statistic,**
    a. two-tailed test, $\pm r_s$; Table 15 (Appendix B, ES10-p827)
    b. one-tailed test, $+r_s$ or $-r_s$; Table 15
                                    (Appendix B, ES10-p827)

3. **the calculated test statistic,**

$$r_s^{\star} = 1 - \frac{6[\sum(d_i)^2]}{n(n^2 - 1)}$$

14.53 a. H_O: There is a no relationship between the two rankings
               H_a: There is a relationship between the two rankings

        b. H_O: The two variables are unrelated
            H_a: The two variables are related

        c. H_O: The is no correlation between the two variables
            H_a: There is positive correlation

        d. H_O: Refrigerator age has no effect on monetary value
            H_a: Refrigerator age has a decreasing effect on monetary
                 value

Table 15, Critical Values of Spearman's Rank Correlation
Coefficient, gives positive critical values based on sample size
and the level of significance. For a two-tailed test, add a plus
and minus sign to the table value. For a one-tailed test, double
the level of significance, then apply a plus or minus sign,
whichever is appropriate.

14.55 a.   b.

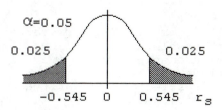

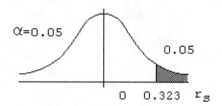

c.

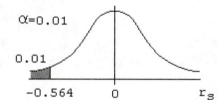

14.57 a. The formula for the rank correlation coefficient:
    0.133

$$r_s = 1 - \frac{6\sum (d_i)^2}{n(n^2 - 1)} : \qquad r_s = 1 - \frac{6(143)}{10(99)} = 1 - .867 = 0.133$$

b.
 Step 1: a. Correlation between the overall rating and the
            street price.
         b. H_o: $\rho_s = 0$
             H_a: $\rho_s > 0$
 Step 2: a. Assume random sample of ordered pairs, one
            ordinal variable and one numerical variable
         b. r_s c. $\alpha = 0.05$
 Step 3: a. n = 10, Σd^2 = 143
         b. r_s* = 0.133
 Step 4: -- using p-value approach ---------------
         a. Using Table 15, Appendix B, ES10-p827
             P > 0.10
             Using computer: **P** = 0.3652
         b. **P** > α

-- 535 --

a. critical region: $r_s \geq 0.564$
b. r_s^* is not in the critical region

Step 5: a. Fail to reject H_o.
      b. There is not sufficient evidence presented
by these data to enable us to conclude that
there is any relationship between overall
performance ratings of 17-inch computer
monitors and their street price, at the
0.05 level of significance.

14.59 Summary of data: $n = 12$, $\Sigma d^2 = 70.5$

$r_s = 1 - [(6)(70.5)/(12)(12^2-1)] = \underline{0.753}$

14.61 Step 1: a. Correlation between undergraduate GPA and GPA at
graduation from a graduate nursing program.
      b. H_o: $\rho_s = 0$
         H_a: $\rho_s > 0$
Step 2: a. Assume random sample of ordered pairs, both
variables are numerical
      b. r_s c. $\alpha = 0.05$
Step 3: a. $n = 10$, $\Sigma d^2 = 43.5$
      b. $r_s = 1 - [(6)(\Sigma d^2)/(n)(n^2-1)]$
         $r_s^* = 1 - [(6)(43.5)/(10)(99)] = 0.736$
         $r_s^* = 0.732$ (using MINITAB)
Step 4: -- using p-value approach ---------------
a. Using Table 15, Appendix B, ES10-p827
         $0.01 < P < 0.025$
Using computer: $P = 0.016$
      b. $P < \alpha$
-- using classical approach -------------
a. critical region: $r_s \geq 0.564$
b. r_s^* is in the critical region

Step 5: a. Reject H_o.
      b. There is sufficient reason to conclude there is
a positive relationship, at the 0.05 level of
significance.

14.63 a.

Lee Co	Florida	U.S.	Lee Rank	FL Rank	US Rank
0.02	0.02	0.026	1	1	1
0.08	0.05	0.065	6	3	2.5
0.05	0.08	0.160	3.5	6	6
0.05	0.05	0.071	3.5	3	4.5
0.25	0.21	0.168	8	8	7
0.03	0.05	0.071	2	3	4.5
0.06	0.06	0.065	5	5	2.5
0.30	0.34	0.359	9	9	8
0.16	0.15	*	7	7	*

b. Step 1: a. Correlation between Lee County job
classification rates and all of Florida.
b. H_o: $\rho_s = 0$
H_a: $\rho_s \neq 0$
Step 2: a. Assume random sample of ordered pairs, both
variables are numerical
b. r_s c. $\alpha = 0.05$
Step 3: a. $n = 9$, $\Sigma d^2 = 16.5$
b. $r_s = 1 - [(6)(\Sigma d^2)/(n)(n^2-1)]$
$r_s^* = 1 - [(6)(16.5)/(9)(80)] = 0.8625$
$r_s^* = 0.860$ (using MINITAB)
Step 4: -- using p-value approach ---------------
a. Using Table 15, Appendix B, ES10-p827
P < 0.01
Using computer: P = 0.003
b. **P** < α
-- using classical approach -------------
a. critical region: $r_s \leq -0.700$ and $r_s \geq 0.700$
b. r_s^* is in the critical region

Step 5: a. Reject H_o.
b. There is sufficient reason to conclude there is
a correlation, at the 0.05 level of
significance.

c. Step 1: a. Correlation between Lee County job
classification rates and all of the US.
b. H_o: $\rho_s = 0$
H_a: $\rho_s \neq 0$
Step 2: a. Assume random sample of ordered pairs, both
variables are numerical
b. r_s c. $\alpha = 0.05$

Step 3: a. $n = 8$, $\Sigma d^2 = 32$

       b. $r_s = 1 - [(6)(\Sigma d^2)/(n)(n^2-1)]$

         $r_s{}^* = 1 - [(6)(32)/(8)(63)] = 0.619$

         $r_s{}^* = 0.612$ (using MINITAB)

Step 4: -- using p-value approach ---------------

       a. Using Table 15, Appendix B, ES10-p827

         P > 0.10

         Using computer: P = 0.107

       b. **P** > α

      -- using classical approach -------------

       a. critical region: $r_s \le -0.738$ and $r_s \ge 0.738$

       b. $r_s{}^*$ is in the critical region

Step 5: a. Fail to reject H_o.

       b. There is sufficient reason to conclude there is
no correlation, at the 0.05 level of
significance.

d. Step 1: a. Correlation between all of Florida's job
classification rates and all of the US.

       b. H_o: $\rho_s = 0$

         H_a: $\rho_s \ne 0$

Step 2: a. Assume random sample of ordered pairs, both
variables are numerical

       b. r_s c. $\alpha = 0.05$

Step 3: a. $n = 8$, $\Sigma d^2 = 11$

       b. $r_s = 1 - [(6)(\Sigma d^2)/(n)(n^2-1)]$

         $r_s{}^* = 1 - [(6)(11)/(8)(63)] = 0.869$

         $r_s{}^* = 0.864$ (using MINITAB)

Step 4: -- using p-value approach ---------------

       a. Using Table 15, Appendix B, ES10-p827

           $0.01 <$ **P** < 0.02

         Using computer: **P** = 0.006

       b. **P** < α

      -- using classical approach -------------

       a. critical region: $r_s \le -0.738$ and $r_s \ge 0.738$

       b. $r_s{}^*$ is in the critical region

Step 5: a. Reject H_o.

       b. There is sufficient reason to conclude there is
a correlation, at the 0.05 level of
significance.

e. Lee Co. and Florida share a significant correlation,
    Florida and the US share a significant correlation,
    however Lee Co. and the US are not correlated. {No
    transitive property!!}

14.65 a.

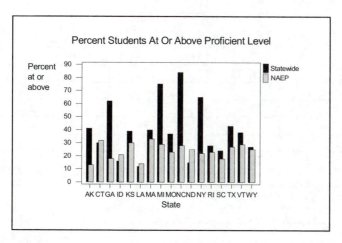

There appears to be very little relationship between
the two sets of percentages. Some of the largest
state percentages are paired with some of the lower
NAEP percentages, while some of the lowest state
percentages are also paired with some of the lowest
NAEP percentages.

b.

	State	NAEP	Rstate	R NAEP	d	dsq	sumdsq
Arkansas	41	13	12	1	-11	121	594
Connecticut	30	32	7	16	9	81	
Georgia	62	18	14	3.5	-11	110.25	
Idaho	16	21	3	5	2	4	
Kansas	39	30	10	15	5	25	
Louisiana	12	14	1	2	1	1	
Massachusetts	40	33	11	17	6	36	
Michigan	75	29	16	13.5	-2.5	6.25	
Missouri	37	23	8	7.5	-0.5	0.25	
New York	65	22	15	6	-9	81	
North Carolina	84	28	17	12	-5	25	
North Dakota	15	25	2	9.5	7.5	56.25	
Rhode Island	28	23	6	7.5	1.5	2.25	
South Carolina	24	18	4	3.5	-0.5	0.25	
Texas	43	27	13	11	-2	4	
Vermont	38	29	9	13.5	4.5	20.25	
Wyoming	27	25	5	9.5	4.5	20.25	

c.

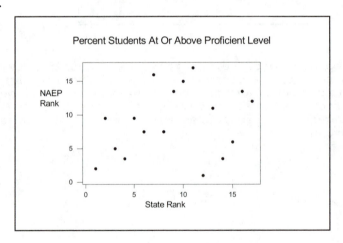

Percent Students At Or Above Proficient Level

d.

Step 1: a. Correlation between statewide assessment and national assessment of education.

b. H_o: $\rho_s = 0$
H_a: $\rho_s \neq 0$

Step 2: a. Assume random sample of ordered pairs, both variables are numerical

b. r_s c. $\alpha = 0.05$

Step 3: a. $n = 17$, $\Sigma d^2 = 594$

b. $r_s = 1 - [(6)(\Sigma d^2)/(n)(n^2-1)]$
$r_s{}^* = 1 - [(6)(594)/(17)(288)] = 0.272$
$r_s{}^* = 0.270$ (using MINITAB)

Step 4: -- using p-value approach ---------------

a. Using Table 15, Appendix B, ES10-p827
P > 0.10
Using computer: $P = 0.294$

b. **P** > α

-- using classical approach -------------

a. critical region: $r_s \leq -0.490$ and $r_s \geq 0.490$

b. $r_s{}^*$ is not in the critical region

Step 5: a. Fail to reject H_o.

b. There is not sufficient reason to conclude there is a correlation between the two sets of percentages, at the 0.05 level of significance.

14.67 Summary of ranks: $n = 12$, $\Sigma x = 78$, $\Sigma y = 78$,
$\Sigma x^2 = 650$, $\Sigma xy = 578.5$, $\Sigma y^2 = 649$

$SS(x) = 650 - (78^2/12) = 143$
$SS(y) = 649 - (78^2/12) = 142$
$SS(xy) = 578.5 - [(78)(78)/12)] = 71.5$

$r_s = 71.5/\sqrt{(143)(142)} = \underline{0.502}$

14.69 a.

```
        Stem-and-leaf of Hydrogen   N  = 52
        Leaf Unit = 0.10
           3      2 078
          10      3 1356688
          18      4 13355689
         (12)     5 001112444589
          22      6 11137
          17      7 2388
          13      8 3378
           9      9 4479
           5     10 56
           3     11 1
           2     12 4
           1     13
           1     14
           1     15 2
```

b. skewed right

c. For n = 52 and $1 - \alpha = 0.95$, the critical value from
    Table 12 is k = 18.
    $x_{k+1} = x_{19} = 5.0$ and $x_{n-k} = x_{34} = 6.3$

    5.0 to 6.3, the 0.95 interval for median M

14.71 a. Using the Sign Test (one median):
         Step 1: a. Median score on exam.
                 b. Ho: Median = 50
                    Ha: Median $\neq$ 50
         Step 2: a. Assume sample is random. Exam score is
                       continuous.
                 b. x = n(least frequent sign)
                 c. $\alpha = 0.05$
         Step 3: a. + = above 50,   - = below 50, 0 = 50; n = 30
                    n(+) = 10, n(0) = 2, n(-) = 20
                 b. x = n(+) = 10
         Step 4: -- using p-value approach ---------------
                 a. $P = 2P(x \leq 10 | n = 30)$;
                    Using Table 12, Appendix B, ES10-p824:
                    $P \approx 0.10$
                    Using computer: P = 0.0987
                 b. $P > \alpha$

                     a. critical region: n(least freq sign)= x ≤ 9
                     b. The test statistic is not in the critical
                            region.

          Step 5: a. Fail to reject Ho.
                     b. The sample evidence is not sufficient to justify
                        the claim that median is different than 50, at
                        the 0.05 level of significance.

     b. Step 1: a. Median score on exam.
                     b. Ho: Median = 50
                        Ha: Median < 50
          Step 2: a. Assume sample is random. Exam score is
                        continuous.
                     b. x = n(least frequent sign)
                     c. α = 0.05
          Step 3: a. + = above 50,   - = below 50;
                            n = 30, n(+) = 10,   n(-) = 20
                     b. x = n(+) = 10
          Step 4: -- using p-value approach ---------------
                     a. P = P(x ≤ 10|n = 30);
                        Using Table 12, Appendix B, ES10-p824:
                        P ≈ 0.05
                        Using computer: P = 0.0494
                     b. P ≤ α
                     -- using classical approach -------------
                     a. critical region: n(least freq sign)= x ≤ 10
                     b. The test statistic is in the critical region.

          Step 5: a. Reject H_O.
                     b. The sample evidence is sufficient to justify the
                        claim that median is less than 50, at the 0.05
                        level of significance.

14.73 Using the Sign Test (dependent samples):
          Step 1: a. Time required to run 220 yd sprint on two
                        tracks.
                     b. H_O: No difference between average times (no
                        faster)
                        H_a: Average time on B is less than on A (B is
                            faster)

```
Step 2:  a. Assume sample is random. Time is continuous.
         b. x = n(least frequent sign)
         c. α = 0.05
Step 3:  a. + = A time is greater;  n = 10
             n(+) = 8, n(0) = 0, n(-) = 2
         b. x = n(-) = 2
Step 4:  -- using p-value approach ---------------
         a. P = P(x ≤ 2|n = 10);
             Using Table 12, Appendix B, ES10-p824:
                 0.05 < P < 0.125
             Using computer: P = 0.0547
         b. P > α
         -- using classical approach -------------
         a. critical region: n(least freq sign)= x ≤ 1
         b. The test statistic is not in the critical
             region.
         ---------------------------------------------
Step 5:  a. Fail to reject H₀.
         b. The evidence is not sufficient to justify the
             claim that track B is faster, at the 0.05 level
             of significance.
```

14.75 Reject for $U \leq 127$

14.77 Using Mann-Whitney U Test (independent samples):
```
Step 1:  a. Line width.
         b. H₀: No difference in line width.
            Hₐ: There is a difference in line width.
Step 2:  a. Independent samples and line widths are
             numerical.
         b. U                 c. α = 0.05
Step 3:  a. ranked data and ranks:  (underlined = normal
             group)
```

27.5	28.0	28.5	29.5	30.5	30.6	30.7	30.9	32.9	35.1
1	2	3	4	5	6	7	8	9	10

$$R_n = 38, \quad R_m = 17$$

```
         b. Uₙ = nₙ.nₘ + [ (nₘ) (nₘ+1)/2] - Rₘ
            Uₙ = (5) (5) + [ (5) (5+1)/2] - 17 = 23
            Uₘ = nₘ.nₙ + [ (nₙ) (nₙ+1)/2] - Rₙ
            Uₘ = (5) (5) + [ (5) (5+1)/2] - 38 = 2; U* = 2
```

Step 4: -- using p-value approach ----------------
     a. $\mathbf{P} = 2P(U \leq 2 \mid n_n = 5, n_m = 5)$;
       Using Table 13, Appendix B, ES10-p825:
       $\mathbf{P} \approx 0.05$
       Using computer: P = 0.0367
     b. $\mathbf{P} \leq \alpha$
    -- using classical approach -------------
     a. U(5,5,0.05) = 2; $U \leq 2$
     b. The test statistic is in the critical region.

Step 5: a. Reject H_O.
     b. There is a significant difference in line width,
       at the 0.05 level of significance.

14.79 a.

w/ ERA, low is best   **w/ BA, high is best**

League	ERA	ERA Rank	League	BA	BA Rank
N	3.74	1	A	0.282	1.5
N	3.75	2	A	0.282	1.5
N	3.81	3	A	0.281	3
N	4.01	4	N	0.278	4
A	4.03	5.5	A	0.276	5
N	4.03	5.5	N	0.275	6
N	4.05	7	N	0.273	7
N	4.09	8	A	0.272	8
N	4.1	9	A	0.27	10.5
A	4.17	10	A	0.27	10.5
A	4.18	11	N	0.27	10.5
N	4.24	12	N	0.27	10.5
A	4.28	13	A	0.268	14
N	4.29	14.5	A	0.268	14
N	4.29	14.5	N	0.268	14
N	4.33	16	N	0.267	16.5
N	4.45	17	N	0.267	16.5
A	4.53	18	A	0.266	18.5
A	4.69	19	A	0.266	18.5

continued

A	4.7	20	N	0.264	20
A	4.76	21	N	0.262	21
A	4.81	22.5	A	0.26	22.5
A	4.81	22.5	N	0.26	22.5
A	4.91	24.5	A	0.259	24
A	4.91	24.5	A	0.258	25
A	4.93	26	N	0.253	26
N	4.98	27	N	0.25	27
A	5.15	28	N	0.249	28.5
N	5.19	29	N	0.249	28.5
N	5.54	30	N	0.248	30

b. <u>Batting Averages</u>:

Step 1: a. Batting averages.
b. H_O: Batting averages in AL are not higher. ($\leq$)
H_a: Batting averages in AL are higher. ($>$)
Step 2: a. Independent samples and batting averages are numerical.
b. U   c. $\alpha = 0.05$
Step 3: a.

R_a = 1.5 + 1.5 + 3.0 + 5.0 + 8.0 + 10.5 + 10.5 + 14.0 + 14.0
              + 18.5 + 18.5 + 22.5 + 24.0 + 25.0 = 176.5

R_n = 4.0 + 6.0 + 7.0 + 10.5 + 10.5 + 14.0 + 16.5 + 16.5 + 20.0 +
           21.0 + 22.5 + 26.0 + 27.0 + 28.5 + 28.5 + 30.0 = 288.5

          b.

$U_a = n_a \times n_n + [(n_n)(n_n + 1)] \div 2 - R_n$
     = 14 × 16 + 16(17) ÷ 2 − 288.5 = 224 + 136 − 288.5
     = 71.5

$U_n = n_a \times n_n + [(n_a)(n_a + 1)] \div 2 - R_a$
     = 14 × 16 + 14(15) ÷ 2 − 176.5 = 224 + 105 − 176.5
     = 152.5

Check: 71.5 + 152.5 = 224 = 14 × 16

          U* = 71.5

Step 4: -- using p-value approach ---------------
  a. $P = P(U < 71.5)$
     Using Table 13, Appendix B, ES10-p825
     $P > 0.05$
  b. $P > \alpha$
  -- using classical approach -------------
  a. critical region: $U \leq 71$
  b. The test statistic is not in the critical
     region.

Step 5: a. Fail to reject Ho
  b. The American League batting average in 2004 was
     not higher than the National League at the 0.05
     level of significance.

Earned Run Averages:

Step 1: a. Earned run averages
  b. H_o: Earned run average for NL is not lower. ($\geq$)
     H_a: Earned run average for NL is lower. ($<$)
Step 2: a. Independent samples and earned run averages are
     numerical.
  b. U   c. $\alpha = 0.05$
Step 3: a.

Ra = 5.5 + 10.0 + 11.0 + 13.0 + 18.0 + 19.0 + 20.0 + 21.0
   + 22.5 + 22.5 + 24.5 + 24.5 + 26.0 + 28.0 = 265.5

Rn = 1.0 + 2.0 + 3.0 + 4.0 + 5.5 + 7.0 + 8.0 + 9.0 + 12.0
   + 14.5 + 14.5 + 16.0 + 17.0 + 27.0 + 29.0 + 30.0 =
   199.5

        b.
Ua = na × nn + [(nn)(nn + 1)] ÷ 2 - Rn
   = 14 × 16 + 16(17) ÷ 2 - 199.5 = 224 + 136 - 199.5
   = 160.5

Un = na × nn + [(na)(na + 1)] ÷ 2 - Ra
   = 14 × 16 + 14(15) ÷ 2 - 265.5 = 224 + 105 - 265.5
   = 63.5

        Check: 160.5 + 63.5 = 224 = 14 × 16

        U* = 63.5

Step 4: -- using p-value approach ---------------
        a. $P = P(U < 63.5)$
           Using Table 13, Appendix B, ES10-p825
           $P < 0.025$
        b. $P > \alpha$
        -- using classical approach -------------
        a. critical region: $U \leq 71$
        b. The test statistic is not in the critical
           region.

Step 5: a. Reject H_O
        b. The National League earned run average in
           2004 is lower than the American League at
           the 0.05 level of significance.

14.81 Using the Runs Test:
        Step 1: a. Randomness in sequence of occurrence of
                   defective and nondefective parts.
                b. H_O: Random order.
                   H_a: Lack of randomness.
        Step 2: a. Each data fits into one of two categories
                b. V   c. $\alpha = 0.05$
        Step 3: a. $n(n) = 20$, $n(d) = 4$
                b. $V* = 9$
        Step 4: -- using p-value approach ---------------
                a. $P = P(V \leq 9)$
                   $P > 0.05$
                b. $P > \alpha$
                -- using classical approach -------------
                a. critical region: $V \leq 4$ and $V \geq 10$
                b. the test statistic is not in the critical region
                ----
        Step 5: a. Fail to reject H_O.
                b. The sample results do not show a significant lack
                   of randomness, at the 0.05 level of significance.

14.83 a. Median = 22.5

Company		Job Growth	Company		Job Growth
1	B	26	11	a	23
2	A	54	12	b	13
3	A	34	13	b	17
4	B	10	14	a	23
5	A	31	15	b	9
6	A	48	16	b	3
7	A	26	17	b	15
8	B	22	18	b	11
9	A	24	19	b	1
10	B	10	20	a	122

Runs above: 6   Runs below: 6

b.

Step 1: a. Randomness of job growth rate percentages.
   b. H_o: The job growth rate percentages are listed
        in a random sequence
      H_a: Lack of randomness.

Step 2: a. Each data fits into one of two categories
   b. V   c. $\alpha = 0.05$

Step 3: a. $n_a = 9$ $n_b = 11$
   b. $V^* = 12$

Step 4: -- using p-value approach ---------------
   a. **P** = $P(V \leq 12)$
      P > 0.05
   b. **P** > α
   -- using classical approach -------------
   a. critical region: $V \leq 6$ and $V \geq 16$
   b. The test statistic is not in the critical region

Step 5: a. Fail to reject H_o.

c. Conclusion: There is not sufficient evidence to reject
the null hypothesis that the job growth rate
percentages are listed in a random sequence at the
0.05 level of significance. Based on this sample
evidence, a higher job growth rate does not imply a
higher rank in attractiveness.

14.85 Using Spearman's Rank Correlation:
Step 1: a. Correlation between two daily high temperatures.
   b. H_o: $\rho_S = 0$ (Independence)
   H_a: $\rho_S > 0$ (Positive correlation)
Step 2: a. Assume random sample of ordered pairs, numerical variables
   b. r_S c. $\alpha = 0.05$
Step 3: a. $n = 18$, $\Sigma d^2 = 116.5$
   b. $r_S = 1 - [(6)(\Sigma d^2)/(n)(n^2-1)]$
   $r_S* = 1 - [(6)(116.5)/(18)(18^2-1)] = 0.880$
Step 4: -- using p-value approach ----------------
   a. Using Table 15, Appendix B, ES10-p827
      P < 0.01
      Using computer: **P** = 0.000
   b. **P** < α
   -- using classical approach -------------
   a. critical regions: $r_S \geq 0.399$
   b. r_S* is in the critical region

Step 5: a. Reject H_o
   b. There is a significant amount of correlation shown between the two sets of temperatures, at the 0.05 level of significance.

14.87 a. $r_{12} = 1 - \dfrac{6(1766)}{25(624)} = 1 - 0.679 = 0.321 = r_{12}$

   $r_{13} = 1 - \dfrac{6(1824)}{25(624)} = 1 - 0.702 = 0.298 = r_{13}$

   $r_{23} = 1 - \dfrac{6(30)}{25(624)} = 1 - 0.012 = 0.988 = r_{23}$

b. The three tests are identical for Steps 1-3:
Step 1: a. Correlation between the rankings.
   b. H_o: $\rho_S = 0$
   H_a: $\rho_S \neq 0$
Step 2: a. Assume random sample of ordered pairs, ordinal variables
   b. r_S c. $\alpha = 0.05$
Step 3: a. $n = 25$
   b. r_S* for each are listed in (a)

For (1) vs. (2), $r_S{}^* = 0.321$

Step 4: -- using p-value approach ---------------
     a. Using Table 15, Appendix B, ES10-p827
        P > 0.10

     b. **P** > α
     -- using classical approach -------------
     a. critical regions: $r_S \leq -0.400$ and $r_S \geq 0.400$
     b. $r_S{}^*$ is not in the critical region

Step 5: a. Fail to reject H_O.
     b. There is not sufficient evidence presented by
        these data to enable us to conclude that the SI
        preseason poll has any relationship to the USA
        Today/ESPN poll, at the 0.05 level of
        significance.

For (1) vs. (3), $r_S{}^* = 0.298$
Step 4: -- using p-value approach ---------------
     a. Using Table 15, Appendix B, ES10-p827
        P > 0.10
     b. P > □
     -- using classical approach -------------
     a. critical regions: $r_S \leq -0.400$ and $r_S \geq 0.400$
     b. $r_S{}^*$ is not in the critical region

Step 5: a. Fail to reject H_O.
     b. There is not sufficient evidence presented by
        these data to enable us to conclude that the SI
        preseason poll has any relationship to the AP
        Top 25 poll, at the 0.05 level of significance.

For (2) vs. (3), $r_S{}^* = 0.988$
Step 4: -- using p-value approach ---------------
     a. Using Table 15, Appendix B, ES10-p827
        P < 0.01

     b. **P** < α
     -- using classical approach -------------
     a. critical regions: $r_S \leq -0.400$ and $r_S \geq 0.400$
     b. $r_S{}^*$ is in the critical region

Step 5: a. Reject H_O.
     b. The USA Today/ESPN poll and the AP Top 25 poll
        have a strong positive relationship, at the 0.05
        level of significance.

INTRODUCTORY CONCEPTS

SUMMATION NOTATION

The Greek capital letter sigma (Σ) is used in mathematics to indicate the summation of a set of addends. Each of these addends must be of the form of the variable following Σ. For example:

1. $\sum x$ means sum the variable x.
2. $\sum (x - 5)$ means sum the set of addends that are each 5 less than the values of each x.

When large quantities of data are collected, it is usually convenient to index the response variable so that at a future time its source will be known. This indexing is shown on the notation by using i (or j or k) and affixing the index of the first and last addend at the bottom and top of the Σ. For example,

$$\sum_{i=1}^{3} x_i$$

means to add all the consecutive values of x's starting with source number 1 and proceeding to source number 3.

▽ ILLUSTRATION 1

Consider the inventory in the following table concerning the number of defective stereo tapes per lot of 100.

Lot Number (I)	1	2	3	4	5	6	7	8	9	10
Number of Defective Tapes per Lot (x)	2	3	2	4	5	6	4	3	3	2

a. Find $\displaystyle\sum_{i=1}^{10} x_i$. b. Find $\displaystyle\sum_{i=4}^{8} x_i$.

Solution

a. $\displaystyle\sum_{i=1}^{10} x_i = x_1 + x_2 + x_3 + x_4 + \cdots + x_{10}$

$= 2 + 3 + 2 + 4 + 5 + 6 + 4 + 3 + 3 + 2 = 34$

b. $\displaystyle\sum_{i=4}^{8} x_i = x_4 + x_5 + x_6 + x_7 + x_8 = 4 + 5 + 6 + 4 + 3 = 22$

$\Delta\Delta$

The index system must be used whenever only part of the available information is to be used. In statistics, however, we will usually use all the available information, and to simplify the formulas we will make an adjustment. This adjustment is actually an agreement that allows us to do away with the index system in situations where all values are used. Thus in our previous illustration, $\displaystyle\sum_{i=1}^{10} x_i$ could have been written simply as $\sum x$.

NOTE The lack of the index indicates that all data are being used.

▽ ILLUSTRATION A-2

Given the following six values for x, 1, 3, 7, 2, 4, 5, find $\sum x$.

Solution

$\displaystyle\sum x = 1 + 3 + 7 + 2 + 4 + 5 = 22$ $\qquad\qquad \Delta\Delta$

Throughout the study and use of statistics you will find many formulas that use the Σ symbol. Care must be taken so that the formulas are not misread. Symbols like Σx^2 and $(\Sigma x)^2$ are quite different. Σx^2 means "square each x value and then add up the squares," while $(\Sigma x)^2$ means "sum the x values and then square the sum."

∇ ILLUSTRATION A-3

Find (a) $\sum x^2$ and (b) $\left(\sum x\right)^2$ for the sample in Illustration A-2.

Solution

a.

x	1	3	7	2	4	5
x^2	1	9	49	4	16	25

$$\sum x^2 = 1 + 9 + 49 + 4 + 16 + 25 = 104$$

b. $\sum x = 22$, as found in Illustration A-2. Thus,

$$\left(\sum x\right)^2 = (22)^2 = 484$$

As you can see, there is quite a difference between $\sum x^2$ and $\left(\sum x\right)^2$.   ΔΔ

Likewise, $\sum xy$ and $\sum x \sum y$ are different. These forms will appear only when there are paired data, as shown in the following illustration.

∇ ILLUSTRATION A-4

Given the five pairs of data shown in the following table, find (a) $\sum xy$ and (b) $\sum x \sum y$.

x	1	6	9	3	4
y	7	8	2	5	10

Solution

a. $\sum xy$ means to sum the products of the corresponding x and y values. Therefore, we have

x	1	6	9	3	4
y	7	8	2	5	10
xy	7	48	18	15	40

$$\sum xy = 7 + 48 + 18 + 15 + 40 = 128$$

b. $\sum x \sum y$ means the product of the two summations, $\sum x$ and $\sum y$. Therefore, we have

$$\sum x = 1 + 6 + 9 + 3 + 4 = 23$$
$$\sum y = 7 + 8 + 2 + 5 + 10 = 32$$
$$\sum x \sum y = (23)(32) = 736 \qquad \Delta\Delta$$

There are three basic rules for algebraic manipulation of the $\sum$ notation.

NOTE c represents any constant value.

RULE 1: $\displaystyle\sum_{i=1}^{n} c = nc$

To prove this rule, we need only write down the meaning of $\displaystyle\sum_{i=1}^{n} c$:

$$\sum_{i=1}^{n} c = \underbrace{c + c + c + \ldots + c}_{n \text{ addends}}$$

Therefore,

$$\sum_{i=1}^{n} c = n \cdot c$$

▽ ILLUSTRATION A-5

Show that $\displaystyle\sum_{i=1}^{5} 4 = (5)(4) = 20$.

Solution

$$\sum_{i=1}^{5} 4 = \underbrace{4_{(\text{when } i=1)} + 4_{(\text{when } i=2)} + 4_{(i=3)} + 4_{(i=4)} + 4_{(i=5)}}$$

five 4s added together

$$= (5)(4) = 20$$

△△

RULE 2:	$\displaystyle\sum_{i=1}^{n} cx_i = c \cdot \sum_{i=1}^{n} x_i$	

To demonstrate the truth of Rule 2, we will need to expand the term $\displaystyle\sum_{i=1}^{n} cx_i$, and then factor our the common term c.

$$\sum_{i=1}^{n} cx_i = cx_1 + cx_2 + cx_3 + \cdots + cx_n$$

$$= c(x_1 + x_2 + x_3 + \cdots + x_n)$$

Therefore,

$$\sum_{i=1}^{n} cx_i = c \cdot \sum_{i=1}^{n} x_i$$

$$\boxed{\text{RULE 3:} \qquad \sum_{i=1}^{n} (x_i + y_i) = \sum_{i=1}^{n} x_i + \sum_{i=1}^{n} y_i}$$

The expansion and regrouping of $\displaystyle\sum_{i=1}^{n} (x_i + y_i)$ is all that is needed to show this rule.

$$\sum_{i=1}^{n} (x_i + y_i) = (x_1 + y_1) + (x_2 + y_2) + \cdots + (x_n + y_n)$$

$$= (x_1 + x_2 + \cdots + x_n) + (y_1 + y_2 + \cdots + y_n)$$

Therefore,

$$\sum_{i=1}^{n} (x_i + y_i) = \sum_{i=1}^{n} x_i + \sum_{i=1}^{n} y_i$$

▽ ILLUSTRATION A-6

Show that $\displaystyle\sum_{i=1}^{3} (2x_i + 6) = 2 \cdot \sum_{i=1}^{3} x_i + 18$.

Solution

$$\sum_{i=1}^{3} (2x_i + 6) = (2x_1 + 6) + (2x_2 + 6) + (2x_3 + 6)$$

$$= (2x_1 + 2x_2 + 2x_3) + (6 + 6 + 6)$$
$$= (2)(x_1 + x_2 + x_3) + (3)(6)$$
$$= 2 \sum_{i=1}^{3} x_i + 18 \qquad\qquad \Delta$$

Δ

▽ ILLUSTRATION A-7

Let $x_1 = 2$, $x_2 = 4$, $x_3 = 6$, $f_1 = 3$, $f_2 = 4$, and $f_3 = 2$. Find $\displaystyle\sum_{i=1}^{3} x_i \cdot \sum_{i=1}^{3} f_i$.

Solution

$$\sum_{i=1}^{3} x_i \cdot \sum_{i=1}^{3} f_i = (x_1 + x_2 + x_3) \cdot (f_1 + f_2 + f_3)$$

$$= (2 + 4 + 6) \cdot (3 + 4 + 2)$$
$$= (12)(9) = 108 \qquad\qquad \triangle\triangle$$

▽ ILLUSTRATION A-8

Using the same values for the x's and f's as in Illustration A-7, find $\Sigma(xf)$.

Solution   Recall that the use of no index numbers means "use all data."

$$\sum (xf) = \sum_{i=1}^{3} (x_i f_i) = (x_1 f_1) + (x_2 f_2) + (x_3 f_3)$$

$$= (2 \cdot 3) + (4 \cdot 4) + (6 \cdot 2) = 6 + 16 + 12 = 34 \quad \triangle\triangle$$

▽△ EXERCISES

A.1   Write each of the following in expanded form (without the summation sign):

a. $\displaystyle\sum_{i=1}^{4} x_i$ b. $\displaystyle\sum_{i=1}^{3} (x_i)^2$ c. $\displaystyle\sum_{i=1}^{5} (x_i + y_i)$

d. $\displaystyle\sum_{i=1}^{5} (x_i + 4)$ e. $\displaystyle\sum_{i=1}^{8} x_i y_i$ f. $\displaystyle\sum_{i=1}^{4} x_i^2 f_i$

A.2   Write each of the following expressions as summations, showing the subscripts and the limits of summation:

      a. $x_1 + x_2 + x_3 + x_4 + x_5 + x_6$

      b. $x_1 y_1 + x_2 y_2 + x_3 y_3 + \cdots x_7 y_7$

      c. $x_1^2 + x_2^2 + \cdots x_9^2$

      d. $(x_1 - 3) + (x_2 - 3) \cdots + (x_n - 3)$

A.3   Show each of the following to be true:

      a. $\displaystyle\sum_{i=1}^{4} (5x_i + 6) = 5 \cdot \sum_{i=1}^{4} x_i + 24$ b. $\displaystyle\sum_{i=1}^{n} (x_i - y_i) = \sum_{i=1}^{n} x_i - \sum_{i=1}^{n} y_i$

A.4   Given $x_1 = 2, x_2 = 7, x_3 = -3, x_4 = 2, x_5 = -1,$ and $x_6 = 1,$ find each of the following:

      a. $\displaystyle\sum_{i=1}^{6} x_i$ b. $\displaystyle\sum_{i=1}^{6} x_i^2$ c. $\displaystyle\left(\sum_{i=1}^{6} x_i\right)^2$

A.5   Given $x_1 = 4, x_2 = -1, x_3 = 5, f_1 = 4, f_2 = 6, f_3 = 2, y_1 = -3,$ $y_2 = 5,$ and $y_3 = 2,$ find each of the following:

      a. $\sum x$ b. $\sum y$ c. $\sum f$ d. $\sum (x - y)$

      e. $\sum x^2$ f. $\left(\sum x\right)^2$ g. $\sum xy$ h. $\sum x \cdot \sum y$

      i. $\sum xf$ j. $\sum x^2 f$ k. $\left(\sum xf\right)^2$

A.6   Suppose that you take out a $12,000 small-business loan. The terms of the loan are that each month for 10 years (120 months) you will pay back $100 plus accrued interest. The accrued interest is calculated by multiplying 0.005 (6 percent/12) times the amount of the loan still outstanding. That is, the first month you pay $\$12,000 \times 0.005$ in accrued interest, the second month $(\$12,000 - 100) \times 0.005$ in interest, the third month $[\$12,000 - (2)(100)] \times 0.005$, and so forth. Express the total amount of interest paid over the life of the loan by using summation notation.

The answers to these exercises can be found in the back of the manual.

USING THE RANDOM NUMBER TABLE

The random number table is a collection of random digits. The term *random* means that each of the 10 digits (0,1,2,3,...,9) has an equal chance of occurrence. The digits, in Table 1 (Appendix B, ES10-pp805-806) can be thought of as single-digit numbers (0-9), as two-digit numbers (00-99), as three-digit numbers (000-999), or as numbers of any desired size. The digits presented in Table 1 are arranged in pairs and grouped into blocks of five rows and five columns. This format is used for convenience. Tables in other books may be arranged differently.

Random numbers are used primarily for one of two reasons: (1) to identify the source element of a population (the source of data) or (2) to simulate an experiment.

∇ ILLUSTRATION 1

A simple random sample of 10 people is to be drawn from a population of 7564 people. Each person will be assigned a number, using the numbers from 0001 to 7564. We will view Table 1 as a collection of four-digit numbers (two columns used together), where the numbers 0001, 0002, 0003, …, 7564 identify the 7564 people. The numbers 0000, 7565, 7566, …, 9999 represent no one in our population; that is, they will be discarded if selected.

Now we are ready to select our 10 people. Turn to Table 1 (Appendix B, ES10-p805). We need to select a starting point and a path to be followed. Perhaps the most common way to locate a starting point is to look away and arbitrarily point to a starting point. The number we located this way was 3909 on page 805. (It is located in the upper left corner of the block that is in the fourth large block from the left and the ninth (last/bottom) large block down.) From here we will proceed down the column, then go to the top of the next set of columns, if necessary. The person identified by number 3909 is the first source of data selected. Proceeding down the column, we find 8869 next. This number is discarded. The number 2501 is next. Therefore, the person identified by 2501 is the second source of data to be selected. Continuing down this column and the top of fourth column on page 806, our sample will be obtained from those people identified by the numbers 3909, 2501, 7485, 0545, 5252, 5612, 0997, 3230, 1051, 2712. (The numbers 8869, 8338, and 9187 were discarded.)

∆∆

∇ ILLUSTRATION 2

Let's use the random number table and simulate 100 tosses of a coin. The simulation is accomplished by assigning numbers to each of the possible outcomes of a particular experiment. The assignment must be done in such a way as to preserve the probabilities. Perhaps the simplest way to make the assignment for the coin toss is to let the even digits (0,2,4,6,8) represent heads and the odd digits (1,3,5,7,9) represent tails. The correct probabilities are maintained: $P(H) = P(0,2,4,6,8) = \dfrac{5}{10} = 0.5$ and

$P(T) = P(1,3,5,7,9) = \dfrac{5}{10} = 0.5$. Once this assignment is complete, we are ready to obtain our sample.

Since the question asked for 100 tosses and there are 50 digits to a "block" in Table 1 (Appendix B, ES10-p805), let's select two blocks as our sample of random one-digit numbers (instead of a column 100 lines long). Let's look away and point to one block on p. 805 and then do the same to select one block from p. 806. We picked the second block down in the first column of blocks on
p. 805 (21 even and 29 odd numbers) and the second block down in the first column of blocks on p. 806 (27 even and 23 odd numbers). Thus we obtain a sample of 48 heads and 52 tails for our 100 simulated tosses.   ∆∆

There are, of course, many ways to use the random number table. You must use your good sense in assigning the numbers to be used and in choosing the path to be followed through the table. One bit of advice is to make the assignments in as simple and easy a method as possible to avoid errors.

∇∆ EXERCISES

1.   A random sample of size 8 is to be selected from a population that contains 75 elements. Describe how the random sample of the 8 objects could be made with the aid of the random number table.

2.   A coin-tossing experiment is to be simulated. Two coins are to be tossed simultaneously and the number of heads appearing is to be recorded for each toss. Ten such tosses are to be observed. Describe two ways to use the random number table to simulate this experiment.

3.   Simulate five rolls of three dice by using the random number table.

The answers to these exercises can be found in the back of the manual.

ROUND-OFF PROCEDURE

When rounding off a number, we use the following procedure.

STEP 1   Identify the position where the round-off is to occur. This is shown by using a vertical line that separates the part of the number to be kept from the part to be discarded. For example,

125.267	to the nearest tenth is written as	125. 2|67
7.8890	to the nearest hundredth is written as	7. 88|90

STEP 2   Step 1 has separated all numbers into one of four cases. (X's will be used as placeholders for number values in front of the vertical line. These X's can represent any number value.)

Case I: XXXX|000...
Case II: XXXX|---(any value from 000...1 to 499...9)
Case III: XXXX|5000...0
Case IV: XXXX|---(any value from 5000...1 to 999...9)

STEP 3   Perform the rounding off.

Case I requires no round-off. It's exactly XXXX.

▽ ILLUSTRATION 1

Round 3.5000 to the nearest tenth.

$$3. 5|000 \text{ becomes } 3.5 \qquad \triangle\triangle$$

Case II requires rounding. We will round down for this case. That is, just drop the part of the number that is behind the vertical line.

ILLUSTRATION 2

Round 37.6124 to the nearest hundredth.

$$37. 61|24 \text{ becomes } 37.61 \qquad \triangle\triangle$$

Case III requires rounding. This is the case that requires special attention. **When a 5 (exactly a 5) is to be rounded off, round to the even digit.** In the long run, half of the time the 5 will be preceded by an even digit $(0,2,4,6,8)$ and you will round down, while the other half of the time the 5 will be preceded by an odd digit $(1,3,5,7,9)$ and you will round up.

∇ ILLUSTRATION 3

Round 87.35 to the nearest tenth.

$$87.\ 3|5\ \text{becomes}\ 87.4$$

Round 93.445 to the nearest hundredth.

$$93.\ 44|5\ \text{becomes}\ 93.44$$

(**Note:** 87.35 is 87.35000... and 93.445 is 93.445000...)   ∆∆

Case IV requires rounding. We will round up for this case. That is, we will drop the part of the number that is behind the vertical line and we will increase the last digit in front of the vertical line by one.

∇ ILLUSTRATION 4

Round 7.889 to the nearest tenth.

$$7.\ 8|89\ \text{becomes}\ 7.9$$ ∆∆

NOTE   Case I, II, and IV describe what is commonly done. Our guidelines for Case III are the only ones that are different from typical procedure.

If the typical round-off rule $(0, 1, 2, 3, 4$ are dropped; $5, 6, 7, 8, 9$ are rounded up) is followed, then $(n + 1)/(2n + 1)$ of the situations are rounded up. (n is the number of different sequences of digits that fall into each of Case II and Case IV.) That is more than half. You (as many others have) may say, "So what?" In today's world that tiny, seemingly insignificant amount becomes very significant when applied repeatedly to large numbers.

∇∆ EXERCISES

1. Round each of the following to the nearest integer:
    a. 12.94   b. 8.762   c. 9.05   d. 156.49
    e. 45.5   f. 42.5   g. 102.51   h. 16.5001

2. Round each of the following to the nearest tenth:
    a. 8.67   b. 42.333   c. 49.666   d. 10.25
    e. 10.35   f. 8.4501   g. 27.35001   h. 5.65
    i. 3.05   j. $\dfrac{1}{4}$

3. Round each of the following to the nearest hundredth:
    a. 17.6666   b. 4.444   c. 54.5454   d. 102.055
    e. 93.225   f. 18.005   g. 18.015   h. 5.555
    i. 44.7450   j. $\dfrac{2}{3}$

The answers to these exercises can be found in the back of the manual.

REVIEW LESSONS

THE COORDINATE-AXIS SYSTEM AND THE EQUATION OF A STRAIGHT LINE

The rectangular coordinate-axis system is a graphic representation of points. Each point represents an ordered pair of values. (Ordered means that when values are paired, one value is always listed first, the other second.) The pair of values represents a horizontal location (the *x*-value, called the abscissa) and a vertical location (the *y*-value, the ordinate) in a fixed reference system. This reference system is a pair of perpendicular real number lines whose point of intersection is the 0 of each line (Figure 1-1, below).

Any point (x, y) is located by finding the point that satisfies both positional values. For example, the point $P(2,3)$ is exactly 2 units to the right of 0 along the horizontal axis and 3 units above 0 along the vertical axis. Figure 1-2 (next page) shows two lines, *A* and *B*. Line *A* represents all the points that are 2 units to the right of 0 along the *x*-(horizontal) axis. Line *B* represents all the points that are 3 units above the 0 along the *y*-(vertical) axis. Point *P* is the one point that satisfies both conditions. Typically we think of locating point *P* by moving along the *x*-axis 2 units in the positive direction and then moving parallel to the *y*-axis 3 units in the positive direction (Figure 1-3, next page).

If either value is negative, we just move in a negative direction a distance equal to the number value.

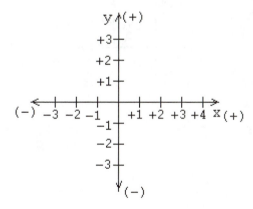

Figure 1-1

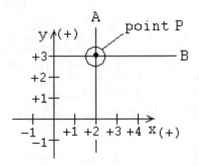

Figure 1-2

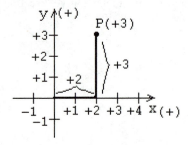

Figure 1-3

∇∆ EXERCISES

1. On a rectangular coordinate axis, drawn on graph paper, locate the following points.

    A(5,2) B(-5,2)   C(-3,-2)
    D(3,0) E(0,-2)   F(-2,5)

The equation of a line on a coordinate-axis system is a statement of fact about the coordinates of all points that lie on that line. This statement may be about one of the variables or about the relationship between the two variables. In Figure 1-2 above, a vertical line was drawn at $x = +2$. A statement that could be made about this line is that every point on it has an x-value of 2. Thus, the equation of this line is $x = 2$. The horizontal line that was drawn on the same graph passed through all the points where the y-values were +3. Therefore the equation of this line is $y = +3$.

All vertical lines will have an equation of $x = a$, where a is the value of the abscissa of every point on that line. Likewise, all horizontal lines will have an equation of $y = b$, where b is the ordinate of every point on that line.

In statistics, straight lines that are neither vertical nor horizontal are of greater interest. Such a line will have an equation that expresses the relationship between the two variables x and y. For example, it might be that y is always one less than the double of x; this would be expressed equationally as $y = 2x - 1$. There are an unlimited number of ordered pairs that fit this relationship; to name a few: (0,-1), (1,1), (2,3), (-2,-5), (1.5,2), (2.13,3.26) (see Figure 1-4).

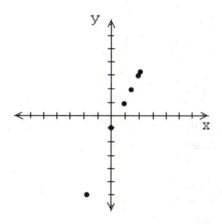

Figure 1-4

Notice that these points fall on a straight line. Many more points could be named that also fall on this same straight line; in fact, all the pairs of values that satisfy $y = 2x - 1$ lie somewhere along it. The converse is also true: all the points that lie on this straight line have coordinates that make the equation $y = 2x - 1$ a true statement. When drawing the line that represents the equational relationship between the coordinates of points, one needs to find only two points of the straight line; however, it is often useful to locate three of four to ensure accuracy.

EXERCISES (CONTINUED)

2. a. Find the missing values in the accompanying chart of ordered pairs, where $x + y = 5$ is the relationship.

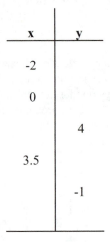

x	y
-2	
0	
	4
3.5	
	-1

   b. On a coordinate axis, locate the same five points and then draw a straight line that passes through all of them.

3. Find five points that belong to the relationship expressed by $y = (3x/2) - 4$; then locate them on the axis system and draw the line representing $y = (3x/2) - 4$.

The form of the equation of a straight line that we are interested in is called the slope-intercept form. Typically in mathematics this slope-intercept form is expressed by $y = mx + b$, where m represents the concept of *slope* and b is the *y-intercept*.

Let's look at the *y*-intercept first. If you will look back at Exercise 3, you will see that the *y*-intercept for $y = (3x/2) - 4 [y = (3x/2) + (- 4)]$ is -4. This value is simply the value of y at the point where the graph of the line intersects the *y*-axis, and all nonvertical lines will have this property. The value of the *y*-intercept may be found on the graph or from the equation. From the graph it is as simple as identifying it, but we will need to have the equation solved for y in order to identify it from the equation. For example, in Exercise 2 we had the equation $x + y = 5$; if we solve for y, we have $y = -x + 5$, and the *y*-intercept is 5 (the same as is found on the graph for Exercise 2).

The *slope* of a straight line is a measure of its inclination. This measure of inclination can be defined as the amount of vertical change that takes place as the value of x increases by exactly one unit. This amount of change may be found anywhere on the line since this value is the same everywhere on a given straight line. If we inspect the graph drawn for Exercise 2(Figure 1-5), we will see that the slope is -1, meaning that each x increase of 1 unit results in a decrease of 1 unit in the *y*-value.

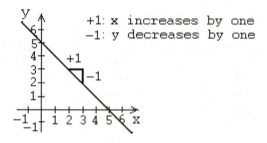

Figure 1-5

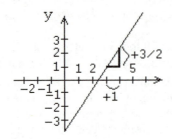

Figure 1-6

An inspection of the graph drawn for Exercise 3 (Figure 1-6) will reveal a slope of +3/2. This means that y increased by 3/2 for every increase of one unit in x.

Illustration: Find the slope of the straight line that passes through the points $(-1,1)$ and $(4,11)$.

Solution: $$m = \frac{\Delta y}{\Delta x} = \frac{11 - 1}{4 - (-1)} = \frac{10}{5} = 2$$

NOTES: 1. $\Delta y = y_2 - y_1$ and $\Delta x = x_2 - x_1$ where (x_1, y_1) and (x_2, y_2) are the two points that the line passes through.
2.   As stated before, these properties are both algebraic and graphic. If you know about them from one source, then the other must agree. Thus, if you have the graph, you should be able to read these values from it, or if you have the equation, you should be able to draw the graph of the line with these given properties.

ILLUSTRATION: Graph $y = 2x + 1$.

SOLUTION: $m = 2$ and $b = 1$. Locate the y-intercept $(y = +1)$ on the y-axis. Then draw a line that has a slope of 2 (Figure 1-7).

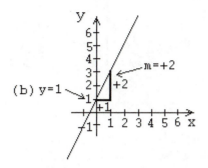

Figure 1-7

Illustration: Find the equation of the line that passes through $(-1,1)$ and $(2,7)$.

Solution: (Algebraically) $m = \dfrac{\Delta y}{\Delta x} = \dfrac{7 - 1}{2 - (-1)} = \dfrac{6}{3} = 2$

$y = 2x + b$, and the line passes through $(2,7)$. Therefore $x = 2$ and $y = 7$ must satisfy (make the statement true) $y = 2x + b$. In order for that to happen, b must be equal to $3[7 = 2(2) + b]$. Therefore the equation of such a line is $y = 2x + 3$.

(Graphically) Draw a graph of a straight line that passes through $(-1,1)$ and $(2,7)$; then read m and b from it (Figure 1-8).

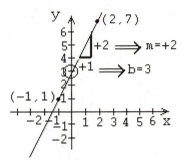

Figure 1-8

Therefore $y = mx + b$ becomes $y = 2x + 3$.

EXERCISES (Continued)

4. Write the equation of a straight line whose slope is 10 and whose y-intercept is -3.

5. Draw the graph of each of the following equations (use graph paper).
    a. $y = x + 2$
    b. $y = -2x + 10$
    c. $y = (1/3)x - 2$

6. Find the equation of the straight line that passes through each of the following pairs of points (a) algebraically and (b) graphically (use graph paper).
    I. $(3,1)$ and $(9,5)$
    II. $(-2,3)$ and $(6,-1)$

The preceding discussion about the equation of a straight line is presented from a mathematical point of view. Mathematicians and statisticians often approach concepts differently. For instance, the statistician typically places the terms of a linear equation in exactly the opposite order from the mathematician's equation.

To the statistician, for example, the equation of the straight line is $y = b + mx$, while to the mathematician it is $y = mx + b$. m and b represent exactly the same concepts in each case — the different order is a matter of emphasis. The mathematician's first interest in an equation is usually the highest-powered term; thus he or she places it first in the sequence. The statistician tends to describe a relationship in as simple a form as possible; thus his or her first interest is usually the lower-powered terms. The equation of the straight line in statistics is $y = b_0 + b_1x$, where b_0 is the y-intercept and b_1 represents the slope.

The answers to the exercises can be found in the back of the manual.

TREE DIAGRAMS

The purpose of this lesson is to learn how to construct and read a tree diagram. A tree diagram is a drawing that schematically represents the various possible outcomes of an experiment. It is called a tree diagram because of the branch concept that it demonstrates.

Let's consider the experiment of tossing one coin one time. We will start the experiment by tossing the coin and will finish it by observing a result (heads or tails)(see Figure 2-1).

Start   **Observation**
                                   Heads
                                   Tails

Figure 2-1

This information is expressed by the tree shown in Figure 2-2.

Figure 2-2

In reading a diagram like this, the single point at the left is simply interpreted as we are ready to start and do not yet know the outcome. The branches starting from this point must represent all of the different possibilities. With one coin there are only two possible outcomes, thus two branches.

∇Δ EXERCISES

1. Draw a tree diagram that shows the possible results from rolling a single die once.

2. Draw a tree diagram showing the possible methods of transportation that could be used to travel to a resort area. The possible choices are car, bus, train, and airplane.

Now let's consider the experiment of tossing a coin and single die at the same. What are the various possibilities? The coin can result in a head (H) or a tail (T) and the die could show a 1, 2, 3, 4, 5, or 6. Thus to show all the possible pairs of outcomes, we must decide which to observe first. This is an arbitrary decision, as the order observation does not affect the possible pairs of results. To construct the tree to represent this experiment we list the above-mentioned possibilities in columns, as shown in Figure 2-3.

```
        Start           Coin            Die

                                        •1
                                         2
                                         3
                         •H•             4
                                         5
                                         6




                         •T•
```

Figure 2-3

From "start" we draw two line segments that represent the possibility of *H* or *T* (Figure 2-4). The top branch means that we might observe a head on the coin. Paired with it is the result of the die, which could be any of the six numbers. We see this represented in Figure 2-5.

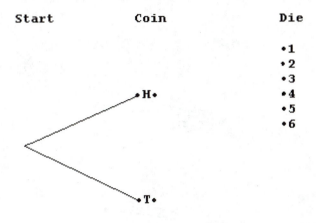

Figure 2-4

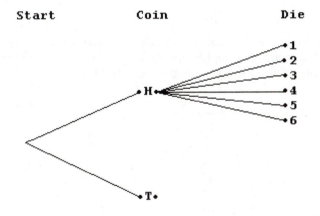

Figure 2-5

However, the outcome could have been *T*, so *T* must have branches going to each of the numbers 1 through 6. To make the diagram easier to read we list these outcomes again and draw another set of branches (Figure 2-6).

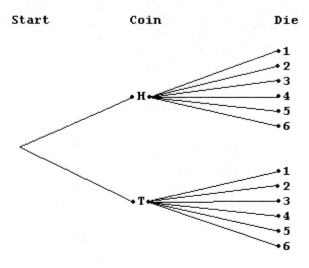

Figure 2-6

Figure 2-6 shows 12 different possible pairs of results — each one of the 12 branches on the tree represents one of these pairs. (A complete branch is a path from the start to an end.) The 12 branches in Figure 2-6, from top to bottom, are H1, H2, H3, H4, H5, H6, T1, T2, T3, T4, T5, T6.

The ordering could have been reversed: the tree diagram in Figure 2-7 shows the die result first and the coin result second.

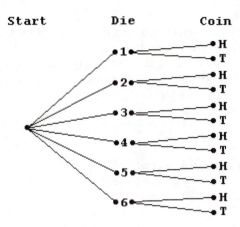

Figure 2-7

Is there any difference in the listing of the 12 possible results? Only the order of observation and the vertical order of the possibilities as shown on the diagram have changed. They are still the same 12 pairs of possibilities. If the experiment contains more than two stages of possible events we may expand this tree as far as needed.

EXERCISES (CONTINUED)

3. a. Draw a tree diagram that represents the possible results from tossing two coins.
    b. Repeat part (a) considering one coin a nickel and the other a penny.
    c. Is the list of possibilities for part (b) any different than it was for part (a)?

4. Draw a tree diagram representing the tossing of the coins.

5. a. Draw a tree diagram representing the possible results that could be obtained when two dice are rolled.
    b. How many branch ends does your tree have?

On occasion the stages of an experiment will be ordered; when this is the case the tree diagram must show ordered sets of branches, as in Figure 2-8, next page. The experiment consists of rolling a die. Then the result of the die will dictate your next trial. If an odd number results, you will toss coin. If a two or a six occurs, you stop. If any other number (a four) occurs, you roll the die again.

Notice that the tree diagram becomes a very convenient "road map" showing all the various possibilities that may occur in an experiment of this nature. Remember that an event is represented by a complete branch (a broken line from the start to an end), and the number of ends of branches is the same as the number of possibilities for the experiment. There are 14 branches in the tree diagram in Figure 2-8. Do you agree?

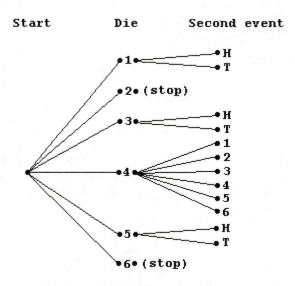

Figure 2-8

EXERCISES (CONTINUED)

6. Students at our college are to be classified as male or female, graduates of public or private high schools, and by the type of curriculum they are enrolled in, liberal arts or career. Draw a tree diagram which shows all of the various possible classifications.

7. There are two scenic routes (A and B) as well as one business route (C) by which you may travel from your home to a nearby city. You are planning to drive to that city by way of one route and come home by a different route.
   a. Draw a tree diagram representing all of your possible choices for going and returning.
   b. How many different trips could you plan?
   c. How many of these trips are scenic in both directions?

The answers to the exercises can be found in the back of the manual.

VENN DIAGRAMS

The Venn diagram is a useful tool for representing sets. It is a pictorial representation that uses geometric configurations to represent *set containers*. For example, a set might be represented by a circle — the circle acts like a "fence" and encloses all of the elements that belong to that particular set. The figure drawn to represent a set must be closed, and the elements are either inside the boundary and belong to that set or they are outside and do not belong to that set. The universal set (sample space or population) is generally represented by a rectangular area, and its subsets are generally circles inside the rectangle. Complements, intersections, and unions of sets then become regions of various shapes as prescribed by the situation. The Venn diagram in Figure 3-1 shows a universal set and a subset *P*.

Figure 3-1

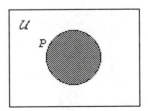

Any element that is represented by a point inside the rectangle is an element of the universal set. Likewise, any element represented by a point inside the circle, P(the shaded area), is a member of set P. The unshaded area of the rectangle then represents $\overline{P}$.

The Venn diagrams in Figure 3-2 show the regions representing $A \cap B$, $A \cup B$, $\left(\overline{A \cap B}\right)$, and $\left(\overline{A \cup B}\right)$. The shaded regions represent the identified sets.

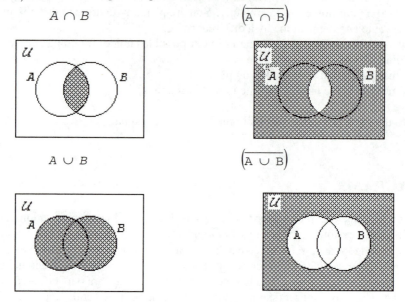

Figure 3-2

When three subsets of the same population are being discussed, three circles can be used to represent all of the various possible situations.

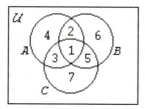

Figure 3-3

In Figure 3-3 the eight regions that are formed by intersecting the three sets have been numbered for convenience. Each of these regions represents the intersection of three sets (sets and/or complements of sets), as shown below.

Region Number	Set Representation
1	$A \cap B \cap C$
2	$A \cap B \cap \overline{C}$
3	$A \cap \overline{B} \cap C$
4	$A \cap \overline{B} \cap \overline{C}$
5	$\overline{A} \cap B \cap C$
6	$\overline{A} \cap B \cap \overline{C}$
7	$\overline{A} \cap \overline{B} \cap C$
8	$\overline{A} \cap \overline{B} \cap \overline{C}$

Region 1 $(A \cap B \cap C)$ might be thought of as the set of elements that belong to A, B, and C. Region 4 represents the set of elements that belong to A, $\overline{B}$ (but not to set B), and $\overline{C}$ (but not to set C). Region 8 represents the set of elements that belong to $\overline{A}$, $\overline{B}$, and $\overline{C}$ (or that do not belong to A, B, or C). The others can be described in similar fashion.

Figure 3-4 shows the union as sets B and C in the shaded areas of all three sets. Notice that $B \cup C$ is composed of regions 1, 2, 3, 5, 6, and 7. (You might note that three of these regions are inside A and three are outside).

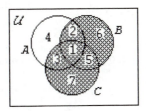

Figure 3-4

∇∆ EXERCISES

1. Shade the regions that represent each of the following sets on a Venn diagram as shown in Figure 3-5.

a. A

b. B

c. $A \cap B$

d. $A \cup B$

e. $\overline{A} \cup B$

f. $\overline{A} \cup \overline{B}$

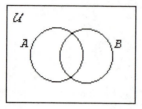

Figure 3-5

2. On a diagram showing three sets, P, Q, and R, shade the regions that represent the following sets.

a. P

b. $P \cap Q$

c. $P \cup R$

d. $\overline{P}$

e. $P \cap Q \cap R$

f. $P \cup \overline{Q}$

g. $P \cup Q \cup R$

h. $P \cup Q \cup \overline{R}$

The answers to these exercises can be found in the back of the manual.

THE USE OF FACTORIAL NOTATION

The factorial notation is a shorthand way to identify the product of a particular set of integers. 5! (five factorial) stands for the product of all positive integers starting with the integer 5 and proceeding downward (in value) until the integer 1 is reached. That is, $5! = 5 \times 4 \times 3 \times 2 \times 1$, which is 120. Likewise, $n!$ symbolizes the product of the integer n multiplied by the next smaller integer $(n - 1)$ multiplied by the next smaller integer $(n - 2)$ and so on, until the last integer, the number 1, is reached.

NOTES: 1.   The number in front of the factorial symbol (!) will always be a positive integer or 0.

2.   The last integer in the sequence is always the integer 1, with one exception: 0! (zero factorial). The value of zero factorial is defined to be 1, that is, $0! = 1$.

1! (one factorial) is the product of a sequence that   starts and ends with the integer 1, thus $1! = 1$.

2! (two factorial) is the product of 2 and 1. That is, $2! = (2)(1) = 2$.

$3! = (3)(2)(1) = 6$

$n! = (n)(n - 1)(n - 2)(n - 3)\ldots(2)(1)$

$(n - 2)! = (n - 2)(n - 3)(n - 4)\ldots(2)(1)$

$(4!)(6!) = (4 \cdot 3 \cdot 2 \cdot 1)(6 \cdot 5 \cdot 4 \cdot 3 \cdot 2 \cdot 1) = (24)(720) = 17280$

$4(6!) = (4)(6!) = 4(6 \cdot 5 \cdot 4 \cdot 3 \cdot 2 \cdot 1) = (4)(720) = 2880$

$$\frac{6!}{4!} = \frac{(6)(5)(4)(3)(2)(1)}{(4)(3)(2)(1)} = (6)(5) = 30$$

∇∆ Exercises

Evaluate each of the following factorials.

1.   4!

2.   6!

3.   8!

4. $(6!)(8!)$

5. $\dfrac{8!}{6!}$

6. $\dfrac{8!}{4!\,4!}$

7. $\dfrac{8!}{6!\,2!}$

8. $2\,\dfrac{8!}{[5!]}$

The answers to these exercises can be found in the back of the manual.

ANSWERS TO INTRODUCTORY CONCEPTS AND REVIEW LESSONS EXERCISES

Summation Notation Exercises

1. (a) $x_1 + x_2 + x_3 + x_4$
   (b) $x_1^2 + x_2^2 + x_3^2$
   (c) $(x_1 + y_1) + (x_2 + y_2) + (x_3 + y_3) + (x_4 + y_4) + (x_5 + y_5)$
   (d) $(x_1 + 4) + (x_2 + 4) + (x_3 + 4) + (x_4 + 4) + (x_5 + 4)$
   (e) $x_1 y_1 + x_2 y_2 + x_3 y_3 + x_4 y_4 + x_5 y_5 + x_6 y_6 + x_7 y_7 + x_8 y_8$
   (f) $x_1^2 f_1 + x_2^2 f_2 + x_3^2 f_3 + x_4^2 f_4$

2. (a) $\sum_{i=1}^{6} x_i$ (b) $\sum_{i=1}^{7} x_i y_i$ (c) $\sum_{i=1}^{9} (x_i)^2$ (d) $\sum_{i=1}^{n} (x_i - 3)$

4. (a) 8   (b) 68   (c) 64

5. (a) 8   (b) 4   (c) 12   (d) 4   (e) 42
   (f) 64 (g) -7   (h) 32   (i) 20   (j) 120
   (k) 400

6. $\sum_{i=1}^{120} [0.005(12,000 - (i - 1)100)]$

Using the Random Number Table Exercises

2. (a)   Use a two-digit number to represent the results obtained. Let the first digit represent one of the coins and the   second digit represent the other coin. Let an even digit indicate heads and an odd digit tails. Observe 10 two-digit numbers from the table. If a 16 is observed, it represents tails and heads on two coins. One head was therefore observed. The probabilities have been preserved.

   (b)   A second way to simulate this experiment is to find the   probabilities associated with the various possible results. The number of heads that can be seen on two coins is 0, 1, or 2. (HH,HT,TH,TT is the sample space.)   P(no heads) = 1/4; P(one head) = 1/2; P(two heads) = 1/4.   Using two-digit numbers, let the numbers 00 to 24 stand for no head appeared, 25 to 74 stand for one head appeared, and 75 to 99 stand for two heads appeared. The probabilities have again been preserved. Observe 10 two-digit numbers.

Round-Off Procedure Exercises

1.   (a) 13   (b) 9   (c) 9   (d) 156   (e) 46
    (f) 42   (g) 103   (h) 17

2.   (a) 8.7   (b) 42.3   (c) 49.7   (d) 10.2   (e) 10.4
    (f) 8.5   (g) 27.4   (h) 5.6   (i) 3.0   (j) 0.2

3.   (a) 17.67   (b) 4.44   (c) 54.55   (d) 102.06   (e) 93.22
    (f) 18.00   (g) 18.02   (h) 5.56   (i) 44.74   (j) 0.67

Review Lessons

The Coordinate-Axis System and the Equation of a Straight Line

1.

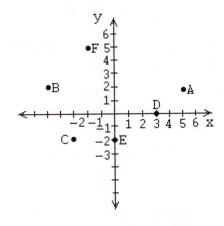

2. a.

Point	x	y
A	-2	7
B	0	5
C	1	4
D	3.5	1.5
E	6	-1

b.

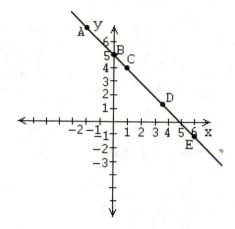

3.   Pick any values of x you wish; $x = $ -2,0,1,2,4 will
     be convenient

$x =$ -2	0	1	2	4
$y =$ -7	-4	-2.5	-1	2

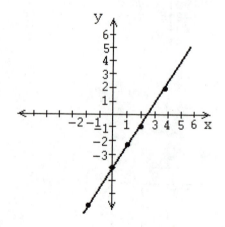

4. $m = 10, b = $ -3, $y = 10x - 3$

5.　a.

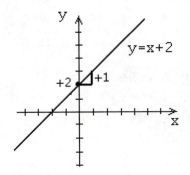

b.

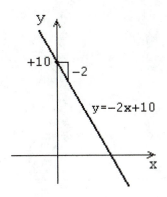

c.

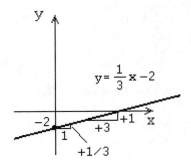

6.   I. (a) Given points $(3,1)$ and $(9,5)$.

$$m = \frac{\Delta y}{\Delta x} = \frac{5-1}{9-3} = \frac{4}{6} = \frac{2}{3}$$

$y = \frac{2}{3}x + b$ and passes through $(3,1)$

$1 = \frac{2}{3}(3) + b$ implies that $b = -1$

Thus $y = \frac{2}{3}x - 1$

I.   (b)

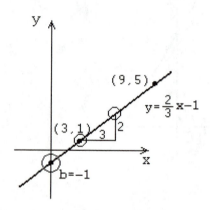

II. (a) Given points $(-2,3)$ and $(6,-1)$.

$$m = \frac{\Delta y}{\Delta x} = \frac{(-1)-(-3)}{6-(-2)} = \frac{-4}{8} = \frac{-1}{2}$$

$y = \frac{-1}{2}x + b$ and passes through $(-2,3)$

$3 = \frac{-1}{2}(-2) + b$ implies that $b = 2$

Thus $y = \frac{-1}{2}x + 2$

(b)

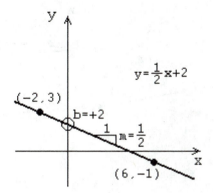

Tree Diagrams

1.

2.

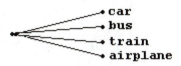

3. a.

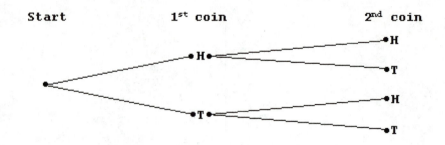

b.

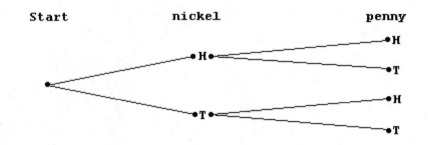

c.   no

4.

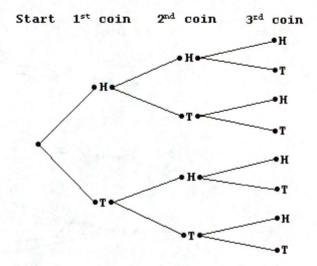

5.　a.

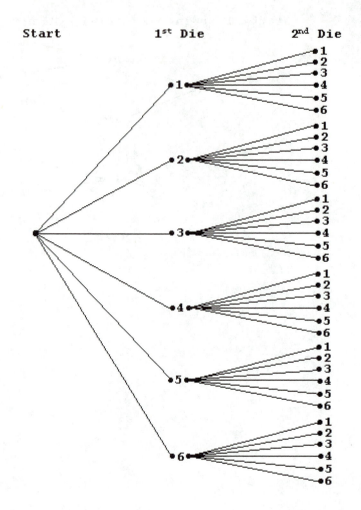

Start	1st Die	2nd Die

b.　36 branch ends

6.

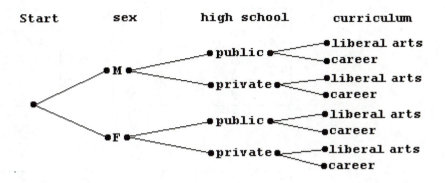

7.   a.

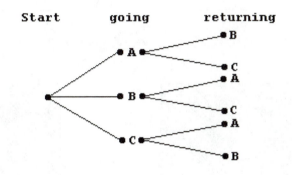

b.   6
c.   2

Venn Diagrams

1. a.

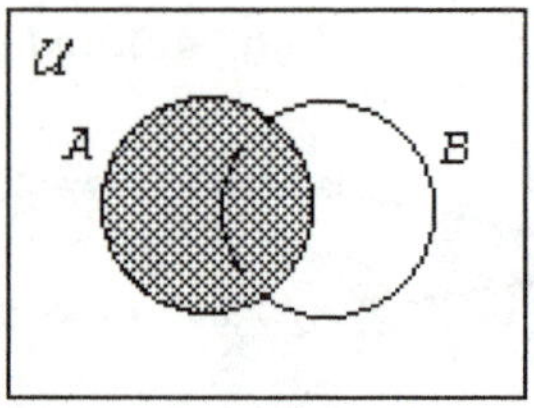

b.

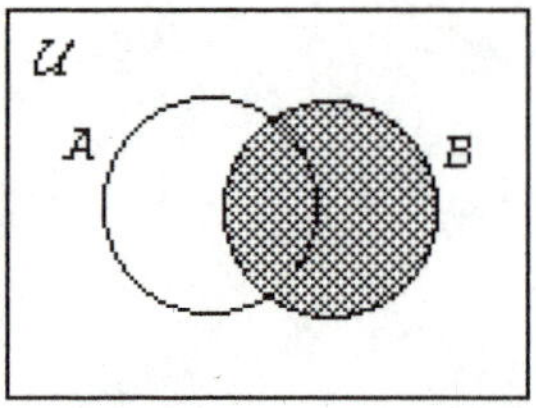

c.

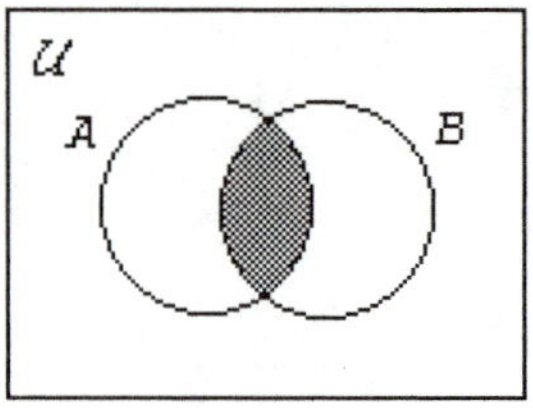

d.

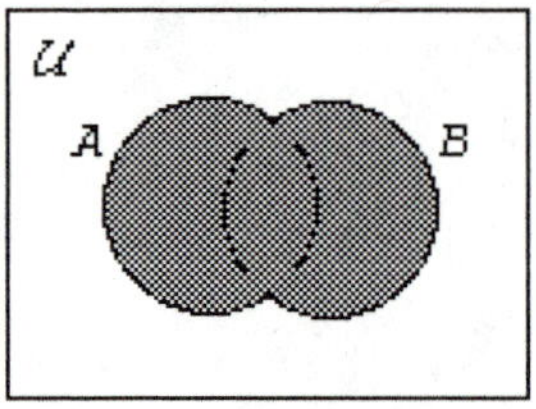

e.

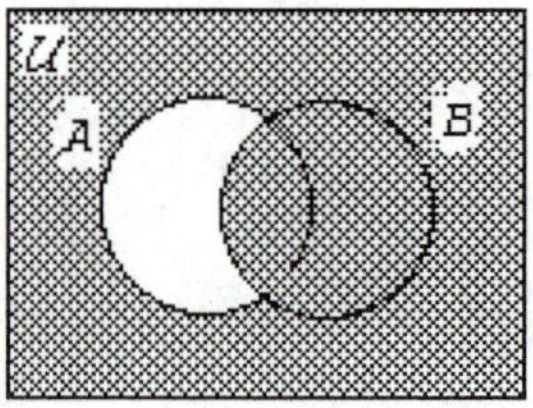

f.

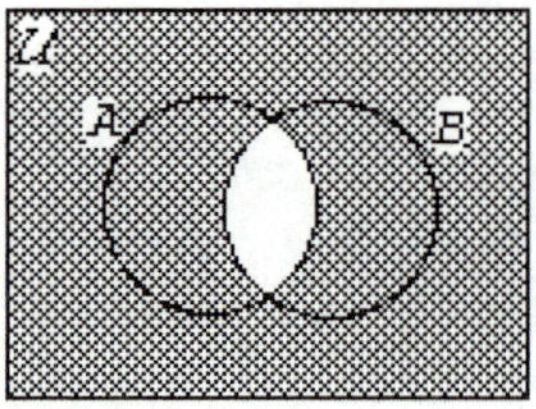

2. a.

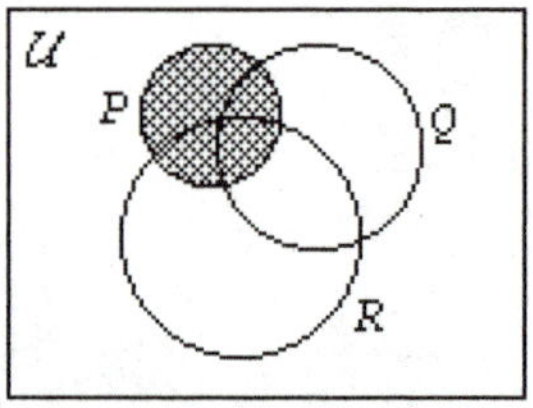

b.

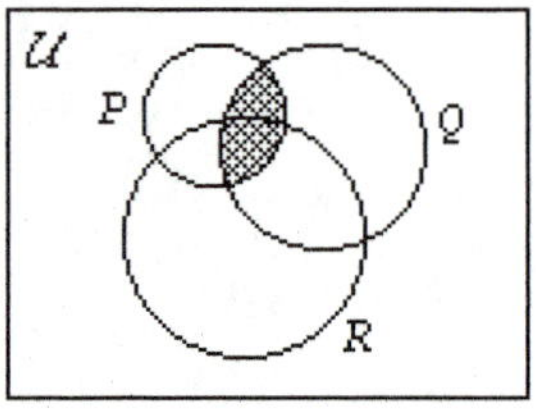

c.

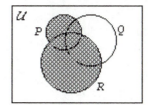

d.

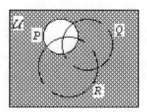

e.

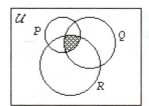

f.

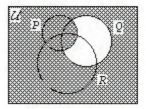

g.

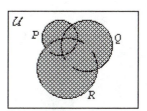

h.

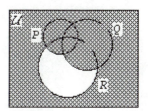

The Use of Factorial Notation

1. $4! = 4 \times 3 \times 2 \times 1 = 24$
2. $6! = 6 \times 5 \times 4 \times 3 \times 2 \times 1 = 720$
3. $8! = 8 \times 7 \times 6 \times 5 \times 4 \times 3 \times 2 \times 1 = 40320$
4. $(6!)(8!) = (720)(40320) = 29{,}030{,}400$
5. $\dfrac{8!}{6!} = \dfrac{8 \times 7 \times (6!)}{6!} = 8 \times 7 = 56$
6. $\dfrac{8!}{4!4!} = \dfrac{8 \times 7 \times 6 \times 5 \times (4!)}{4 \times 3 \times 2 \times 1 \times (4!)} = 2 \times 7 \times 5 = 70$
7. $\dfrac{8!}{6!2!} = \dfrac{8 \times 7 \times (6!)}{(6!) \times 2 \times 1} = 4 \times 7 = 28$
8. $2\left(\dfrac{8!}{5!}\right) = 2\left(\dfrac{8 \times 7 \times 6 \times (5!)}{5!}\right) = 2 \times 8 \times 7 \times 6 = 672$